PRINCIPLES OF ENERGY CONVERSION

PRINCIPLES
OF
ENERGY
CONVERSION

Archie W. Culp, Jr.

Associate Professor of Mechanical Engineering
University of Missouri-Rolla

McGraw-Hill Book Company

New York   St. Louis   San Francisco   Auckland   Bogotá   Düsseldorf
Johannesburg   London   Madrid   Mexico   Montreal   New Delhi
Panama   Paris   São Paulo   Singapore   Sydney   Tokyo   Toronto

PRINCIPLES OF ENERGY CONVERSION

1234567890 FGRFGR 7832109

Library of Congress Cataloging in Publication Data

Culp, Archie W
    Principles of energy conversion.

    Includes bibliographical references and index.
    1. Power (Mechanics) 2. Power resources.
3. Energy conservation. I. Title.
TJ163.9.C84   621   78-14760
ISBN 0-07-014892-9

This book was set in Times Roman.
The editor was Frank J. Cerra;
the cover was designed by Albert M. Cetta;
the production supervisor was Donna Piligra.
Fairfield Graphics was printer and binder.

To
DonaFay
and
Craig, Brian, and Eric

CONTENTS

Appendixes

Index

PREFACE

The area of energy conversion is such a broad area and covers so many disciplines that it is a difficult course to teach. Moreover, so much research is being carried out in this area that the field is constantly changing. The area is further clouded by political decisions and regulations. Writing a textbook on energy conversion is like trying to write an objective political or religious treatise on any highly controversial subject. Each "expert" involved in the energy program has a pet energy source and/or conversion system that he or she feels should be exploited.

The author has tried to be relatively objective in presenting the basic theory associated with almost all the energy-conversion systems. A relatively detailed description is presented for those systems that currently produce the bulk of humanity's fuel energy. This will undoubtedly generate some controversy because solar energy and some of the other exotic and promising systems are not extensively covered in this text, although the basic principles are covered. The author does not intend to imply that solar energy or some of these other systems are not important energy sources; it just means that at this time in history, solar energy and the other systems produce very little of our fuel energy, and more attention and detail are given to those systems that do produce the bulk of our fuel energy. It is felt that such an approach better prepares the student for a possible career in the energy area.

One can find several excellent textbooks on nuclear engineering and reactor systems and also good books on fossil fuel systems and direct energy conversion. Unfortunately, there are very few books that cover all aspects of energy conversion, describing systems as well as covering the theory. This book attempts to bridge this gap.

Most of the material presented here has been given in a mechanical engineering course on energy conversion that was developed by the author during the last five years. This course is one which is required for all M.E. students and replaced an established course in steam-power systems along with a modern physics course in the old M.E. curriculum.

This book is intended for use by senior-level or early graduate-level students in the various engineering disciplines associated with energy production, conversion, and utilization. These disciplines include the conventional engineering fields of chemical, electrical, mechanical, and nuclear engineering. It is expected that students will have a reasonable background in basic classical thermodynamics.

Because of the wide scope of the material presented in this text, it is difficult to cover the book in one three-semester-hour course. If it is used for such a course, the instructor must be somewhat selective of the material coverage. Some of the material may be a duplication of material presented in other courses, depending on the background of the student. If so, at the discretion of the instructor that material can be reviewed very quickly or omitted altogether.

Chapters 1 and 2 should be covered rather thoroughly as the basic background and fundamental conversion principles are presented in these chapters. Chapters 3, 7, and 8 deal with the production of thermal, mechanical, and electrical energy, respectively, and should have the next order of priority. Chapter 6 gives the student an insight into some of the environmental problems associated with energy generation along with some of the possible solutions. Chapter 4 presents a more detailed description of the conventional fossil-fuel combustion systems and their components. Chapter 5 should be covered by those people who want to go into some of the details of nuclear reactor design and operation. Chapter 9 presents some general background information on energy-storage systems.

There are a number of example problems included in the text, but many of the problems can be worked simply by using dimensional analysis. Students should be urged to write down the dimensions associated with each of the variables along with those of the appropriate conversion factors. The dimensions should be cancelled just like numbers when working the problems. This procedure has been employed in most of the example problems except that none of the dimensions are actually cancelled.

If the instructor discovers any mistakes in the text, either typographical or theoretical, and there are undoubtedly some of both, or if the instructor feels that certain areas should be covered in greater detail, then he or she should contact the author. In fact, the author would be receptive to any comments, good, bad, or indifferent, about the book.

The author would like to express his appreciation to his family and friends for their moral support during preparation of the manuscript. He would also like to thank Dr. H. J. Sauer, Jr., for his assistance in locating some of the technical information. Finally, the author is indebted to the Halliburton Oil Company whose unrestricted grant to the M.E. Department for faculty development permitted him to work full time on the preparation of the manuscript during the summer of 1977.

Archie W. Culp, Jr.

ENERGY CLASSIFICATION, SOURCES, AND UTILIZATION

1.1 INTRODUCTION

Throughout the history of the human race, major advances in civilization have been accompanied by an increased consumption of energy. Today, energy consumption appears to be directly related to the level of living of the populace and the degree of industrialization of the country. Those countries that have had abundant supplies of energy available to them have realized substantially high rates of industrial growth and a corresponding increase in the gross national product. In many instances, the availability of low-cost energy has led to the inefficient utilization of the energy and in some instances with disastrous ecological effects. However, it is obvious that in order to raise the level of living of the majority of the world's population, the present energy consumption must be greatly expanded. Figure 1.1 shows the relationship, for the various countries of the world, between the per capita energy consumption and the level of living as measured by the value of the per capita gross national product.

Recently, some of the countries that have large supplies of low-cost energy are using those supplies as potential political and economic weapons to achieve political ends that could not be realized by normal diplomatic means. Because of this "energy blackmail," the people of the energy-dependent countries of the

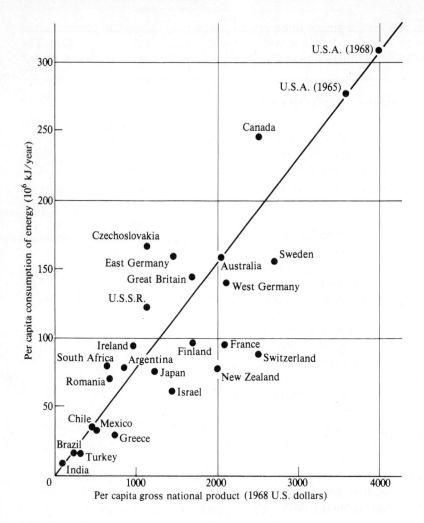

Figure 1.1 Dependence of the gross national product on energy consumption, 1968.

world are becoming increasingly aware of the importance of the conversion, conservation, and development of new energy sources.

It is the responsibility of the scientist, power engineer, and technician to locate, develop, and exploit these new sources. In order to accomplish this, it is imperative that these people have an intimate knowledge of the various energy forms, sources, conversion techniques, and conservation methods, along with their limitations and inherent problems.

In the first half of the twentieth century, energy sources were exploited with the primary consideration given to economics—low cost. Today, the power engineer must be concerned with the three "E's"—energy, economy, and ecology.

Thus, the modern engineer must try to develop systems that produce large quantities of energy at low cost with minimal impact on the environment. The proper balancing of these three " E's " is a major technological challenge.

1.2 MASS-ENERGY DEPENDENCE

An early statement of the first law of thermodynamics stated, in effect, that energy must be conserved in any process. A related postulate stated that mass can neither be created nor destroyed. In 1922, however, Albert Einstein hypothesized that energy and mass are actually related by the following relationship:

$$E = mc^2 \tag{1.1}$$

where E is the energy release in joules, m is the actual mass converted into energy in kilograms, and c is the velocity of light (3×10^8 m/s). This equation actually represents a reversible process, but the important thing is that the sum of mass and energy must be conserved in any energy-conversion process.

When Eq. (1.1) is used, it becomes evident that a small amount of mass, completely annihilated, produces huge quantities of energy. A 600,000-kW$_e$ (the subscript " e " indicates that this is electrical energy) coal-fired power plant, operating continuously, consumes about 220 tons of coal per hour or about 2,000,000 tons of coal per year. A 600,000-kW$_e$ nuclear power plant, operating continuously, consumes about 1 ton of uranium fuel per year. The actual fuel mass that is converted into energy in both of these systems is around 640 g or less than $1\frac{1}{2}$ lb per year.

When energy is produced or released, as in a chemical or nuclear reaction, there must be a corresponding decrease in mass accompanying the process. In this text, no distinction is made between the actual energy or the mass converted into energy. When reference is made to chemical or nuclear energy, it will actually refer to that portion of the total mass of the reactants which can be converted into other energy forms in some type of conversion process.

1.3 ENERGY TYPES AND CLASSIFICATIONS

There are two general types of energy—transitional energy and stored energy. Transitional energy is energy in motion, and as such can move across system boundaries. Stored-energy forms, as implied, are energy forms that exist as mass, position in a force field, etc. These stored forms can usually be easily converted into some form of transitional energy.

While there is no generally accepted method or system of energy classification, this text will divide the different energy forms into six major groups or classifications. These six groups or categories are: mechanical energy, electrical energy, electromagnetic energy, chemical energy, nuclear energy, and thermal energy.

In thermodynamics, mechanical energy is defined as energy which can be used to raise a weight. The common system of units for mechanical energy in the United States is the foot pound for energy and horsepower for the unit of power. In this text, Standard International (SI) units are generally employed. In this system, the unit of energy is the joule (or wattsecond) and the unit of power is the watt.

The transitional form of mechanical energy is called work. Mechanical energy may be stored as either potential energy or as kinetic energy. Potential energy is the energy that a given quantity of material possesses as the result of its position in a force field. This includes gravitational-field energy, the energy associated with a compressed fluid, the energy associated with the position of ferromagnetic substances in a magnetic field, and the energy associated with elastic strain, as in springs and torsion bars. Kinetic energy is the energy associated with a given mass of material due to its relative motion with another body. The flywheel is an example of a system that stores mechanical energy in the form of kinetic energy. Mechanical energy is a very useful form of energy and can be easily and efficiently converted into other energy forms.

Electrical energy is that class of energy associated with the flow or accumulation of electrons. This form of energy is commonly reported in units of power and time, such as watthours or kilowatthours. The transitional form of electrical energy is electron flow, usually through a conductor of some kind. Electrical energy may be stored as either electrostatic-field energy or as inductive-field energy. Electrostatic-field energy is the energy associated with the electric field produced by the accumulation of charge (electrons) on the plates of a capacitor. Inductive-field energy, which is sometimes called electromagnetic-field energy, is the energy associated with the magnetic field established by the flow of electrons through an induction coil. Electrical energy, like mechanical energy, is a very useful energy form because it can be easily and efficiently converted into other energy forms.

Electromagnetic energy is that form of energy associated with electromagnetic radiation. Radiation energy is usually reported in the very small energy units of electronvolts (eV) or million-electronvolts (MeV). This energy unit is also commonly used in the evaluation of nuclear energy.

Electromagnetic radiation is a form of pure energy in that there is no mass associated with it. This radiation exists as only transitional energy traveling at the speed of light, c. The velocity of a wave c is equal to the product of the frequency v, in cycles per second or hertz, and the wavelength λ, in meters, of the radiation. The energy E of these waves is directly proportional to the frequency v of the radiation and is given by the following relationship:

$$E = hv = \frac{hc}{\lambda} \tag{1.2}$$

where E is the energy in joules, h is Planck's constant (6.626×10^{-34} J·s), v is the frequency, and λ is the wavelength. The more energetic electromagnetic waves have short wavelengths and high frequencies.

There are several different classes of electromagnetic radiation depending on the wavelength (energy) or the source of the radiation. Gamma radiation is the most energetic form of electromagnetic energy and most of them emanate from the atomic nucleus. The next most energetic form of radiation are x-rays, and are produced as the result of the exitation of orbiting electrons. Thermal radiation is electromagnetic radiation that is produced as the result of atomic vibration. This band of electromagnetic energy is very broad and includes the high-temperature or ultraviolet radiation, the narrow band of visible radiation, and the band of low-temperature or infrared radiation. Microwave and millimeter-wave radiation is the next most energetic form of radiation and is used in radar and microwave cookers. The least energetic form of electromagnetic radiation is radiowave radiation. The electromagnetic-radiation spectrum is shown in Fig. 1.2.

Chemical energy is energy that is released as the result of electron interactions in which two or more atoms and/or molecules combine to produce a more stable chemical compound. Chemical energy exists as only a stored-energy form. If energy is released in a chemical reaction, it is called an exothermic reaction. This energy release is commonly reported in units of calories or British thermal units (Btu) per unit mass of fuel reactant. In some chemical reactions, energy is absorbed and these reactions are called endothermic reactions. The most important source of fuel energy for the human race is the exothermic chemical reaction called combustion. The combustion reaction involves the oxidation of fossil fuels.

Nuclear energy is another energy form that exists only as stored energy which is released as the result of particle interactions with or within the atomic nucleus. This energy is released as the result of the product particles assuming a

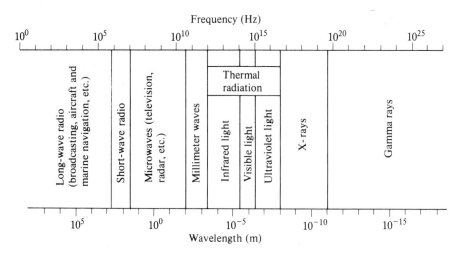

Figure 1.2 The electromagnetic energy spectrum.

more stable configuration. This energy release is usually reported in units of million-electronvolts per reaction. There are three general types of nuclear reactions, including radioactive decay, fission, and fusion. The radioactive-decay process is one in which only one unstable nucleus, a radioisotope, randomly decays to a more stable configuration, with the release of particles and energy. The fission reaction, which is the principal process in the nuclear reactor, occurs when a heavy-mass nucleus absorbs a neutron and the resulting excited compound nucleus splits into two or more nuclei with the release of energy. In the fusion reaction, two light-mass nuclei combine to produce more stable configurations with the release of energy.

The annihilation reaction is commonly listed as a nuclear reaction, but it is actually a separate reaction that is not necessarily associated with the atomic nucleus. This reaction is the ultimate energy-conversion reaction in that all the reactant mass is converted into energy. In the annihilation reaction, matter and antimatter combine and are converted into electromagnetic energy. This reaction is the only reaction in which atomic particles, as such, are completely destroyed, but the only known naturally occurring reaction of this type involves subatomic particles and it is not an important reaction.

The last major energy classification is thermal energy and this energy is associated with atomic and molecular vibration. Thermal energy is a basic energy form in that all other energy forms can be completely converted into thermal energy but the conversion of thermal energy into other forms is severely limited by the second law of thermodynamics. The transitional form of thermal energy is heat and it is commonly expressed in units of calories or British thermal units. Thermal energy can be stored in almost any media as either sensible heat or latent heat. Sensible-heat storage is accompanied by an increase in temperature while latent-heat storage is an isothermal process associated with a change of phase.

1.4 ENERGY SOURCES

Energy sources may be grouped into two general categories—celestial or income energy, which is the energy reaching the earth from outer space, and capital energy, which is energy that already exists on or in the earth. Income energy includes solar and lunar energy while capital energy sources include geothermal and atomic energy sources.

Celestial energy sources actually include all possible sources that provide energy to the earth from outer space. This includes electromagnetic, gravitational, and particle energy from stars, planets, and the moon as well as the potential energy of meteorites entering the earth's atmosphere. The only useful celestial energy sources are the electromagnetic energy of the earth's sun, called direct solar energy, and the gravitational energy of the earth's moon, producing tidal flows. The utilization of celestial energy sources is very attractive because they are continuing or nondepletable sources of energy and because they are relatively pollution free—a very important consideration.

The direct solar energy also generates some indirect nondepletable sources of energy. The solar heating, combined with the earth's rotation, produces some large convection currents in the form of wind in the atmosphere and ocean currents in the seas. The absorption of solar energy also generates relatively large thermal gradients in the ocean that have the potential of producing power. In addition, the evaporation of surface water generates clouds, which, when condensed into rain at higher altitudes, provides the source for hydroelectric or water power. The wind also generates large ocean waves that have the potential of generating energy.

The other major source of celestial or income energy is lunar energy, which is primarily the gravitational energy of the moon. The gravitational energy of the moon is manifested primarily in the form of tidal flows that have a variation of from a few inches to about 25 or 30 ft in the Passamaquoddy Bay, which is a part of the Bay of Fundy located between Maine, U.S.A., and New Brunswick, Canada.

There have been several proposals to harness tidal power to produce electricity, including one to build an 800- to 14,000-MW_e tidal-electric system in Passamaquoddy Bay. Such a system involves the construction of a dam across the entrances to the tidal basin and letting the water flow back and forth through a number of reversible water turbines in the dam.

Two tidal-electric systems have been constructed. The Russians built a small 2-MW_e tidal-electric plant at Kislaya Guba, about 600 mi north of Murmansk. The French built a 240-MW_e tidal-electric system on the Rance Estuary off the channel-island coast of France. This 24-turbine system is shown in Fig. 1.3, and this plant is also used as a pumped storage system. During times of low power demand, the generator-motor units are reversed and they pump seawater into the estuary which is subsequently released to the sea during times of peak power demand.

The total potential of all the tidal-power systems of the world has been estimated to be 64,000 MW_e. While this is a very large block of power, it is comparatively small when compared to the 356,800-MW_e electrical generation capacity of the United States in 1970. Although the utilization of tidal power does not provide a solution to the world's energy requirements, this source is nondepletable and the energy is essentially pollution free.

The major source of capital energy employed presently is atomic energy. The term atomic energy, as used here, refers to energy released as the result of any kind of reaction with atoms—including both chemical and nuclear energy. Both nuclear and chemical energy have been discussed to some extent earlier, and the conversion of these energy forms will be presented in greater detail in subsequent chapters.

The final major source of fuel energy available is geothermal energy. This

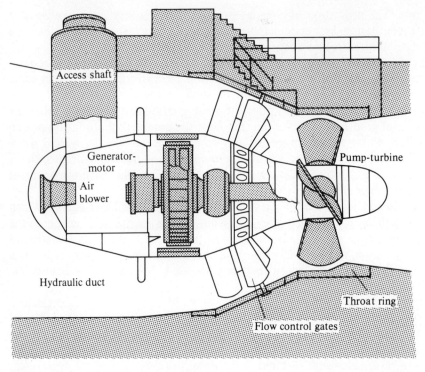

Figure 1.3 A pump-turbine unit of the La Rance tidal-electric station. The system is composed of 24 of these 10-MW$_e$ units. (*Courtesy of the Editors of Power, the McGraw-Hill Magazine of Energy Systems Engineering.*)

source is essentially thermal energy trapped beneath and within the solid crust of the earth. This energy is manifested as steam, hot water, and/or hot rock and is released naturally in the form of fumaroles, geysers, hot springs, and volcanic eruptions. Although there are tremendous reserves of thermal energy trapped beneath the earth's crust, it has not been possible to drill through the crust, despite several attempts. Consequently, the only useful geothermal reserves are those located in pockets trapped within the crust, and such pockets are normally found near active fault lines.

The utilization of geothermal energy is not a new technology as the first geothermal steam well was drilled at Larderello, Italy, in 1904 and the present capacity of that plant is 370 MW$_e$. The Pacific Gas and Electric Company operates a 400-MW$_e$ geothermal power complex at Geyserville, California. A typical geothermal power system is shown in Fig. 1.4.

Many people are promoting geothermal power as a major source of pollution-free energy. Upon closer examination, however, geothermal energy may not be as pollution free as is promoted. Air pollution at a geothermal

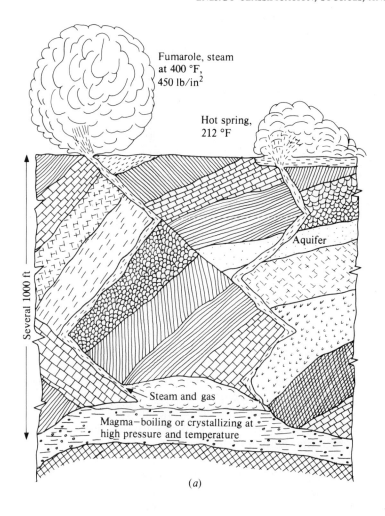

Fumarole, steam at 400 °F, 450 lb/in²

Hot spring, 212 °F

Aquifer

Several 1000 ft

Steam and gas

Magma—boiling or crystallizing at high pressure and temperature

(a)

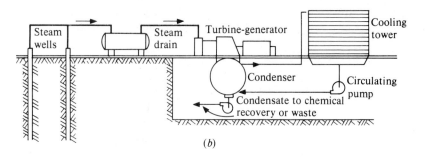

Steam wells

Steam drain

Turbine-generator

Cooling tower

Condenser

Circulating pump

Condensate to chemical recovery or waste

(b)

Figure 1.4 Schematic diagram of a typical geothermal deposit and power plant. *(Summers, 1971.)*

Table 1.1 Geothermal power projects

Country	Fields	Year commercial production started	Present capacity MW_e	Type of source
China	Kwantang Province	1958	???	???
Iceland	Namafjall	1969	3.0	Hot water
Italy	Larderello	1912	375.0	Steam
	Monte Amiata	1959	25.5	Steam
Japan	Matsukawa	1966	20.0	Steam
	Otake	1967	13.0	Steam
Mexico	Cerro Prieto	1973	75.0	Hot water
New Zealand	Wairakei	1960	170.0	Steam
U.S.S.R.	Pauzhetsk	1967	13.0	Hot water
United States	The Geysers	1960	395.0	Steam

installation may be a significant problem because of the emission of heavy radioactive gases and hydrogen sulfide (H_2S), a poisonous gas. Because of the relatively poor steam conditions, a geothermal power plant usually pumps three times as much thermal energy into the environment for a given unit of electricity as a conventional fossil unit. This is called "thermal pollution." Hot-water geothermal sources have such high mineral content that disposal of the cooled water is a problem. Other serious problems associated with the utilization of geothermal energy include possible land subsidence and increased seismic activity, particularly if water is injected into hot rock to recover the thermal energy.

A summary of the current status of geothermal power plants and the expected developments in the near future are presented in Table 1.1. While there are vast quantities of geothermal energy trapped beneath the earth's crust in the hot mantle and molten core, the recoverable geothermal energy trapped within the crust is fairly limited. Moreover, these pockets, like fossil fuels, are normally depleted when the energy is removed. The total estimated recoverable energy from the world's major geothermal areas is about 3,000,000 MW_{th}·years.

1.5 ENERGY UTILIZATION

Ever since the dawn of history, the human race has used more and more energy with the discovery of new energy sources and the development of new and better conversion methods. The earliest source of energy was muscle power—first their own and later that of work animals. Some time early in evolution, people learned to produce energy from the combustion of carbohydrates (plants and wood). Around 3000 B.C. people learned to utilize the wind to drive ships and during the dark ages they applied the wind power to drive windmills.

Table 1.2 Production and consumption of energy in the United States

Year	Total 10¹⁵ Btu	Production by type of fuel, %					Consumption by economic sector, %				
		Oil	Coal	Natural gas	Hydro	Nuclear	Commercial/ residential†	Industrial	Transportation	Electricity generation	Other
1950	34.0	39.7	38.0	18.1	4.2	0.0	22.3	36.2	25.3	14.7	1.4
1955	39.7	44.1	29.1	23.3	3.5	0.0	21.6	35.2	24.8	16.6	1.8
1960	44.6	45.0	22.8	28.5	3.7	<0.1	22.8	32.9	24.3	18.5	1.5
1965	53.3	43.6	22.3	30.2	3.8	0.1	22.2	32.2	23.8	20.8	1.0
1970	67.1	44.0	18.9	32.8	4.0	0.3	20.8	30.1	24.2	24.2	0.7
1971	68.7	44.5	17.5	33.2	4.2	0.6	20.7	29.1	24.8	25.1	0.3
1972	71.9	45.8	17.3	32.0	4.1	0.8	20.3	28.5	25.1	25.8	0.3
1973	74.7	46.6	17.8	30.4	4.0	1.2	19.1	28.6	25.3	26.6	0.4
1974	73.0	45.8	17.9	30.2	4.5	1.6	19.1	27.9	25.3	27.4	0.3
1975	70.6	46.2	18.2	28.4	4.6	2.6	19.2	25.4	26.3	28.7	0.4
1976	74.2	47.2	18.4	27.4	4.1	2.8	20.2	25.4	25.6	28.8	0.0

† Does not include electricity.
Sources: Bureau of Mines, 1950–1975; FEA, 1976.

Water power was first utilized around the birth of Christ, but it was not until the late eighteenth century that thermal energy was utilized as a large-scale source of mechanical energy.

It is interesting to examine the primary sources of fuel energy in the United States during the last 120 years or so to see how the primary sources of fuel energy have changed. In 1850, over 90 percent of the fuel energy came from the combustion of wood and wood products. Sixty years later, in 1910, coal provided about 80 percent of the fuel energy and wood had dropped to 10 percent. Sixty years later, in 1970, 75 percent of the fuel energy came from the combustion of petroleum and natural gas and 20 percent was provided by coal. During 1970, the United States consumed approximately 73×10^{18} J (69×10^{15} Btu). The energy sources and utilization for the United States are presented in Table 1.2, and these data are shown graphically for the last 50 years or so in Fig. 1.5. A flow diagram showing the sources and utilization of energy during 1970 is presented in Fig. 1.6.

It is interesting to speculate as to what the primary source of fuel energy will be in the year 2030, approximately 50 years from now. Will it be fossil fuels, fission reactors, solar energy, fusion reactors, or some other energy source? One can find many "energy experts" who will predict that any of these four sources plus some others will predominate by 2030. Projections of energy sources and consumption in the United States for the future are presented in Fig. 1.7.

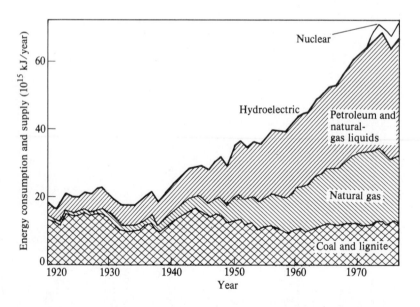

Figure 1.5 U.S. energy supply and consumption, 1920–1976. (*U.S. Bureau of Mines and the Federal Energy Agency.*)

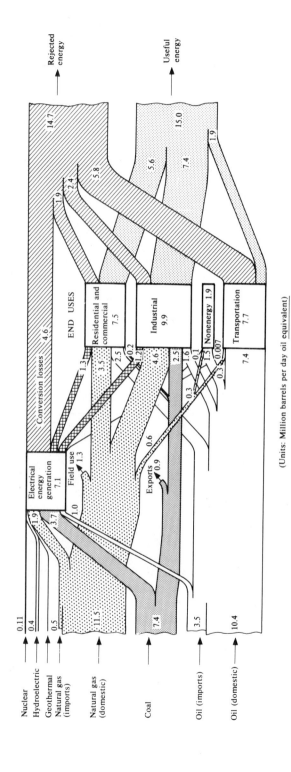

(Units: **Million barrels per day oil equivalent**)

Figure 1.6 The sources and utilization of fuel energy in the United States, 1970. (*From charts presented to the Congressional Joint Committee on Atomic Energy at the May 3, 1973 meeting.*)

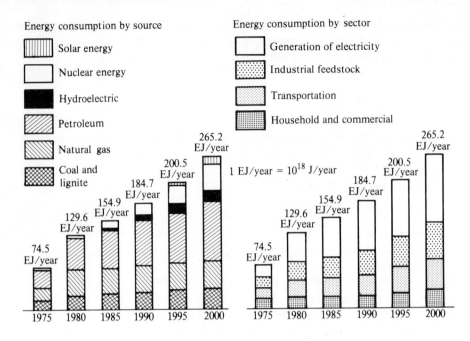

Figure 1.7 Estimates of the projected energy source and consumption in the United States, A.D. 1970–2000.

1.6 ELECTRICAL POWER GENERATION AND CONSUMPTION

The electrical power generation in the United States for the year 1970 totaled 1.62×10^{12} kW$_e$·h (5.5×10^{15} Btu$_e$) and the total installed capacity was 356,800 MW$_e$. This means that the average load for the electrical grid was about 50 percent of the maximum possible load. The prime sources for the production of electricity were as follows: 6.3 percent from oil, 22.0 percent from natural gas, 52.5 percent from coal and lignite, 18.2 percent from hydroelectric plants, and 1.0 percent from nuclear power. Barring any significant technical breakthroughs in the utilization of other energy sources or politically motivated moratoriums, the primary sources of energy for the generation of electricity is expected to be coal and lignite, nuclear fission reactors, and hydroelectric systems, for the balance of the twentieth century.

By 1975, the installed nuclear capacity had risen to around 8 percent and over half of the new electrical generation ordered in 1975 was nuclear. As of December 1975, there were a total of 58 nuclear reactors in operation in the United States with a total capacity of 39,600 MW$_e$. At the same time another 159 reactors were either under construction or on order. The estimated capacity of nuclear power plants was expected to be over 100,000 MW$_e$ in the United States by 1980 and this figure was expected to double by 1985, again barring any legal moratoriums or technical difficulties.

1.7 GROWTH RATES

It is interesting to examine the consequences of a quantity that is constantly increasing, whether it is energy consumption, population, gross national product, personal income, etc. If a quantity such as power P increases at the same frac-tional rate i each year, the rate of change of the quantity with time becomes

$$\frac{dP}{dt} = Pi \tag{1.3}$$

Setting the initial power at P_0 at some arbitrary time, $t = 0$, Eq. (1.3) can be integrated to yield

$$\int_{P_0}^{P} \frac{dP'}{P'} = \ln\left(\frac{P}{P_0}\right) = \int_{0}^{t} i \, dt' = it \tag{1.4}$$

or
$$P = P_0 e^{it} \tag{1.5}$$

A plot of power versus time is shown in Fig. 1.8. To make this plot more general, dimensionless parameters, P/P_0 and t/t_d, are used where t_d is the doubling time.

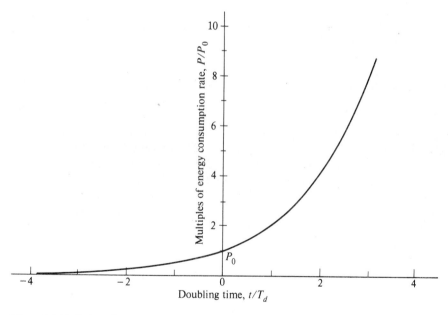

Figure 1.8 Variation of energy-consumption rate for a constant rate of growth.

t_d is the time required for the power to increase by a factor of two. Thus, when $t = t_d$, $P = 2P_0$ and

$$t_d = \frac{\ln 2}{i} = \frac{0.693}{i} \tag{1.6}$$

The electrical energy consumption in the United States has historically averaged about 7 percent per year, which means that the electrical generating capacity has doubled every 0.693/0.07 or 9.9 years.

The total energy consumed during a given period of time can be found by integrating Eq. (1.5) over the time interval. It is convenient to assume that the exponential growth rate has occurred from time $t = -\infty$. The total energy consumed E_0 from time $t = -\infty$ to time $t = t_1$ is obtained as follows:

$$E_0 = \int_{-\infty}^{t_1} P_0 e^{it} \, dt = \frac{P_0}{i} e^{it_1} \tag{1.7}$$

Let E_1 be equal to the total energy consumed between $t = t_1$ and $t = t_2$. Then

$$E_1 = \int_{t_1}^{t_2} P_0 e^{it} \, dt = \frac{P_0}{i} \left(e^{it_2} - e^{it_1} \right) = \frac{P_0}{i} e^{it_1} \left(e^{i(t_2 - t_1)} - 1 \right)$$

$$= E_0 \left(e^{i(t_2 - t_1)} - 1 \right) \tag{1.8}$$

If the time interval $(t_2 - t_1)$ in Eq. (1.8) is set equal to the doubling time t_d, the exponential term in (1.8) becomes numerically equal to 2 and this means that $E_1 = E_0$. Thus, in any given doubling period, the total energy consumed in that period is equal to the total energy consumption in all the time prior to the doubling period. This result leads to some sobering if not frightening prospects and conclusions. If new fuel discoveries extend our present fuel reserves by an order of magnitude, the new reserves will last only 3.32 doubling times unless the rate of increase in the growth rate is lowered.

Example 1.1 Using the 1970 value of U.S. energy consumption rate of 70.8×10^{18} J/year, and assuming a 4 percent growth rate per year, find the doubling time and estimate the energy consumption rate in the year 2000.

SOLUTION Given: $i = 0.04$ per year, $P_0 = 70.8 \times 10^{18}$ J/year, $t = 30$ year

Doubling time $= t_d = (\ln 2)/i = 0.693/0.04 = 17.33$ years

Power at A.D. 2000 $= P = P_0 e^{it} = 70.8 \times 10^{18} e^{(0.04)(30)}$

$$= 235.1 \times 10^{18} \text{ J/year}$$

Example 1.2 Using the consumption rate and growth rate of Example 1.1 along with the U.S. coal reserves of 70×10^{21} J, estimate how long the coal would last as the energy source if it supplies all the fuel energy.

SOLUTION Given: $i = 0.04$ per year, $P_0 = 70.8 \times 10^{18}$ J/year, $t_1 = 0$,

$$E_1 = 70 \times 10^{21} \text{ J}$$

$$E_1 = 70 \times 10^{21} = (P_0/i)e^{it_1}(e^{i(t_2 - t_1)} - 1) = (70.8 \times 10^{18}/0.04)(e^{0.04t_2} - 1)$$
$$t_2 = (1/0.04) \ln [1 + (70 \times 10^{21})(0.04)/(70.8 \times 10^{18})] = 92.56 \text{ years}$$

1.8 ENERGY RESERVES

The energy reserves of the earth can be divided into four general categories. They include renewable or nondepletable sources, fossil fuels, fissionable and fertile isotopes, and the fusionable isotopes. In some instances, particularly for shale oil and uranium, the reserves are strongly dependent on the current market price of the raw fuel. As the cost of energy increases, it becomes profitable to mine the low-grade ores.

A good example of how the price effects the available reserves is given by uranium-235, the only naturally occurring fissionable isotope. In Table 1.3, the U-235 reserves are listed as 13.7×10^{21} J, which corresponds to the 1973 price of $8 per ton of U_3O_8. If the uranium price had been $30 a ton, the available reserves would have increased to 22.0×10^{21} J, and if the price had been $500 a ton, the available reserves would have increased to $30,880 \times 10^{21}$ J. While a sixtyfold increase in the price of uranium feedstock would certainly increase the cost of nuclear power, it is estimated that the total increase in the cost of nuclear power would be less than a factor of four since the ore cost is a small part of the overall cost.

Table 1.3 lists some of the energy reserves of the earth. The values listed in this table came from a number of different sources, but in each case the highest or most optimistic values are listed. These values change constantly as new discoveries are made and as the world energy consumption increases.

Care must be taken when comparing the reserves of the various energy sources from Table 1.3. The values listed in the table are pure energy values and most are utilized by converting them to thermal energy. If the final energy form desired is either mechanical or electrical energy, some of the sources, such as tidal, water, and wind power, can be converted to these energy forms with a much higher conversion efficiency than the other sources.

Figure 1.9 depicts several logarithmic energy scales along with some of the more important energy terms. The energy scale ranges from very small values such as the kinetic energy of atoms at 20°C through 56 orders of magnitude to the daily energy output of the sun. There is also a mass-energy conversion scale.

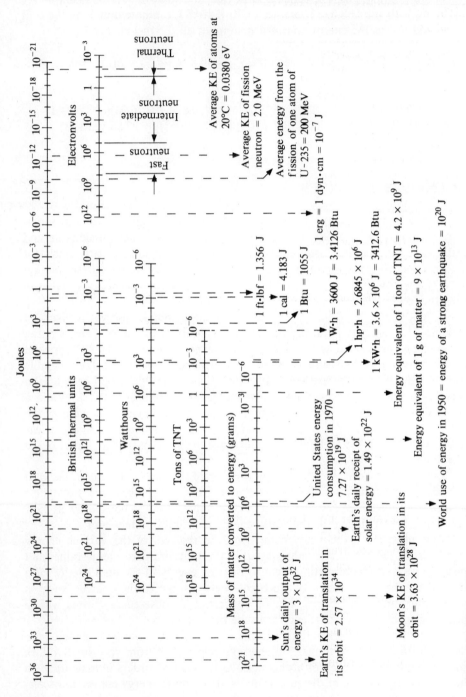

Figure 1.9 Energy scales.

Table 1.3 Estimated energy reserves of the world

Source	Amount	Type
Total available tidal power[a,b]	6.7×10^{10} W	Mechanical
Total available water power[a,b]	300.0×10^{10} W	Mechanical
Wind power in the United States[a,b]	970.0×10^{10} W	Mechanical
Recoverable geothermal energy	0.4×10^{21} J	Thermal
Shale-oil reserves[c,d]	1.2×10^{21} J	Chemical
Tar-sand oil reserves[c,d]	1.8×10^{21} J	Chemical
Natural gas reserves[c]	9.5×10^{21} J	Chemical
Petroleum reserves[c]	11.7×10^{21} J	Chemical
Uranium-235 reserves[d,e]	13.7×10^{21} J	Nuclear
Solar power in the United States[a]	$187,000.0 \times 10^{10}$ W	Electromagnetic
Thorium $(^{233}_{92}\text{U})$[f]	71.7×10^{21} J	Nuclear
Coal and lignite[c]	200.0×10^{21} J	Chemical
Uranium-238 $(^{239}_{94}\text{Pu})$[f]	$1,800.0 \times 10^{21}$ J	Nuclear
Deuterium $(^{2}_{1}\text{H})$-tritium $(^{3}_{1}\text{H})$[g]	$6,100.0 \times 10^{21}$ J	Nuclear
Deuterium $(^{2}_{1}\text{H})$-deuterium $(^{2}_{1}\text{H})$[g]	$6,000,000.0 \times 10^{21}$ J	Nuclear

[a] These reserves are nondepletable reserves and are actually power sources.

[b] These reserves can be converted directly to mechanical energy while most of the rest of the reserves must normally be converted into thermal energy.

[c] These reserves are classed as fossil fuels.

[d] These reserves are strongly dependent on the price of energy.

[e] This is the only naturally occurring fissionable fuel isotope.

[f] Utilization of these reserves depends on the development of the breeder fission reactor and subsequent fissioning of the isotope shown in parenthesis.

[g] Utilization of these reserves depends on the development of the fusion reactor.

1.9 ENERGY ECONOMICS

Energy costs from a given power system can be divided into two general categories—capital costs and operational costs. The capital costs are that part of the overall power cost which is constant and must be paid regardless of whether the plant operates or not. Consequently, the capital cost decreases per unit energy output as the plant output is increased. The capital costs include the cost of the land, the construction cost, certain taxes, insurance, and interest on the investment. The largest portion of these costs is normally the investment and interest charges.

If the original cost of the energy converter is A dollars, the interest that this investment can earn over the lifetime of the system can be determined from the compound interest formula. When an investment of A dollars is made at interest rate i, which is compounded n times a year, the total worth of the investment A_T at the end of an operating period of t_{op} years is

$$A_T = A\left(1 + \frac{i}{n}\right)^{nt_{op}} \qquad (1.9)$$

If the power payments are made to the company in equal increments of S dollars over the operating period, these funds can be invested and subsequently used to pay off the original investment plus interest at the end of the operating period. The total sum that can be accumulated, which must equal A_T, by investing S dollars m times a year at a yearly interest rate of j, and which is compounded m times a year, is found from the annuity relationship:

$$A_T = \frac{(1 + j/m)^{mt_{op}} - 1}{j/m} S \qquad (1.10)$$

Using Eqs. (1.9) and (1.10), one can determine the amount of money that should be collected each $1/m$ year to compensate for the capital investment. This amount, S, can then be divided by the average amount of energy produced each $1/m$ period to determine that part of the unit energy cost attributable to the capital investment. This cost is usually reported in units of cents per kilowatthour.

Some energy-conversion devices, such as the typical electrical power generating station, require many years of construction before they start producing energy. Since money is spent during the entire construction period, the initial investment cost, A, is considerably higher than the actual cash outlay during the construction period. If the construction cost is paid in discrete uniform increments of R dollars during construction, the actual capital investment at the time of startup becomes

$$A = \frac{(1 + i/p)^{pt_c} - 1}{i/p} R \qquad (1.11)$$

where R is the incremental construction cost that is paid p times a year over the construction period of t_c years and i is the interest rate. The quantity R can be determined from the following equation:

$$R = \frac{(\text{cost per kW}_e)(\text{rated power, kW}_e)}{pt_c} \qquad (1.12)$$

In evaluating the unit power cost associated with the capital investment from Eq. (1.10), it was assumed that the salvage value of the plant at the end of the operating period essentially paid for the dismantling of the system. If the salvage value of the plant is appreciable, this sum should be subtracted from the value of A_T before solving for S.

The operational power costs include all the expenses incurred during operation of the plant, including personnel wages and fringe benefits, fuel costs, maintenance, and certain taxes. Since these costs are essentially directly proportional to the energy output of the plant, that part of the unit power cost associated with operation is independent of the energy output of the plant. Moreover, since these costs are incurred as the system is operated and the company is promptly reimbursed for the energy, there is normally no interest charge associated with these expenses. The principal operating cost is the fuel cost and this can be easily calculated.

The energy cost to the utility of the fuel needed to operate the power plant is commonly contracted in terms of X cents per million Btu or its equivalent. If this fuel cost, X, is known, the average unit power cost associated with the fuel is

$$\text{Fuel unit power cost} = 0.003413X/\eta_{th} \quad \text{cents/kW}_e \cdot \text{h} \tag{1.13}$$

where η_{th} is the overall thermal efficiency of the power plant.

The operational costs associated with labor is somewhat dependent on the power output of the plant if the system produces any energy at all. The labor cost, in cents per kilowatthour, can be evaluated from the following equation:

$$\text{Labor unit power cost} = \frac{1.14 \times 10^{-5} N(\text{AS})}{P_{max}(\text{LF})} \tag{1.14}$$

where N is the number of people working at the plant, AS is the average annual salary, in dollars, P_{max} is the maximum electrical power output of the plant, in megawatts, and LF is the average load factor of the plant or the ratio of average to maximum power.

Example 1.3 Assume that a solar-powered home heating system can be built for $8000 to supply all the heating requirements of the house for 20 years. If the interest on the money is 8 percent, compounded annually, what is the effective cost of heating the house? Assume that the salvage value of the system just compensates for the maintenance and operating costs over the operational period.

SOLUTION Given: $A = \$8000$, $i = j = 0.08$ per year, $m = n = 1$ year,

$$t_{op} = 20 \text{ year}, \, t_c = 0$$

$$A_T = A\left(1 + \frac{i}{n}\right)^{nt_{op}} = 8000(1 + 0.08)^{20} = \$37{,}287.66$$

$$A_T = \frac{(1 + j/m)^{mt_{op}} - 1}{j/m} S = \frac{(1.08)^{20} - 1}{0.08} S = 45.762S = \$37{,}287.66$$

S = yearly heating cost (all capital costs) = $814.82 per year

Example 1.4 Calculate the unit capital and fuel power costs associated with a 1-gigawatt (1000 MW$_e$) nuclear power plant if the nominal cost is $800 per kilowatt, the interest is compounded quarterly, and the annual interest rate is 8 percent. Assume that the overall thermal efficiency of the plant is 33 percent, the load factor is 70 percent, with a construction period of 10 years, an operating period of 40 years, a fuel cost of 25 cents per million Btu, and a construction pay period every quarter. Neglect the salvage value of the plant.

SOLUTION Given: Unit cost = $800 per kW$_e$, $P_{max} = 10^6$ kW$_e$,

$$m = n = p = 4.0 \text{ per year}$$

$t_c = 10$ year, $t_{op} = 40$ year, $i = j = 0.08$ per year, $\square/\eta_{th} = 0.33$, LF $= 0.7$

Assume the average salary of the workers is $20,000 per year.

Capital cost:

$$\frac{\text{Construction payment}}{\text{Quarter}} = R = \frac{(\text{unit cost})(P_{max})}{pt_c}$$

$$= \frac{(\$800 \text{ per kW}_e)(10^6 \text{ kW}_e)}{(4 \text{ per yr})(10 \text{ year})}$$

$$= \$20,000,000 \text{ per quarter}$$

Value of the facility at the end of construction = A:

$$A = \frac{(1 + j/m)^{mt_c} - 1}{j/m} R = \frac{(1 + 0.08/4)^{40} - 1}{0.02} (2 \times 10^7)$$

$$= 1.208 \times 10^9 \text{ dollars}$$

Total capital cost = $A_T = A\left(1 + \frac{i}{n}\right)^{nt_{op}} = 1.208 \times 10^9 (1.02)^{160}$

$$= 2.8714 \times 10^{10} \text{ dollars}$$

$$A_T = \frac{(1 + i/m)^{mt_{op}} - 1}{i/m} S = \frac{(1.02)^{160} - 1}{0.02} S$$

$$= 1138.5S$$

Quarterly power payments needed to meet the capital costs = S:

$$S = \frac{A_T}{1138.5} = \frac{2.8714 \times 10^{10}}{1138.5} = \$2.5221 \times 10^7 \text{ per quarter}$$

Energy output per quarter $= E_Q = (P_{max})(LF)(\text{time per quarter})$

$$= \frac{(10^6 \text{ kW}_e)(0.7)(8766 \text{ h/year})}{4 \text{ quarters/year}}$$

$$= 1.534 \times 10^9 \text{ kW}_e \cdot \text{h/quarter}$$

Capital cost per $\text{kW}_e \cdot \text{h} = \dfrac{S}{E_Q} = \dfrac{(\$2.5221 \times 10^7 \text{ per } Q)(100 \text{ cents/\$})}{1.534 \times 10^9 \text{ kW}_e \cdot \text{h}/Q}$

$$= 1.644 \text{ cents/kW}_e \cdot \text{h} = 16.44 \text{ mils/kW}_e \cdot \text{h}$$

Operating costs:

Fuel cost $= 0.003413 X/\eta_{th} = \dfrac{(0.003413 \text{ MBtu/kW} \cdot \text{h})(25 \text{ cents/MBtu})}{0.33}$

$$= 0.259 \text{ cents/kW}_e \cdot \text{h} = 2.59 \text{ mils/kW}_e \cdot \text{h}$$

Labor cost: Assume that it takes 30 people to operate the plant.

Unit labor cost $= \dfrac{1.14 \times 10^{-5} N(AS)}{P_{max}(LF)}$

$$= \frac{1.14 \times 10^{-5}(30)(20{,}000)}{(10^3)(0.7)}$$

$$= 0.00977 \text{ cents/kW}_e \cdot \text{h} = 0.0977 \text{ mils/kW}_e \cdot \text{hr}$$

Total costs (excluding taxes, profit, etc.) $= 16.44 + 2.59 + 0.10$

$$= 19.13 \text{ mils/kW}_e \cdot \text{h}$$

Coal-fired power plants are characterized by relatively low capital costs but have high fuel costs with the prospects of even higher fuel costs. Nuclear power plants are characterized by relatively high capital cost and low fuel costs. Thus, nuclear power plants are particularly well suited for operation as base-loaded power systems, which operate most of the time.

Example 1.4 gives an insight into the amount of capital required to build a typical modern power plant. This example is fairly representative for a modern nuclear station except that some of the values may be somewhat low. Some estimated power costs for large power stations scheduled to go "on line" in 1981 and 1986 are presented in Table 1.4.

Table 1.4 Power costs

(a) Estimated capital costs for 1000-MW$_e$ power systems for operation in 1981 ($/kW$_e$)

Costs	Type of power plant		
	Nuclear (LWR)	Coal	Oil
Direct costs			
Land	1	1	1
Structures and site facilities	45–60	30–40	25–35
Reactor or boiler plant equipment	80–85	75–85	65–75
Turbine plant equipment	85–95	65–75	65–75
Electric plant equipment	30–35	15–20	15–20
Miscellaneous plant equipment	5–6	55–60	4–5
Contingency and spare parts allowance	20–25	20–25	15–20
Subtotal	266–307	261–306	190–231
Indirect costs			
Professional services	45–50	25–30	20–25
Miscellaneous costs	25–30	30–35	20–25
Interest during construction (at 7 percent/year)	85–95	75–85	55–65
Subtotal	155–175	130–150	95–115
Total plant cost (no escalation)	421–482	391–456	285–346
Escalation during construction			
At 4 percent	75–88	73–83	51–63
At 6 percent	116–137	105–121	74–90
At 8 percent	161–189	138–160	100–121

Construction time for nuclear plant assumed to be 7.5 and 6.0 years for fossil systems. All systems assumed to use natural-draft cooling towers and coal system employs sulfur dioxide (SO_2) scrubbers (included in cost of miscellaneous plant equipment).

(b) Estimated power costs for a 2000-MW$_e$ power system scheduled for operation in 1986

Capital cost estimates ($/kW$_e$)

Capital costs	Nuclear (LWR)	Coal-fired with SO_2 scrubbers	Oil-fired
Plant cost (no escalation) in 1977 dollars	580	480	315
Escalation at 7 percent per year	260	260	175
Interest during construction (at 9 percent)	260	170	110
Total cost on completion ($/kW$_e$)	1100	910	600

Table 1.4—*Continued*

Power generation costs

Costs	Nuclear (LWR)		Coal-fired with SO$_2$ scrubbers		Oil-fired	
	$/kW$_e$	cents/kW·h	$/kW$_e$	cents/kW·h	$/kW$_e$	cents/kW·h
Capital costs	1100	2.69	910	2.23	600	1.46
Fuel costs		1.33		2.12		4.96
Operation and maintenance		0.33		0.44		0.19
Total costs		4.35		4.79		6.61

Fixed charge rate of 15 percent and plant capacity factor of 70 percent (from Bechtel).

1.10 ENERGY AND POWER UNITS

Care must be exercised when performing calculations not to confuse energy units with power units. Power is an energy rate ($P = dE/dt$) and energy is equal to the integral of power over a given time interval. Standard International (SI) units are commonly used throughout the text, although other units are occasionally employed.

The SI unit of energy is the joule (J), but some of the other energy units are electronvolts (eV), million-electronvolts (MeV), calories (cal), British thermal units (Btu), and foot pound-force (ft·lbf). In addition, energy is commonly expressed in terms of power and time units, such as the watthour (W·h), kilowatthour (kW·h), horsepower-hour (hp·h), etc. Subscripts such as "e" and "th" are used throughout the text to identify the energy (or power) terms as electrical or thermal quantities, respectively.

The SI unit of energy is the watt (W), and this unit and multiples of it are commonly employed, such as kilowatts (kW), megawatts (MW), gigawatts (GW), and terawatts (TW). Occasionally, the English power unit, horsepower (hp), will be employed. Power units can also be expressed in terms of energy rates such as joules per second (J/s), British thermal units per hour (Btu/h), etc.

The SI unit of mass is the kilogram (kg). Other mass units used in this text are the pound-mass (lbm), the atomic mass unit (amu), and multiples of grams.

1.11 ENERGY-CONVERSION MATRIX

An energy-conversion matrix is presented in Fig. 1.10 that lists the various processes, reactions, and systems used in the conversion of one form of energy to another. Those processes, reactions, and systems which are discussed at length in this text are indicated in the matrix with an asterisk.

From → To ↓	Mechanical energy				
			Potential energy		
	Work	Kinetic energy	Gravitational	Elastic strain	Compressed fluid
Mechanical energy — Work	****	Impulse and momentum changes (turbine blades)†	Lowering weight	Springs	Air motors Turbines
Kinetic energy	Flywheel† Linear acceleration	****	Dropping weight	Catapult	Nozzle
Potential energy — Gravitational energy	Raise weight	Pitot tube and manometer Rockets	****	Spring-weight systems	Upward nozzles Hydraulic jack
Elastic strain energy	Spring compression and extension Elastic deformation	Pitot tube and gage	Spring-weight systems	****	????
Fluid compression	Mechanical pump	Diffuser	Piston in cylinder	????	****
Electrical energy	Dc generator† Alternator†	Magnetohydrodynamic generator (MHD)† Electrogasdynamic generator (EGD)† Electrokinetic converter	Hydroelectric generators† Liquid-drop generators†	Piezioelectricity	Electrogasdynamic generator†
Electromagnetic energy	????	Phosphorescence Bremsstrahlung Cerenkov radiation	????	????	????
Chemical energy	????	Dissociation by radiolysis	????	????	????
Nuclear energy	????	Charged-particle reactions	????	????	????
Thermal energy — Heat	Friction	Impact (inelastic) Friction	Friction	Plastic flow	Unrestrained expansion
Internal energy	Compression Friction	Diffuser	????	????	????

† These systems are discussed in the text.

Figure 1.10 Energy-conversion matrix.

Electrical energy	Electromagnetic energy	Chemical energy	Nuclear energy	Thermal energy	
				Heat	Internal energy
Electric motors† Electromagnets Electrostriction	????	Muscles	????	Heat-engine cycles†	Expansion processes
Particle accelerator Electromagnetic pump†	Compton scattering Radiometer	Firearms Rockets	Particle emission	Pulse jet Turbo jet†	Nozzle
Electromagnets	????	????	????	Hot-air balloons Evaporation (clouds)	????
????	????	????	????	Thermal stress Bimetallic strips	????
Electromagnetic pump†	????	Combustion	????	Fluid heating	????
****	Photoelectricity Solar cells† Radio antenna	Fuel cells† Batteries†	Nuclear battery†	Thermoelectrics† Thermionics† Thermomagnetism Ferroelectricity Nernst effect	????
Lasers and masers Radio transmitter Electroluminescense	****	Chemiluminescense (fireflies)	Phosphors X-ray emission Gamma emission Annihilation	Thermal radiation	Thermal radiation
Electrolysis Battery charging	Photosynthesis Photochemistry	****	Ionization Radiation catalysis	Endothermic reactions	????
????	Gamma reactions	Phosphors	****	????	????
Joule heating	Absorption	Exothermic reactions Combustion†	Fission† Radioactive decay† Fusion†	****	Conduction Convection
????	????	????	Radiation damage	Sensible heat† Latent heat†	****

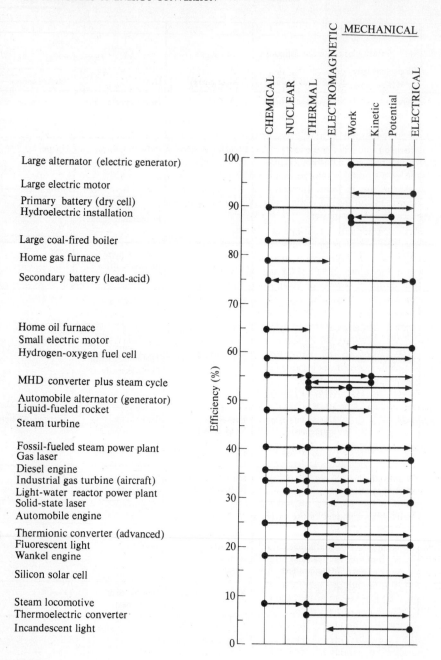

Figure 1.11 Some typical conversion efficiencies.

PROBLEMS

1.1 A 2400-MW_e power plant has the following power demand for a given day.

12–5 A.M.: 850 MW_e	9–12 A.M.: 2150 MW_e	5–6 P.M.: 2250 MW_e
5–7 A.M.: 1250 MW_e	12–1 P.M.: 2040 MW_e	6–8 P.M.: 1850 MW_e
7–8 A.M.: 1840 MW_e	1–4 P.M.: 2350 MW_e	8–10 P.M.: 1500 MW_e
8–9 A.M.: 1960 MW_e	4–5 P.M.: 2400 MW_e	10–12 P.M.: 1150 MW_e

Find the total energy output, in MW_e·days, kW_e·h, MeV, J, and Btu_e. Assume that the plant burns coal with a heating value (energy content) of 29,310 kJ/kg (12,600 Btu/lbm) with an overall thermal efficiency of 37 percent, and find the total mass of coal, in tons, consumed during the day's operation and the maximum coal rate, in tons per hour, required for proper operation of the plant. Determine the heat rate for the unit, in British thermal units per kilowatthour (Btu_{th}/kW_e·h), and find the load factor for the system during this period of operation. The load factor is defined as the ratio of average to maximum energy output.

1.2 A 1-GW power plant with a heat rate of 9400 Btu_{th}/kW_e·h operates at full power for a period of one day. Find the mass of coal, in tons, and the volume of oil, in barrels, required to provide sufficient energy for each day's operation at full power. Assume that the energy value (the higher heating value) of the coal is 27,910 kJ/kg and that of the oil is 141,000 kJ/gal. There are 42 gal of petroleum in a barrel of petroleum. Also, determine the actual mass, in grams, of matter that is actually converted into energy for each day of full-power operation.

1.3 In 1973, the production of fossil fuels in the United States (including Alaska) was as follows: 22,647,549 million standard cubic feet (ft_s^3) of natural gas; 634,423,000 barrels (42-gal barrels) of natural-gas liquids; 3,360,903,000 barrels of petroleum; and 598,569,000 tons of coal. If the higher heating value (HHV) of the fuels are as listed below, find the energy in joules (keep in mind that some of these fuels are used in the production of plastics and petrochemicals) that could be realized from their combustion:

Coal:	HHV = 29,070 kJ/kg
Petroleum:	HHV = 141,000 kJ/gal
Natural-gas liquids:	HHV = 130,000 kJ/gal
Natural gas:	HHV = 1060 kJ/ft_s^3

1.4 If the world's reserves of fossil fuels (from Table 1.3) is uniformly consumed over a 10-year period, find the ratio of the fossil-fuel energy release to the amount of solar energy absorbed by the earth for the same time period. In the determination of the amount of solar energy received by the earth, use the "solar constant" of 1395 W/m^2 as the incident energy flux at the earth's orbital position and assume that the earth receives solar energy as a disk and that 30 percent of the incident energy is reflected. The diameter of the earth is 1.2756×10^7 m and the distance from the earth to the sun is about 1.49×10^{11} m.

1.5 A common form of residential electrical rate is as follows:
First 10 kW_e·h/month or less at $1.50
Next 40 kW_e·h/month at 7 cents/kW_e·h
Next 50 kW_e·h/month at 5 cents/kW_e·h
Next 100 kW_e·h/month at 3 cents/kW_e·h
All kW_e·h in excess of 200 kW_e·h at 2.5 cents/kW_e·h
(a) Calculate the monthly bill for 8 kW_e·h of consumption.
(b) Calculate the monthly bill for 90 kW_e·h of consumption.
(c) Calculate the monthly bill for 600 kW_e·h of consumption.

1.6 An industrial concern purchases energy under the following rate:
For energy purchased monthly, the equivalent of
First 30 h at the maximum demand at 5 cents/kW_e·h
Next 100 h at the maximum demand at 2 cents/kW_e·h
Next 200 h at the maximum demand at 1 cent/kW_e·h
All energy in excess of the foregoing blocks at 0.8 cents/kW_e·h

The total monthly charge shall not be less than $3000.

(a) Calculate the monthly bill for energy consumed of 420,000 $kW_e \cdot h$ with a maximum demand of 2000 kW_e.

(b) Calculate the minimum bill for the same total energy usage.

1.7 The atomic bomb dropped on Hiroshima, Japan, had a yield of about 5 kilotons of TNT. Determine the electrical power output of a conventional power plant, in megawatts, that could be operated with the same energy equivalent if it ran at constant power for a period of one year with an overall thermal efficiency of 40 percent.

1.8 A 1-GW power plant with an overall thermal efficiency of 38 percent has a load factor of 80 percent for an operating period of 345 days with 20 days/year allotted for scheduled outages; find the annual fuel cost and the annual income. The cost of coal is 90 cents/million Btu and the company charges an average of 3 cents/$kW_e \cdot h$ output. Also find the mass that is actually converted into energy per year, in pounds mass per year and in kilograms per year.

1.9 Determine the tons of coal that could be saved per year if all the world's water and tidal power (from Table 1.3) could be harnessed. It is assumed that 80 percent of this power could be converted into electricity but that only 35 percent of the coal's energy could be transformed into electricity. Assume that the higher heating value (HHV) of the coal is 28,000 kJ/kg.

1.10 A 600-MW_e power plant with a heat rate of 9275 $Btu_{th}/kW_e \cdot h$ is operated at a load factor of 75 percent for an operating period of 335 days with 30 days allotted for scheduled outages. Find the annual fuel cost, in dollars per year, the area of coal field that must be mined, in acres, and the total income, in dollars per year, for the plant. Assume that the cost of the coal is $1.20 per million British thermal units and that the company receives 3.5 cents/$kW_e \cdot h$ of energy output. The density of coal is 1.2 kg/liter, the HHV equals 12,600 Btu/lbm, and the coal seam is 2 m thick. Also find the mass of coal that is actually converted into energy, in pounds-mass per year.

1.11 A reservoir has a water surface of 1200 acres and an available storage of 4000 million gal. The watershed contributing to and including the reservoir covers 35 mi^2. The dam of the reservoir contains a hydroelectric turbine generator with an available head of 40 ft and a mechanical-electrical efficiency of 83 percent. If the runoff for the area amounts to 42 in/year and the evaporation from the surface of the reservoir is 8 in/year, compute the average electrical output of the generator, in kilowatts, assuming that all the overflow goes through the turbine.

1.12 A proposed solution for the supply of electrical energy to a house involves the use of a windmill to pump water from a depth of 30 m with a reversible water pump-turbine. The pumped water is then stored in a tank for use when there is no wind. The windmill shaft is not only connected to the water pump-turbine but it is also directly connected to an electrical generator. If the average household uses 120 $kW_e \cdot h$ of energy per day, determine the size of tank, in gallons, required to store water for a 3-day period of no wind. Assume that the mechanical-electrical conversion efficiency of the water turbine and generator unit is 60 percent. Also determine the dimensions of the cylindrical storage tank if the tank diameter is equal to its height.

1.13 The cost of mining and transporting coal are roughly independent of the heating value of the coal. Consider that the coal in the ground is valued at 40 cents/million kJ, that mining costs $8 per ton, and that transportation costs 6 cents/ton-mile. If the price of other delivered coals are $1.20 per million kilojoules, find:

(a) The radius from the mine, in miles, that a coal of 32,560 kJ/kg can be delivered and sold for zero profit or loss

(b) The same radius for a 23,200-kJ/kg coal

(c) The minimum heating value of a coal that could be used locally

1.14 The fuel cost for an 800-MW_e steam power plant is 90 cents/million kJ. Find the total yearly fuel cost for the plant if the overall thermal efficiency is 36 percent and the load factor is 75 percent. Also determine the savings in fuel cost if the overall thermal efficiency of the power plant can be increased by 1 percent to 37 percent.

1.15 The fuel cost for a 2400-MW_e steam power plant is 90 cents/million kJ. Find the total yearly fuel cost for the plant if it has a thermal efficiency of 34 percent and a load factor of 65 percent. Also determine the savings in fuel cost if the overall thermal efficiency of the power plant can be increased by 1 percent to 35 percent.

1.16 It is proposed to power an automobile with a hybrid mechanical-electrical drive system. The prime mover is a 50-hp, external-combustion steam engine with a thermal efficiency of 30 percent. The steam engine drives an alternator with an overall mechanical-electrical efficiency of 70 percent. Power from the alternator is either stored in a lead-acid battery system or is fed directly to electric traction motors that have an electromechanical efficiency of 90 percent. The storage-battery efficiency, including charging, discharging, and storage losses, is around 70 percent. Find the power delivered to the drive wheels when the storage batteries are bypassed and calculate the power input, in watts and British thermal units per hour, as well as the overall efficiency of the system when the batteries are used.

1.17 Using the "solar constant" of 1395 W/m^2 outside the earth's atmosphere and the fact that the earth is 1.49×10^{11} m from the sun, calculate the power of the sun, in gigawatts, if it radiates isotropically. Also determine the rate, in tons per second, at which the sun is converting mass into energy. Find the solar power, in gigawatts, received by the earth if the earth receives energy as a disk with a diameter of 1.2756×10^7 m.

1.18 Determine the area of solar cells required to drive an electric car if the overall conversion efficiency of the propulsion system, including the electromagnetic-electric-mechanical conversion is 11 percent. Assume that the car requires 30 hp and that the average gross solar input is 650 W/m^2. If the system could store energy while sitting in the parking lot, etc., and is storing energy at the rate of 4 h for each hour of operation, find the area required. Assume that the storage efficiency is 60 percent.

1.19 Some people estimate that the electrical energy requirements of the United States in the year 2000 will be 10^{10} $MW_e \cdot h$. Determine the size, in miles, and the cost of a square array of solar cells that will produce the energy. Assume that the cells have a conversion efficiency of 15 percent and that the location chosen for the installation receives an average solar energy flux of 700 W/m^2 for 10 h/day and 300 days/year. Assume that the cost of the array including support and storage material is $90 a square meter. Also determine the length of a dual-lane (four 12-ft lanes) highway, in miles, that could be constructed from the same area as the array.

1.20 Estimate the power required in the United States during A.D. 2030, in gigawatts, and the energy consumed, in gigajoules, from 1970 to 2030, if the consumption rate increases at an average rate of 3.3 percent per year. Start with the 1970 consumption rate.

1.21 Using the data in Table 1.2 determine the average rate of growth of energy consumption from 1950 to 1976.

1.22 If the capital cost of a 70-MW_e gas-turbine power system is $150 per kilowatt, find the cost of energy, in mils per kilowatthour, associated with the capital investment. Assume that the plant has a load factor of 35 percent and a life of 15 years with a salvage value of 35 percent. Also, assume that the initial investment could have been invested at 8 percent interest, compounded quarterly, and that the power payments can be invested quarterly at 6 percent compound (quarterly) interest. If the plant efficiency is 35 percent and the fuel costs are $2.60 per million kilojoules, find the unit power cost associated with the fuel and capital costs, in mils per kilowatthour.

1.23 A proposed solar heating system costs $6000 per home and lasts for 20 years. The system is financed by a loan with an annual compound interest rate of 10 percent. If the salvage value of the system is essentially zero, find the effective yearly heating cost, in dollars per year, if the yearly savings could be invested at 6 percent interest, compounded annually.

REFERENCES

Hammond, A. L.: Energy Options: Challenge for the Future, *Science*, vol. 177, pp. 875–876, 1972.

Hammond, A. L.: Geothermal Energy: An Emerging Major Resource, *Science*, vol. 177, pp. 978–980, 1972.

Hammond, A. L.: Solar Energy: The Largest Resource, *Science*, vol. 177, pp. 1088–1090, 1972.

Hollander, J. M., and M. K. Simmons (eds.): "Annual Review of Energy," vol. I, Annual Reviews, Inc., Palo Alto, Calif., 1976.

Hubbert, M. K.: *Scientific American*, vol. 224(3), p. 61, 1971.

Koenig, J. B.: Worldwide Status of Geothermal Resources Development, in P. Kruger and C. Otto (eds.), "Geothermal Energy," Stanford University Press, Stanford, Calif., 1973.

Krenz, J. H.: "Energy Conversion and Utilization," Allyn and Bacon, Inc., Boston, Mass., 1976.

Roberts, R.: Energy Sources and Conversion Techniques, *American Scientist*, vol. 61, pp. 66–75, 1973.

Shepard, M. L., F. H. Cocks, J. B. Chaddock, and C. M. Harmon: "Introduction to Energy Technology," Ann Arbor Science Publishers, Inc., Ann Arbor, Mich., 1976.

Summers, C. M.: *Scientific American*, vol. 224(3), p. 151, 1971.

PRINCIPAL FUELS FOR ENERGY CONVERSION

2.1 INTRODUCTION

This chapter reviews the properties and some conversion principles for those sources of energy currently used for fuel energy. Specifically, three general categories of fuels are studied—fossil fuels, nuclear fuels, and solar energy. The energy-conversion processes for the radioactive decay, the fusion, and solar energy systems are also presented in this chapter, but the fossil-fuel and nuclear fission systems are discussed in detail in later chapters.

2.2 FOSSIL FUELS

2.2.1 Background

The three general classes of fossil fuels are coal, oil, and natural gas. Other fuels, such as shale oil, tar-sand oil, and fossil-fuel derivatives are somewhat different, but they are still considered to be fossil fuels and are commonly lumped under one of the three main fossil-fuel categories.

All of the fossil fuels were produced from the fossilization of carbohydrate compounds. These compounds, with a chemical formula of $C_x(H_2O)_y$, were produced by living plants in the photosynthesis process as they converted direct solar energy into chemical energy. Most of the fossil fuels were produced during the Carboniferous Age of the Paleozoic Era of the earth, some 325 million years ago. After the plants died, the carbohydrates were converted by pressure and heat, in the absence of oxygen, into hydrocarbon compounds with a general chemical formula of C_xH_y. Since all fossil fuels are composed of hydrocarbon compounds, a general review of hydrocarbon chemistry would appear to be warranted.

2.2.2 Hydrocarbon Chemistry

Although hydrocarbon compounds are composed of only carbon and hydrogen, in some of the more complex molecules the same number of carbon and hydrogen atoms can be arranged in different structures to produce compounds with strikingly different chemical and physical properties. The hydrogen atom has only one electron and therefore needs to share an additional electron to fill the innermost or K shell. This means that hydrogen has a chemical valence of ± 1 and that it will share one bond with another atom in an organic molecule. Carbon atoms have six electrons and, as such, the inner K shell is completely filled with two electrons, leaving four electrons in the L shell. Since eight electrons are needed to fill the L shell, carbon has a chemical valence of ± 4 and will share four electrons or bonds with the other atoms in the molecule.

There are three major groups of hydrocarbon compounds—the aliphatic hydrocarbons, the alicyclic hydrocarbons, and the aromatic hydrocarbons. The aliphatic hydrocarbons are chain compounds and most of the fossil-fuel compounds fall into this major group. The other two major hydrocarbon groups are ring hydrocarbons. The terms "saturated" and "unsaturated" are sometimes applied to hydrocarbon compounds. Saturated hydrocarbons are those compounds in which there are only single bonds between any of the carbon atoms. Unsaturated hydrocarbons have at least two carbon atoms that share multiple bonds.

The aliphatic hydrocarbons or chain hydrocarbons are further divided into three subgroups—the alkane, the alkene, and the alkyne hydrocarbons. The alkane hydrocarbons, also called the paraffin series, are the saturated group of chain hydrocarbons. The general chemical formula for this group is C_nH_{2n+2}. Many of the common fuel compounds fall into this subgroup and some of the typical compounds are listed below with their chemical formulas:

Methane, CH_4 Pentane, C_5H_{12} Nonane, C_9H_{20}

Ethane, C_2H_6 Hexane, C_6H_{14} Decane, $C_{10}H_{22}$

Propane, C_3H_8 Heptane, C_7H_{16} $\vdots$

Butane, C_4H_{10} Octane, C_8H_{18} Hexadecane, $C_{16}H_{34}$

Some of the compounds listed above are readily recognizable as the prime components of common fuels. Methane and ethane comprise most of natural gas, propane and butane make up liquified petroleum gas, and octane is a common compound in gasoline. As the number of carbon atoms in the alkane molecule increases, the hydrogen fraction decreases and the hydrocarbon becomes less volatile. The first four compounds are gases at room temperature and pressure and the balance of those listed are liquids. Some longer-chained molecules are solids.

As indicated earlier, the exact structure of the hydrocarbon molecule strongly influences its chemical and physical properties. If the prefix "*n*," which stands for normal, appears in front of the name of the hydrocarbon, it means that all the carbon atoms are connected in one long chain. The prefix "iso" in front of the hydrocarbon name means that there are carbon-atom branches, usually methyl groups (CH_3^-), connected to the main chain. The *n*-octane and isooctane molecules are shown below:

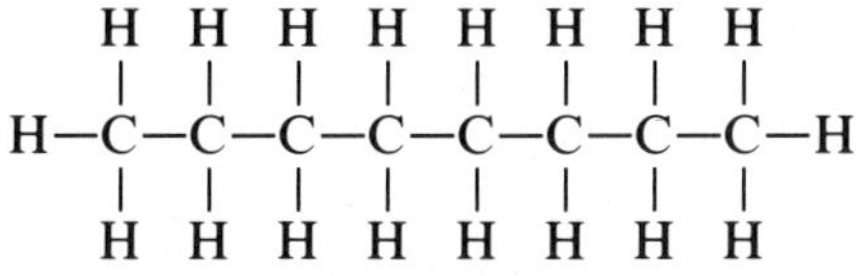

n-Octane

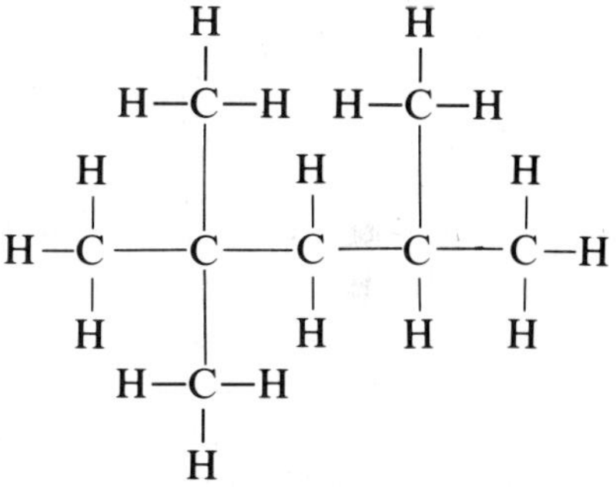

An isooctane
(2,2,4-trimethylpentane)

The isooctane shown here is called 2,2,4-trimethylpentane because there are three methyl groups (trimethyl) attached to the basic pentane chain (pentane) at the second (2), second (2), and fourth (4) carbon-atom positions. Both of these molecules have the same basic chemical formula, C_8H_{18}, but they have very different chemical and physical properties.

The alkene and alkyne subgroups of aliphatic hydrocarbons are unsaturated hydrocarbon compounds. The alkene hydrocarbons, also called the olefin series, have one double carbon-atom bond in the chain. The general formula for this group is C_nH_{2n}, and some typical compounds are ethylene (C_2H_4), propylene (C_3H_6), butene (C_4H_8), pentene (C_5H_{10}), and hexene (C_6H_{12}). The alkyne hydrocarbons, also called the acetylene series, have one triple carbon-atom bond in the hydrocarbon chain. The general formula for this group is $C_nH_{2(n-1)}$, and some typical compounds are acetylene (C_2H_2) and ethylacetylene (C_4H_6). Two unsaturated aliphatic hydrocarbons are shown below:

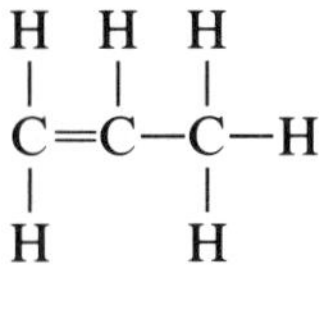

Propylene

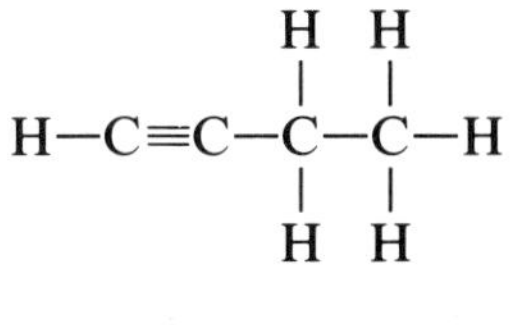

Ethylacetylene

The other two major hydrocarbon groups, the alicyclic and aromatic compounds, are called "ring" hydrocarbons because the molecules are composed of carbon-atom rings. The alicyclic hydrocarbons are saturated rings and have a general chemical formula which is the same as that for the alkene group of aliphatic hydrocarbons, C_nH_{2n}. The names are simply the alkane group names preceded by the prefix "cyclo." Thus, there are cyclopropane (C_3H_6), cyclobutane (C_4H_8), cyclopentane (C_5H_{10}), etc. A typical alicyclic compound is shown as follows:

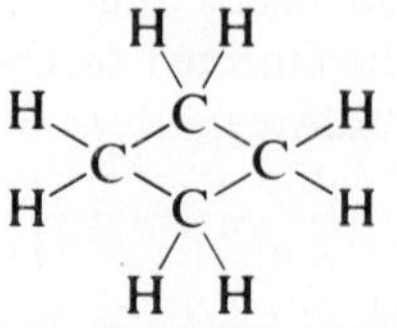

Cyclobutane

The aromatic hydrocarbons are composed of the basic "benzene" ring or rings. This ring is a six-atom carbon ring with double bonds between every other carbon atom (three double bonds in the simple ring). The general formula for this group is C_nH_{2n-6} for single-ringed molecules and C_nH_{2n-12} for the double-ringed molecules. Some typical aromatic compounds are benzene (C_6H_6), toluene (C_7H_8), xylene (C_8H_{10}), and naphthalene ($C_{10}H_8$). These compounds are constructed by adding methyl groups to the basic ring or rings. The basic rings are shown below:

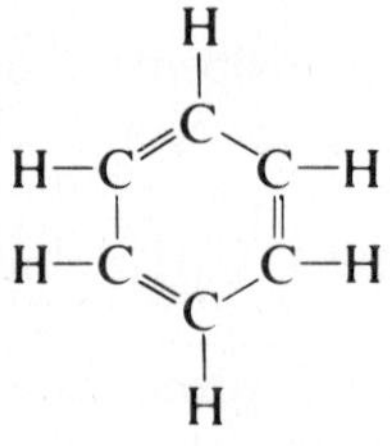

Benzene 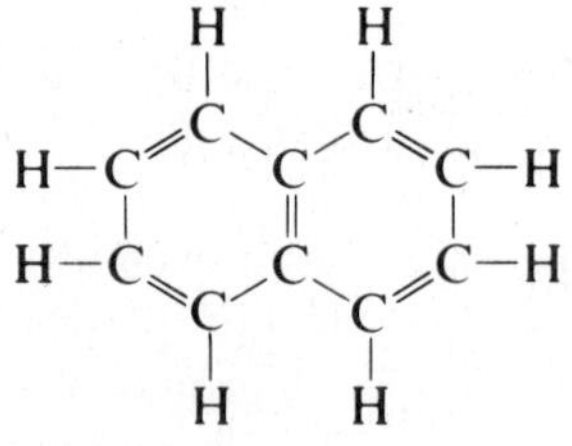   Naphthalene

2.2.3 Standard Fuels

There are a number of basic hydrocarbon compounds that are used as fuel standards for internal-combustion engines. The spark-ignition, internal-combustion engine fuels are rated according to the octane number. The compression-ignition, internal-combustion engine fuels are rated according to the cetane number.

The 100-octane fuel standard is 2,2,4-trimethylpentane (the isooctane shown previously), while the 0-octane fuel standard is n-heptane. The octane number of an unknown fuel is determined in a cooperative fuels research engine (CFR engine). This engine is a single-cylinder engine with a compression ratio that can be adjusted from about 4 to 1 to about 14 to 1. The unknown fuel is burned in the engine and the compression ratio is slowly increased until a certain "knock" or detonation reading is obtained from a vibration detector. Blends of the standard fuels are then burned at the same compression ratio until approximately the same "knock" reading is obtained. The percentage by volume of 100-octane fuel in the blend is the octane number of the unknown fuel. The octane ratings of most "regular" gasolines range from 85 to 95.

Some premium gasolines have octane numbers greater than 100. Octane numbers in excess of 100 can be achieved by using lighter hydrocarbons and/or by putting additives, such as tetraethyl lead (TEL), in the basic fuel. These fuels are sometimes rated according to the "performance number" instead of octane number. A fuel with a performance number of 5 has the same combustion characteristics as a mixture of 1 gal of 100 octane and 5 cm^3 of tetraethyl lead.

The 100-cetane standard for diesel fuel is n-hexadecane ($C_{16}H_{34}$), sometimes called n-cetane. The 0-cetane standard is alpha-methylnaphthalene ($C_{11}H_{10}$), which is similar to the naphthalene molecule shown on the previous page except that one of the hydrogen atoms in the alpha position (one of the two top or bottom atoms in the figure) is replaced with a methyl group. The cetane rating of an unknown diesel fuel is equal to the percentage by volume of n-hexadecane in a mixture of the standard fuels that has the same combustion characteristics in a CFR diesel engine as that of the unknown fuel. The cetane ratings of most diesel-engine fuels range between 30 and 60.

2.2.4 Coal

Coal composition and rank Coal, the most abundant fossil fuel, is thought to be fossilized vegetation. It is estimated that at least 20 ft of compacted vegetation was necessary to produce a 1-ft-thick seam of coal. This compacted vegetation, in the absence of air and under the influence of high pressure and temperature, is subsequently converted into peat, a very low-grade fuel, then into brown coal, then into lignite, then into subbituminous coal, then into bituminous coal, and finally into anthracitic coal. As the aging process progresses, the coal becomes harder, the hydrogen and oxygen content decrease, the moisture content usually decreases, and the carbon content increases, as indicated in Fig. 2.1.

Coal is normally found in seams in the earth's crust. The average seam thickness in the United States is around 1.65 m, although there is a 36-m seam in Lincoln County, Wyoming. A 130-m seam of coal was found in Manchuria.

There are several different systems for classifying coal but the American Society for Testing Materials (ASTM) has a classification method that ranks coal into four major categories with subdivisions in each class. According to this system, the four major classes are, starting with the oldest, anthracitic coals, bituminous coals, subbituminous coals, and lignitic coals. Under this classification system, the coals are ranked according to certain properties. The ASTM classification system and the classification parameters are presented in Table 2.1.

Table 2.1 Classification of coals by rank[a] (ASTM D 388)

Class	Group	Fixed carbon limits, % (dry, mineral-matter-free basis)		Volatile matter limits, % (dry, mineral-matter-free basis)		Calorific value limits, Btu/lb (moist,[b] mineral-matter-free basis)		Agglomerating character
		Equal or greater than	Less than	Greater than	Equal or less than	Equal or greater than	Less than	
I. Anthracitic	1. Metaanthracite	98	......	...	2	......	......	Nonagglomerating
	2. Anthracite	92	98	2	8	......	......	
	3. Semianthracite[c]	86	92	8	14	......	......	
II. Bituminous	1. Low-volatile bituminous coal	78	86	14	22	......	......	
	2. Medium-volatile bituminous coal	69	78	22	31	......	......	
	3. High-volatile A bituminous coal	...	69	31	...	14,000[d]	......	Commonly agglomerating[e]
	4. High-volatile B bituminous coal	...	...	...	...	13,000[d]	14,000	
	5. High-volatile C bituminous coal	...	...	...	...	11,500	13,000	
						10,500[e]	11,500	Agglomerating
III. Subbituminous	1. Subbituminous A coal	...	...	...	...	10,500	11,500	
	2. Subbituminous B coal	...	...	...	...	9,500	10,500	
	3. Subbituminous C coal	...	...	...	...	8,300	9,500	Nonagglomerating
IV. Lignitic	1. Lignite A	...	...	...	...	6,300	8,300	
	2. Lignite B	...	...	...	...	......	6,300	

[a] This classification does not include a few coals, principally nonbanded varieties, which have unusual physical and chemical properties and which come within the limits of fixed carbon or calorific value of the high-volatile bituminous and subbituminous ranks. All of these coals either contain less than 48 percent dry, mineral-matter-free fixed carbon or have more than 15,500 Btu/lb on the moist, mineral-matter-free basis.

[b] Moist refers to coal containing its natural inherent moisture but not including visible water on the surface of the coal.

[c] If agglomerating, classify in low-volatile group of the bituminous class.

[d] Coals having 69 percent or more fixed carbon on the dry, mineral-matter-free basis shall be classified according to fixed carbon, regardless of calorific value.

[e] It is recognized that there may be nonagglomerating varieties in these groups of the bituminous class, and there are notable exceptions in the high-volatile C bituminous group.

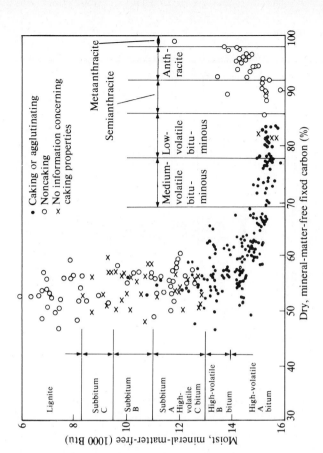

$$\text{Dry, Mm-free FC} = \frac{\text{FC} - 0.15S}{100 - (M + 1.08A + 0.55S)} \times 100 \quad \%$$

Dry, Mm-free VM = 100 − dry, Mm-free FC %

$$\text{Moist, Mm-free Btu} = \frac{\text{Btu} - 50S}{100 - (1.08A + 0.55S)} \times 100 \quad \text{per lb}$$

where Mm = mineral matter
    Btu = heating value per lb
    FC = fixed carbon, %
    VM = volatile matter, %
    M = bed moisture, %
    A = ash, %
    S = sulfur, %

All are for coal on a moist basis.

39

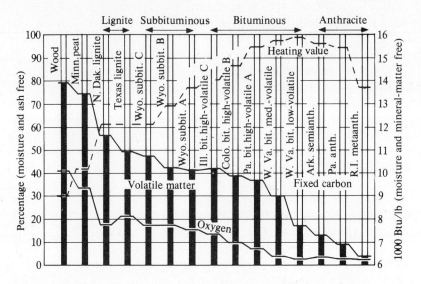

Figure 2.1 Variation of coal composition and properties with age. *(De Lorenzi, 1952.)*

Coal analyses The two basic coal analyses are the proximate analysis and the ultimate analysis. Both of these analyses give mass or gravimetric fractions of the components in the coal and either analysis can be reported in several different ways. In any coal seam, there are two components that can show a significant variation throughout the seam. These components are moisture and ash. The ash fraction varies because ash is essentially the inorganic matter deposited with the organic material during the compaction process. The moisture content of the coal varies significantly, depending on the exposure to groundwater before mining and upon exposure to the weather during transportation and storage before it is burned.

Since the moisture and ash fractions can vary widely for a given coal, it is common practice to report coal analyses (both ultimate and proximate) on a moisture-free (dry), ash-free basis. However, for combustion and coal-handling calculations, these analyses must be converted to an as-burned or as-received basis, which includes both the moisture and ash fractions in the coal.

The proximate analysis is the simplest coal analysis and gives the mass fractions of fixed carbon (FC), volatile matter (VM), moisture (M), and ash (A) in the coal. This analysis can be determined by simply weighing, heating, and burning a small sample of the coal. A powdered coal sample is carefully weighed and then heated to 110°C (230°F) for 20 min. The sample is then weighed again and the mass loss divided by the original mass gives the mass fraction of moisture in the sample. The remaining sample is then heated to 954°C (1750°F) in a closed container for 7 min, after which the sample is reweighed. The resulting mass loss divided by the original mass is equal to the mass fraction of volatile matter in the sample. The sample is then heated to 732°C (1350°F) in an open crucible until it is completely burned. The residue is then weighed and the final weight divided by the original weight is the ash fraction. The mass fraction of fixed carbon is obtained by subtracting the moisture, volatile matter, and ash fractions from unity. In addition to the FC, the VM, the M, and the A, most proximate analyses list separately the sulfur mass fraction (S) and the higher heating value (HHV) of the coal.

The ultimate coal analysis is a laboratory analysis that lists the mass fractions of carbon (C), hydrogen (H_2), oxygen (O_2), sulfur (S), and nitrogen (N_2) in the coal along with the higher heating value (HHV). Most ultimate analyses list the moisture M and ash A separately, but some analyses include the moisture as part of the hydrogen and oxygen mass fractions. The ultimate analysis is required to determine the combustion-air requirements for a given combustion system and this, in turn, is used to size the draft system for the furnace. These calculations should be based on the as-burned ultimate coal analysis, if possible.

Coal analyses for some typical American coals are presented in App. C, along with the higher heating values. All of these analyses are reported on a moisture-free, ash-free basis along with the expected range of moisture and ash fractions of the as-mined coal. Once the moisture and ash fractions have been determined, the other mass fractions and the higher heating value of the coal can be determined from the following equations:

As-burned mass fraction = (dry, ash-free mass fraction)$(1 - M - A)$ (2.1)

As-burned higher heating value = (dry, ash-free HHV)$(1 - M - A)$ (2.2)

As was pointed out earlier, the ASTM coal classification system depends on the coal properties—specifically, the dry, mineral-matter-free fixed carbon (or volatile matter) and/or the moist, mineral-matter-free higher heating value (Btu). The appropriate formulas for determining these values are presented in Table 2.1, along with the classification parameters. When using these equations, the as-mined mass percentages should be used instead of mass fractions.

The ultimate coal analysis is much harder to obtain than the proximate analysis. Since the ultimate analysis is normally of greater interest to the power engineer, it is sometimes necessary to estimate the ultimate analysis from an experimentally determined proximate analysis. A method of making such an estimate is presented in a Power's data sheet (a McGraw-Hill publication), presented in App. D.

Coal properties There are a number of coal properties that should be considered when selecting a coal for a given application. Among these properties are the sulfur content, the burning characteristics of the coal, the weatherability of the coal, the ash-softening temperature, the grindability of the coal, and the energy content of the coal.

One very important consideration in the selection of a coal is the sulfur content. While sulfur is one of the combustible elements in the coal and generates energy, the primary combustion product, sulfur dioxide (SO_2), is a major atmospheric pollutant. It is difficult and expensive to either remove the sulfur before the coal is burned or to remove the sulfur dioxide from the combustion products. Consequently, it is important that the coal have a low sulfur content, at least 1 percent or less.

When choosing a coal for a particular combustion system, one must be concerned as to how the coal burns. If the coal is burned in a stationary bed with little agitation (e.g., a chain-grate stoker furnace), the coal should be a free-burning coal, not a caking coal. A free-burning coal tends to break apart as it burns thereby exposing the unburned coal to the combustion air. This facilitates the complete combustion of the coal. A caking coal produces a fused coal mass as it burns so that much of the fixed carbon is not burned. These coals are commonly employed in the production of coke and if they are to burn effectively, the coal bed must be mechanically agitated to break up the fused coal masses. A high value of the free-swelling index for a given coal normally indicates that the coal is a free-burning coal.

The weatherability of a coal is a measure of the ability of a coal to withstand exposure to the environmental elements without excessive crumbling. All large coal-fired power stations normally store coal reserves in a large pile near the power plant. As the coal is received by unit train or barge, it is spread in thin layers and packed down with large earth-moving machines to get as much air out of the pile as possible. This reduces the risk of a fire starting by spontaneous combustion. A typical coal pile is shown in Fig. 2.2. If the coal crumbles severely

Figure 2.2 A typical coal pile containing about 200,000 tons of coal. *(From "Steam/Its Generation and Use," 1972.)*

during storage, the small particles will be washed away during rainstorms, producing a monetary and energy loss as well as polluting the watershed.

Another important property that should be considered when picking a coal for a power plant is the grindability index of the coal. This is particularly true for the common pulverized-coal power systems where the coal is ground into a powder finer than face powder. The grindability index is inversely proportional to the power required to grind the coal to a certain fineness. Thus, if coal A has a grindability index of 100 while coal B has an index of 50, coal B will require twice the grinding power of coal A. An index of 100 has been arbitrarily assigned to a low-volatile bituminous (class II, group 1) coal. The anthracitic coals, because of their hardness, and the lignitic coals, because of their plasticity, have low-grindability indices.

The ash-softening temperature is an important consideration in the choice of coals for a particular power plant. The ash-softening temperature is the temperature where the ash becomes very plastic, somewhat below the melting point of the ash. This temperature is determined by heating ash cones similar to those shown in Fig. 2.3. Some furnaces discharge the ash from the firebox in the form of a molten slag and a coal with a low ash-softening temperature is desirable for these systems. A coal with a high ash-softening temperature is desirable for those systems that handle the ash as a solid. If a coal with a low ash-softening temperature is used in a stoker furnace, large troublesome fused ash masses, called clinkers, may be produced.

The energy content or heating value of a coal is a very important property. The heating value represents the amount of chemical energy in a given mass or volume of fuel. In most American textbooks and periodicals, it has been common practice to report the heating values of coal in units of British thermal units per pound-mass (Btu/lbm), but this text uses the units of kilojoules per kilogram (kJ/kg).

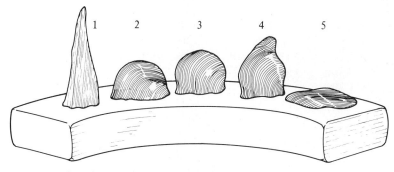

Cone 1, initial deformation temperature
Cones 2 and 3, softening temperature
Cone 4 has almost reached softening temperature
Cone 5, fluid temperature

Figure 2.3 Fusibility of ash cones in determining the ash-softening temperature. (*De Lorenzi, 1952.*)

Reference has been made in previous pages to the higher heating value or HHV. There are actually two heating values—a higher or gross heating value and a lower or net heating value. The difference between these values is essentially the latent heat of vaporization of the water vapor present in the exhaust products when the fuel is burned with dry air. In an actual combustion system, this includes the water present in the as-burned fuel (the moisture) and the water produced from the combustion of hydrogen, but it does not include any moisture that is introduced by the combustion air. Since the latent heat of vaporization of water at 1 lb/in^2 abs (the approximate partial pressure of the water vapor in the exhaust) is around 2400 kJ/kg, the difference between the higher and lower heating values is approximated by the following equation which is applicable to any fuel on a mass basis:

$$\text{HHV} - \text{LHV} = 2400(M + 9H_2) \quad \text{kJ/kg} \tag{2.3}$$

In Eq. (2.3), M and H_2 are the moisture and hydrogen mass fractions of the fuel.

Most combustion tables will always list the higher heating values for coal because the moisture content varies so widely and this would change the lower heating value for each moisture value. The higher heating values, on a dry, ash-free basis, are listed in App. C. These values should be converted to an as-burned basis, as given by Eq. (2.2) when determining the fuel rate for a given power plant. If an experimental value of the higher heating value is unavailable, the higher heating value, in kilojoules per kilogram, for better grades of coal can be estimated from the ultimate analysis and the use of Dulong's formula:

$$\text{HHV} = 33{,}950\text{C} + 144{,}200\left(\text{H}_2 - \frac{\text{O}_2}{8}\right) + 9400\text{S} \quad \text{kJ/kg} \tag{2.4}$$

All values in the above equation are mass fractions and the resulting higher heating value is for the same basis (dry, ash-free, as-burned, etc.) as the ultimate analysis that is used in the equation.

Example 2.1 Determine the as-received proximate and ultimate analyses, the estimated lower heating value from the listed higher heating value, the higher heating value as determined by Dulong's formula, and determine the ASTM classification (group and class) of Stark County, North Dakota, with $A = 8$ and $M = 39$.

SOLUTION Given: Stark County coal with $M = 39$ percent and $A = 8$ percent.
From App. C,
Dry, ash-free proximate coal analysis:

VM = 54.0%	S = 2.8%
FC = 46.0%	

100.0%　　HHV = 28,922 kJ/kg = 12,435 Btu/lbm

Dry, ash-free ultimate analysis:

$$C = 72.4\%, \; H_2 = 4.7\%, \; O_2 = 18.6\%, \; N_2 = 1.5\%, \; S = 2.8\%$$

To convert to an as-received coal basis, the correction factor or multiplier is $(1 - M - A) = (1.00 - 0.39 - 0.08) = 0.53$. The as-received proximate analysis becomes

VM = 0.53 × 54.0 = 28.62%		S = 0.53 × 2.8 = 1.48%
FC = 0.53 × 46.0 = 24.38%		
M	= 39.00%	HHV = 0.53 × 28,922 = 15,329 kJ/kg
A	= 8.00%	= 0.53 × 12,435 = 6,591 Btu/lbm

$$\overline{ 100.00\%}$$

The as-received ultimate analysis becomes

$$
\begin{aligned}
C \; &= 0.53 \times 72.4 = \; 38.37\% \\
H_2 &= 0.53 \times 4.7 = \; 2.49\% \qquad \text{HHV} = 15,329 \text{ kJ/kg}\\
O_2 &= 0.53 \times 18.6 = \; 9.86\% \qquad \phantom{\text{HHV}} = 6,591 \text{ Btu/lbm}\\
N_2 &= 0.53 \times 1.5 = \; 0.80\% \\
S \; &= 0.53 \times 2.8 = \; 1.48\% \\
M \; & = \; 39.00\% \\
A \; & = \; 8.00\%
\end{aligned}
$$

$$\overline{100.00\%}$$

Lower heating value:

$$LHV = HHV - 2400(M + 9H_2) = 15,329 - 2400(0.39 + 9 \times 0.0249)$$

$$= 13,855 \text{ kJ/kg} = 5957 \text{ Btu/lbm}$$

Estimated value of the higher heating value using Dulong's formula:

$$HHV = 33,950C + 144,200\left(H_2 - \frac{O_2}{8}\right) + 9400S$$

$$= 33,950(0.3837) + 144,200(0.0249 - 0.0986/8)$$

$$+ 9400(0.0148)$$

HHV (Dulong's formula) $= 14,979$ kJ/kg $= 6440$ Btu/lbm

Coal classification according to the ASTM method:

$$\text{Dry, Mm-free FC} = \frac{100(FC - 0.15S)}{100 - M - 1.08A - 0.55S}$$

$$= \frac{100(24.38 - 0.15 \times 1.48)}{100 - 39.0 - 1.08 \times 8.0 - 0.55 \times 1.48}$$

$$= 46.87\%$$

Since this value is less than 69 percent, the coal cannot be classed according to the dry, mineral-matter-free value of fixed carbon.

$$\text{Moist, Mm-free Btu} = \frac{100(\text{Btu} - 50\text{S})}{100 - 1.08A - 0.55\text{S}}$$

$$= \frac{100(6591 - 50 \times 1.48)}{100 - 1.08 \times 8.0 - 0.55 \times 1.48}$$

$$= 7197 \text{ Btu/lbm}$$

Since the moist, mineral-matter-free higher heating value is between 6300 and 8300 Btu/lbm, this coal is ranked as a class IV, group 1 coal—lignite A.

2.2.5 Petroleum

Formation and classification Whereas coal is thought to be fossilized vegetation, petroleum is thought to be partially decomposed marine life. Petroleum or crude oil is normally found in large domes of porous rock. Crude oils are normally ranked into three categories, depending on the type of residue left after the lighter fractions have been distilled from the crude. Under this system, the petroleum is classified as paraffin-based crudes, as asphalt-based crudes, or as mixed-based crudes.

Although crude oil is a composition of many organic compounds, the ultimate analyses of all crudes are fairly constant. The carbon mass fraction ranges from 84 to 87 percent, the hydrogen mass fraction ranges from 11 to 16 percent, the sum of the oxygen and nitrogen mass fractions range from 0 to 7 percent, and the sulfur mass fraction ranges from 0 to 4 percent.

Shale oil is not exactly the same as petroleum but it is composed of an oil-like compound, called kerogen. The United States has tremendous reserves of shale oil in the mountains of Colorado and Wyoming. One ton of high-grade oil shale yields about 26 gal of shale oil. Kerogen can be used to produce essentially the same products that can be obtained from crude oil. The principal problem associated with the development of this energy source is the economical recovery of the oil with minimum environmental impact. Unfortunately, if the shale is mined and the crushed ore is heated to distill the oil, the residue expands so that the residue cannot be put back in the same excavation. Some companies are working on an "in-situ" process in which the oil can be recovered from the ore without mining it. Development of this huge energy source presents a major technological challenge.

There are six commercial grades of fuel oil, although fuel oil No. 3 is no longer commercially available. Fuel oil No. 1 is the lightest, least viscous fuel oil which is made for vaporizing burners. Fuel oil No. 2 is the general-purpose domestic heating oil. Fuel oil No. 4 is a relatively light commercial-grade heating oil and is the heaviest grade of fuel oil that can be pumped without heating at moderate temperatures. Fuel oil No. 5 is a heavy, viscous, commercial-grade fuel oil and fuel oil No. 6 or "bunker-C" oil is the heaviest and most viscous of the fuel oils. Both No. 5 and No. 6 fuel oils require heating before they can be pumped. The average properties of these fuel oils are given in App. E.

Properties of petroleum products The important properties of petroleum and petroleum derivatives are the heating value, the specific gravity, the flash point, and the pour point. The heating value, usually the higher heating value, is reported in units of either kilojoules per kilogram (or British thermal units per pound-mass) or kilojoules per liter (or British thermal units per gallon). The heating value of petroleum and petroleum products is given as a function of the specific gravity of the product in Fig. 2.4. The heating value, on a unit mass basis, of petroleum derivatives increases as the specific gravity of the product decreases or as the °API (API stands for the American Petroleum Institute) and °Bé (°Baumé) increase.

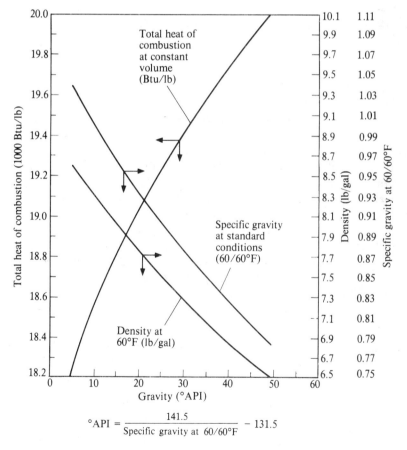

$$°API = \frac{141.5}{\text{Specific gravity at } 60/60°F} - 131.5$$

Figure 2.4 Properties of petroleum derivatives. (*From "Steam/Its Generation and Use," 1972.*)

The specific gravity of any liquid is the density of the liquid divided by the density of water at 60°F (15.6°C). The specific gravity of petroleum and petroleum products is usually expressed in units of °Bé or °API. The relationship between the specific gravity s and these units are as follows:

$$\text{Specific gravity} = s = \frac{140}{130 + °\text{Bé}} \tag{2.5}$$

$$\text{Specific gravity} = s = \frac{141.5}{131.5 + °\text{API}} \tag{2.6}$$

It will be noted in both of these equations that water has a specific gravity of 10° and as the °Bé and °API increase the specific gravity of the liquid decreases.

The flash point of a liquid fuel is the minimum fluid temperature at which the vapors coming from the fluid surface will just ignite. At a slightly higher temperature, called the fire point, the vapors will support combustion. Care should be taken to assure that the maximum oil temperature does not exceed the flash point of the product.

The pour point of a petroleum product is the lowest temperature at which an oil or oil product will flow under standard conditions. It is determined by finding the maximum temperature at which the surface of an oil sample in a standard test tube does not move for 5 s when the test tube is rotated to the horizontal position. The pour point is equal to this temperature plus five Fahrenheit degrees.

Oil has a number of advantages over coal when it is burned. Oil is cleaner and easier to handle, store, and transport. It is easier to burn than coal and has little ash. There are some problems encountered in the combustion of fuel oil, however. Although there is little ash produced, it is difficult to remove. Some crudes have a significant sulfur fraction and although it can be removed from the oil before it is burned, it is an expensive process. Another problem element found in some crudes is vanadium. Vanadium oxidizes during combustion to vanadium pentaoxide (VaO_5) and this compound causes rapid corrosion of ferrous materials found in most boilers.

2.2.6 Gaseous Fuels

General Almost all gaseous fuels are either fossil fuels or byproducts of fossil fuels. These fuels can be divided into three general groups including natural gas, manufactured gas, and byproduct gas. Some of the compositions and other properties of gaseous fuels are presented in App. F.

The composition of fuel gases is commonly expressed in terms of the mole or volume fractions of the gaseous components. The analysis can also be expressed in terms of the elemental mass fractions. For mixtures of ideal gases, the molar and volume fractions are the same. The heating value of any fuel gas is commonly expressed in units of energy per unit volume, such as kilojoules per cubic meter or British thermal units per cubic foot, but this value is directly proportional to the absolute pressure and inversely proportional to the absolute temperature. The heating value can also be expressed in terms of energy per unit mass (kilojoules per kilogram) and this quantity is independent of the pressure and temperature.

The volumetric heating value of a mixture of fuel gases is equal to the sum of the products of the individual component volume or mole fraction and the corresponding volumetric heating value of the component. If the volumetric heating values of the gas components at some reference temperature, T_r, and reference pressure, P_r, are known, the volumetric heating value of the gas mixture, HHV_v, is obtained from the following equation:

$$(HHV_v \text{ of mixture})_{P_r, T_r} = \sum_i (HHV_{v, i})_{P_r, T_r}(V_i) \tag{2.7}$$

where $HHV_{v, i}$ and V_i are the volumetric higher heating value and the volumetric fraction of the ith gaseous component, respectively. The higher heating values of a number of combustible compounds are tabulated in App. G. The following equation can be used to convert the volumetric higher heating value at the reference pressure and temperature to some other pressure and temperature:

$$(HHV_v)_{P, T} = (HHV_v)_{P_r, T_r} \frac{P}{P_r} \frac{T_r}{T} \tag{2.8}$$

The pressures and temperatures in Eq. (2.8) must be absolute values.

A volumetric heating value HHV_v at some temperature T and pressure P can be converted to a gravimetric heating value HHV_m by multiplying the volumetric value by the specific volume v of the gas at the same pressure and temperature:

$$HHV_m = (HHV_v)_{P, T}(v)_{P, T} \tag{2.9}$$

The specific volume of the gas mixture can be determined from the molecular weight of the gas (MW) and the ideal-gas equation of state, as follows:

$$v = \frac{V}{m} = \frac{RT}{P} = \frac{R_u T}{P(MW)} \tag{2.10}$$

where R_u is the universal gas constant.

Example 2.2 Calculate the higher heating value (kilojoules per cubic meter and kilojoules per kilogram) at 10°C and three atmospheres for a gas mixture with the following composition: 94.3% CH_4, 4.2% C_2H_6, and 1.5% CO_2.

SOLUTION Given: Mole fractions of the gas components.
At 20°C and 1 atm (from App. G):

$$(HHV_v)_{CH_4} = 37,204 \text{ kJ/m}^3 \qquad (HHV_v)_{C_2H_6} = 65,727 \text{ kJ/m}^3$$

$$(HHV_v)_{CO_2} = 0$$

Molecular weight of gas mixture $= 0.943(16) + 0.042(30) + 0.015(44)$

$$= 17.01 \text{ kg/kg·mol}$$

At 20°C and 1 atm:

$$(HHV_v)_{mixture} = 0.943(HHV_v)_{CH4} + 0.042(HHV_v)_{C2H6}$$
$$+ 0.015(HHV_v)_{CO2}$$
$$= 0.943(37,204) + 0.042(65,727) = 37,844 \text{ kJ/m}^3$$

At 10°C and 3 atm:

$$(HHV_v)_{10°C, \, 3 \text{ atm}} = 37,844 \frac{P \, T_r}{P_r \, T} = 37,844(3/1)(293/283)$$
$$= 117,544 \text{ kJ/m}^3$$

Specific volume of gas mixture $= v = \dfrac{R_u T}{P(MW)}$

where $P = 3 \text{ atm} = 1.013 \text{ bar/atm} (3 \text{ atm}) = 3.039 \text{ bar}$
$T = 10 + 273 = 283 \text{ K}$
$R_u = 0.08315 \text{ bar} \cdot \text{m}^3/(\text{kg} \cdot \text{mol}) \text{ (K)}$
$v = (0.08315 \text{ bar} \cdot \text{m}^3/\text{kg} \cdot \text{mol} \cdot \text{K})(283 \text{ K})/(3.039 \text{ bar})$
$\quad \times (17.01 \text{ kg/kg} \cdot \text{mol})$
$v = 0.4552 \text{ m}^3/\text{kg}$

$$HHV_m = v(HHV_v) = 0.4552 \text{ m}^3/\text{kg}(117,544 \text{ kJ/m}^3) = 53,506 \text{ kJ/kg}$$
$$= 23,008 \text{ Btu/lbm}$$

Typical fuel gases Natural gas is the only true fossil-fuel gas and is usually trapped in limestone casings on the top of petroleum reservoirs. Reservoir pressures may run as high as 350 to 700 bar (5000 to 10,000 lb/in²). Natural gas is primarily composed of methane with small fractions of other gases. The composition of some typical natural gases is presented in App. F.

Natural gas has the highest gravimetric heating value of all fossil fuels, about 55,800 kJ/kg or 24,000 Btu/lbm. The volumetric heating value of natural gas is about 37,000 kJ/m³ or 1000 Btu/ft³ at 1 atm and 20°C (68°F). Natural gas is commonly sold in units of "therms," where 1 "therm" is equal to 100,000 Btu. At the present time, interstate natural-gas rates are regulated at a fixed low level, which, according to some people, discourages gas production. Of the three principal types of fossil fuels, natural gas has the lowest fuel reserves.

The combustion of natural gas has several advantages compared to the combustion of oil and coal. It is probably the easiest fuel to burn and it mixes well with air. It burns cleanly with little ash. Natural gas can be easily and cheaply transported in pipelines and overseas gas is sometimes converted to liquified natural gas (LNG) at −127°C and shipped in cryogenic tankers. If there is any disadvantage associated with the use of natural gas as a source of fuel energy, it would be that it is difficult to store large quantities of energy in the form of natural gas. Some gas companies are actually injecting high-pressure gas into large underground cavities including domed, sealed aquefers. As the gas is pumped into an aquefer, it displaces the groundwater.

There are a number of manufactured fuel gases, including liquified petroleum gas (LPG), water gas, carbureted water gas, synthetic or substitute natural gas (SNG), and producer gas. There are a number of other fuel gases that are produced as a byproduct of some other process. These gases include coke-oven gas, sewage gas, and blast-furnace gas. The composition and properties of some of these gases are presented in App. F.

Liquified petroleum gas, sometimes called refinery gas or simply LPG, is composed of the light distillates of petroleum—primarily propane and butane. Because of the higher molecular weight and density of this gas, it has a higher volumetric heating value than natural gas. Actually, LPG is heavier than air which probably makes it more dangerous to handle than natural gas. This fuel gas is usually transported and stored under pressures ranging from 4 to 20 bars, depending on the atmospheric temperature.

Water gas is a manufactured fuel gas that is produced by alternately passing steam and air through a bed of incandescent coke. The steam reacts with the hot coke to produce hydrogen and carbon monoxide. Oil vapors are sometimes added to the water gas to improve the heating value of the product gas. The resulting fuel gas is called carbureted water gas.

There are many proposed processes for producing a "high-Btu-fuel gas" from coal. Such a fuel gas is commonly called synthetic natural gas, substitute natural gas, or simply SNG. Theoretically, these processes will permit the utilization of large quantities of high-sulfur coal by converting most of the energy into a cheap, clean fuel gas. Solid hydrocarbon compounds, such as those found in coal, have a low hydrogen-to-carbon ratio compared to the gaseous hydrocarbons. Thus, any scheme for the conversion of coal into gas requires the addition of hydrogen to the coal.

In the hydrogenation process, pressurized hydrogen at 900°C (1650°F) is reacted with the coal to produce a number of light hydrocarbon compounds, particularly methane. This process, as is the case with any SNG process, is not presently economically competitive with the other fuel gases. Some of the various SNG conversion processes under development along with some "low-Btu-fuel gases" are listed in Table 2.2.

Producer gas is a fuel gas formed by burning low-grade coal seams in the ground or in situ, with insufficient air for complete combustion. Only enough air is added to maintain the bed at a high enough temperature to drive off some of the hydrogen and to oxidize some of the carbon to carbon monoxide. While the resulting fuel gas is a low-quality fuel gas, it does permit the utilization of thin-seam, low-grade coals that cannot be economically mined. This method has been extensively used in the U.S.S.R. and has been tried on an experimental basis in the United States.

Blast-furnace gas is a low-quality fuel gas that is a byproduct of the steel industry. It is produced by burning coal with insufficient air. The resulting exhaust gas is used to provide a reducing atmosphere over the molten metal to inhibit oxidation of the melt. Although the gas has a heating value which is only one-tenth that of natural gas, there is so much of this gas generated in the

Table 2.2 Characteristics of coal-to-SNG processes

Process	Type of process†	Chemical reactor parameters		Status
		Temp, °C	Pressure, bars abs	
Koppers-Totzek	G	1480–1820	1+	Commercial
Wellman-Galusha	G	540–650	1+	Commercial
Winkler	G	820–1010	1+	Commercial
Texaco	G	1480–1820	70	Pilot plant
Ash agglomerating	G	870–980	16	Pilot plant
COGAS	G	870–930	1+	Pilot plant
Lurgi	G/HD	620–870	30	Commercial
Hygas	G/H	650–980	75	Pilot plant
CO₂ acceptor	G/HD	860–1040	10	Pilot plant
BI-GAS	G/HD	1650–930	70	Pilot plant
Synthane	G/H	980	70	Pilot plant

† *G*—gasification process; *H*—hydrogasification process; HD—hydrodevolatilization process.

In the gasification process, the coal is reacted with steam to produce a gas that can subsequently be converted to methane. In the hydrogasification process, high-pressure hydrogen is added to the coal for the direct production of methane gas.

operation of a blast furnace that it is economically feasible to recover and burn it. Blast-furnace gas is composed primarily of nitrogen, carbon monoxide, and carbon dioxide. The only combustible component of this gas is the carbon monoxide.

Sewage gas has been used as a heating fuel in several cities in Eastern United States. Currently, most of the interest in sewage gas involves the utilization of animal and vegetable wastes, particularly the waste from large cattle-feed lots, to generate the gas. Sewage gas is composed of essentially pure methane which is produced in the decay process.

2.3 NUCLEAR FUELS

2.3.1 Source of Nuclear Energy

The mass-to-energy conversion of chemical energy in a given chemical reaction is too small to be detected. In nuclear reactions, however, the energy release per reaction is high enough that the mass conversion can actually be detected. Consequently, it is possible to calculate the energy release per reaction from a mass balance of the reactants and products without having to rely on an experimental determination of the energy release.

In any energy-conversion reaction, the sum of mass and energy must be conserved along with momentum. While these laws are true in any conversion process, they are particularly valuable in determining the total energy release as well as the energy distribution among the products of the nuclear reactions.

The atomic number Z is equal to the number of protons (positively charged ions) existing in the atomic nucleus. In a nonionized atom, the atomic number is also equal to the number of electrons (much smaller, negatively charged ions) that orbit the nucleus. All atoms with the same atomic number are members of the same chemical element and these atoms have essentially the same chemical behavior.

The nucleus of an atom is comprised of protons and neutrons where the neutron is a neutrally charged particle with the approximate mass of a proton. Both the protons and the neutrons are called nucleons and the total number of nucleons in a given atom is called the atomic mass number A.

As indicated earlier, atoms with the same atomic number Z are members of the same chemical element. If these atoms have a different atomic mass number A they are said to be isotopes of the element. The different isotopes of the element differ only in the number of neutrons in the nucleus. These atoms exhibit essentially the same chemical behavior but vastly different nuclear behaviors.

It is common practice to identify a given atom or nucleus with the chemical symbol preceded by a superscript of the atomic mass number and preceded by a subscript of the atomic number. Thus, $^{1}_{1}\text{H}$ represents a hydrogen atom or nucleus with one proton, one nucleon, and no neutrons; $^{235}_{92}\text{U}$, on the other hand, represents an uranium atom or nucleus with 92 protons, 235 nucleons, and 143 neutrons. Another common uranium isotope, $^{238}_{92}\text{U}$, has 92 protons, 238 nucleons, and 146 neutrons.

The mass of an atomic nucleus is less than the mass of the individual particles or nucleons that comprise it. This mass difference, called the mass defect, is what holds the nucleus together and prevents the coulomb forces of the positive charges (protons) in the nucleus from tearing it apart. This mass defect is, in effect, "negative mass glue." In order to break a nucleus into its individual nucleons, a minimum amount of energy that is equivalent to the mass defect would have to be added to the nucleus. The mass defect of a given nucleus can be calculated as follows:

$$\text{Mass defect} = Zm_p + (A - Z)m_n - \text{nuclear mass} \qquad (2.11)$$

where m_p and m_n are the mass of a proton and neutron, respectively. Unfortunately, it is very difficult to measure nuclear masses directly, but atomic masses (nuclear mass plus the mass of the orbiting electrons) can be accurately measured. Consequently, the mass defect is usually determined from the following equation:

$$\text{Mass defect} = Zm_H + (A - Z)m_n - \text{atomic mass} \qquad (2.12a)$$

where m_H is the mass of the light-hydrogen ($^{1}_{1}\text{H}$) atom.

Atomic masses are very small and they are usually reported in units called atomic mass units or amu, where one of these units is approximately equal to the mass of a proton or a neutron. One amu is equal to the reciprocal of Avagadro's number in grams, or 1.66×10^{-24} grams. The energy equivalent of 1 amu is 931 MeV. A partial list of isotopes along with their atomic masses is presented in App. I. The masses in App. I have been normalized so that the carbon-12 isotope has an atomic mass of exactly 12.00000 amu. Some of the older compilations of atomic masses were normalized so that the oxygen-16 had an atomic mass of 16.00000 amu. Care should be exercised when using the atomic masses from more than one table to be sure that they are normalized to the same value.

The atomic mass of the hydrogen-1 isotope is 1.007825 amu and the mass of a neutron is 1.008665 amu. Using these values, Eq. (2.12a) reduces to

$$\text{Mass defect} = 1.007825Z + 1.008665(A - Z) - \text{atomic mass} \qquad (2.12b)$$

The mass defect increases as the atomic mass number increases because there are more particles in the nucleus. The energy equivalent of the mass defect is called the total binding energy and this energy, in million-electrovolts, can be found from the following relationship:

$$\text{Total binding energy} = 931 \ (\text{mass defect}) \qquad (2.13)$$

The total binding energy is the absolute minimum amount of energy required to break up a nucleus into its individual components or nucleons.

While the mass defect or binding energy increases with the atomic mass, the average mass defect or binding energy per nucleon rapidly increases and then slowly decreases as the atomic mass is increased. A plot of the average binding energy per nucleon, which is obtained by dividing the total binding energy from Eq. (2.13) by the atomic mass number A, is shown in Fig. 2.5.

It will be noted in Fig. 2.5 that the average binding energy per nucleon is a maximum of 8 to 9 MeV for intermediate-mass nuclei. This indicates that these intermediate-mass nuclei are the most stable nuclei. This means that excess binding energy is released in any nuclear reaction in which a heavy-mass nucleus is broken into intermediate-mass nuclei, as in the fission reaction. Also, excess binding energy is released in any nuclear reaction in which two light-mass nuclei are combined to form a heavier nucleus, as in the fusion reaction. The energy released in any radioactive-decay process is also excess binding energy as the nucleus decays to a more stable configuration.

The average binding energy per nucleon is theoretically the minimum energy required to remove the average proton or neutron from a nucleus. It is also the average excitation energy of a nucleus after it absorbs a proton or neutron. This excitation energy is usually released in the form of gamma radiation, called capture gamma rays. In some heavy-mass nuclei, the excitation energy produced by neutron absorption is sufficient to initiate the fission process.

In the fission, fusion, and the radioactive-decay processes, the total number of nucleons (the sum of the atomic mass numbers) and the total charge (the sum of the atomic numbers) is conserved along with the sum of mass and energy and

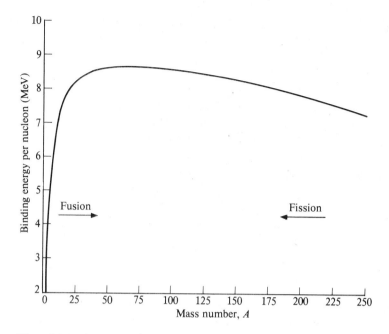

Figure 2.5 Variation of the binding energy per nucleon with the atomic mass. (*From "Steam/Its Generation and Use," 1972.*)

the momentum. Thus, no particles, as such, are converted into energy in these reactions; only excess binding energy is released. This energy is normally released in the form of the kinetic energy of the product particles and/or in the form of electromagnetic radiation (gamma rays). The only reaction in which particles are completely converted into energy is the annihilation reaction. In this reversible reaction, matter and antimatter react to generate electromagnetic radiation.

Example 2.3 Calculate the average binding energy per nucleon for the following isotopes: (*a*) heavy hydrogen, ^2_1H, (*b*) a nickel isotope, $^{59}_{28}Ni$, and (*c*) an uranium isotope, $^{235}_{92}U$.

SOLUTION

(*a*) For hydrogen-2, the atomic mass is 2.0141 amu (from App. I).
   Mass defect $= 1.007825(1) + 1.008665(2 - 1) - 2.0141 = 0.00239$ amu
   Total binding energy $= 931$(mass defect) $= 931(0.00239) = 2.225$ MeV

$$\text{Average binding energy per nucleon} = \frac{\text{total binding energy}}{A}$$

$$= \frac{2.225}{2} = 1.1125 \text{ MeV}$$

(*b*) For nickel-59, the atomic mass is 58.9342 amu (from App. I).

Mass defect $= 1.007825(28) + 1.008665(59 - 28) - 58.9342$

$= 0.55352$ amu

Total binding energy $= 931(0.55352) = 515.33$ MeV

Average binding energy per nucleon $= \dfrac{515.33}{59} = 8.734$ MeV

(*c*) For uranium-235, the atomic mass is 235.0439 amu (from App. I).

Mass defect $= 1.007825(92) + 1.008665(235 - 92) - 235.0439$

$= 1.91510$ amu

Total binding energy $= 931(1.9151) = 1782.95$ MeV

Average binding energy per nucleon $= \dfrac{1782.95}{235} = 7.587$ MeV

2.3.2 Radioactive Decay

Mechanics of the decay process In the radioactive-decay process, a radioisotope spontaneously assumes a more stable configuration with the release of excess binding energy and usually with the release of a light particle. There are four basic types of radioactive decay, including alpha decay, beta decay, positron decay, and K-capture decay. These decay processes may or may not be accompanied with the release of gamma (decay gamma) radiation. Usually, most of the decay energy is released in the form of the kinetic energy of the product particles, with the balance carried away by the gamma radiation.

In any radioactive-decay process, the sum of mass and energy must be conserved and the energy distribution among the product particles can be determined from the conservation of momentum. Suppose that a given radioactive-decay process produces a heavy-mass nucleus M and a light-mass nucleus m and that the two nuclei have respective velocities of U and u. The total kinetic energy of the two particles can be determined from the difference between the mass of the parent nucleus and the sum of the product masses and then subtracting the energy of the decay-gamma radiation:

$$\text{Total kinetic energy of the products} = KE_T = \frac{MU^2}{2} + \frac{mu^2}{2} \quad (2.14)$$

$$KE_T = 931[(\text{mass of parent nucleus, amu}) - (\text{mass of product particles, amu})]$$

$$- (\text{total decay gamma energy, MeV}) \quad (2.15)$$

If the momentum of the gamma radiation is neglected, and the initial momentum of the parent nucleus is assumed to be zero, the product MU must equal the product mu in order to conserve momentum. Substituting this relationship into Eq. (2.14) gives

$$KE_T = \frac{MU^2}{2}\left(1 + \frac{M}{m}\right) = \frac{mu^2}{2}\left(1 + \frac{m}{M}\right) \tag{2.16}$$

Solving for the individual kinetic energy of the product particles gives

$$\text{Kinetic energy of the heavy-mass product} = \frac{MU^2}{2} = \frac{KE_T}{1 + M/m} \tag{2.17}$$

$$\text{Kinetic energy of the light-mass product} = \frac{mu^2}{2} = \frac{KE_T}{1 + m/M} \tag{2.18}$$

Since the mass of the light particle m is usually very much smaller than the mass of the heavy product M, Eqs. (2.17) and (2.18) indicate that the light-mass nucleus carries off almost all the kinetic energy.

Example 2.4 Uranium-235 atoms undergo alpha (helium-4) decay with the emission of a 0.17-MeV gamma ray. Find the kinetic energy of the product nucleus and the alpha particle.

SOLUTION

Decay reaction: $\quad ^{235}_{92}U \longrightarrow ^{231}_{90}Th + ^{4}_{2}He + 0.17$ MeV gamma

From App. I, the atomic masses are

$$m_{U\text{-}235} = 235.0439 \text{ amu} \qquad m_{Th\text{-}231} = 231.0347 \text{ amu}$$

$$m_{He\text{-}4} = 4.0026 \text{ amu}$$

$$\text{Total KE} = 931(m_{235} - m_{231} - m_4) - 0.17$$

$$= 931(235.0439 + 231.0347 - 4.0026) - 0.17 = 5.9746 \text{ MeV}$$

$$\text{KE of Th-231 nucleus} = \frac{KE_T}{1 + M/m} = \frac{5.9746}{1 + 231/4} = 0.1017 \text{ MeV}$$

$$\text{KE of alpha particle} = \frac{KE_T}{1 + m/M} = \frac{5.9746}{1 + 4/231} = 5.8729 \text{ MeV}$$

Types of radioactive decay One of the important radioactive-decay processes with respect to power production is the alpha-decay process with or without the production of gamma radiation. There are a total of about 150 radioisotopes that emit alpha particles. A typical alpha-decay process is given below:

$$^{235}_{92}U \longrightarrow ^{231}_{90}Th + ^{4}_{2}\alpha + \gamma_d \qquad (T_{1/2} = 7.1 \times 10^8 \text{ years})$$

The alpha particle, α, is actually a helium-4 nucleus and γ_d is gamma radiation produced in the decay process (a decay gamma ray or rays). In any alpha-decay process, the product or daughter nucleus (Th-231 in the above reaction) has an atomic mass number which is four less than the parent nucleus (U-235 in the above reaction) and the atomic number of the product nucleus is decreased by two. The half-life, designated by $T_{1/2}$, is the time required for one-half of the radioactive atoms to decay and is constant for a particular radioisotope. Thus, 1 kg of U-235 will decay to $\frac{1}{2}$ kg 710,000,000 years later. The resulting product nucleus of any radioactive-decay process may be stable or it may be radioactive.

The alpha-decay process is normally limited to heavy-mass radioisotopes and a few very light-mass radioisotopes. The alpha particles are essentially mono-energetic (constant kinetic energy) and have very high kinetic energy (4 to 6 MeV). Alpha particles have very low penetrating power and do not present a biological hazard unless they are injested.

The most common type of radioactive decay is the beta-decay process with or without decay gamma radiation. There are approximately 450 beta-emitting radioisotopes. A typical beta-decay process, commonly used in medical radiation therapy, is shown below:

$$\,^{60}_{27}\text{Co} \longrightarrow \,^{60}_{28}\text{Ni} + \,^{0}_{1}\beta + \gamma_d + v \qquad (T_{1/2} = 5.3 \text{ years})$$

The beta particle, β, is essentially an electron and the v is a strange particle called a neutrino that is produced in the beta-decay process. The average kinetic energy of the beta particle is about one-third the total kinetic energy with the balance of the energy being carried away by the neutrino. The neutrino has no mass but it does have spin, and the kinetic energy of this particle cannot be recovered. In any beta-decay reaction, the atomic number of the product nucleus is increased by one over that of the parent nucleus and the atomic mass number is unchanged because the electron mass is very small (1/1847 times the proton mass). The beta-decay process is another reaction that is useful in the production of thermal energy.

One radioactive-decay process that is not useful in the production of radio-isotope power is the positron-decay process. Approximately 150 radioisotopes emit positrons. An example of a positron-decay process is shown below:

$$\,^{57}_{27}\text{Co} \longrightarrow \,^{57}_{26}\text{Fe} + \,^{0}_{+1}\beta + \gamma_d \qquad (T_{1/2} = 270 \text{ days})$$

The atomic mass number of the product nucleus is always unchanged with respect to the parent nucleus while the atomic number decreases by one in a positron reaction.

The positron is an antielectron and as such it rapidly reacts with an ordinary electron and both particles are destroyed in an annihilation reaction. The products of this reaction are two annihilation gamma rays, γ_a, each having the rest-mass energy of an electron (0.51 MeV). Both rays go in opposite directions in order to conserve momentum. This is the only known reaction occurring in nature in which mass is completely converted to energy. This annihilation reaction is shown in Fig. 2.6. This particular reaction happens to be reversible in that

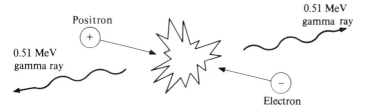

Figure 2.6 The annihilation of an electron and a positron.

if a high-energy gamma ray (the energy must exceed 1.02 MeV) passes near a nucleus, it can be converted into an electron and a positron in the gamma-ray reaction called pair production.

The fourth common type of radioactive decay is the *K*-capture process. Approximately 350 radioisotopes decay by means of the *K*-capture process. In this decay process, the nucleus absorbs or captures one of the two nearest orbiting electrons from the inner or *K* shell as a proton in the nucleus is converted into a neutron. An example of the *K*-capture reaction is given below:

$$\ce{^{71}_{32}Ge} + {_{-1}^{0}e} \longrightarrow \ce{^{71}_{31}Ga} + \gamma_d \qquad (T_{1/2} = 11.4 \text{ days})$$

In the above reaction, the resulting product nucleus (Ga-71) has the same atomic mass number as the parent nucleus (Ge-71) but the atomic number is decreased by one. Some of the remaining orbiting electrons fall into the vacated positions thereby producing a series of x-rays as the electrons assume lower potential energies.

Sources of radioisotopes The two basic sources of radioisotopes are the naturally occurring isotopes and manufactured radioisotopes. The naturally occurring radioisotopes include those produced by cosmic radiation and the long-lived isotopes and their products. The manufactured radioisotopes are the activation products produced from stable nuclei and the radioactive products from the fission reaction (fission products).

The long-lived radioisotopes and their products have been present since the formation of the earth and consist of the primary products, U-235, U-238, and Th-232. A number of short-lived radioisotopes are produced as the result of the decay of these long-lived isotopes. Radium-226, the radioisotope isolated by Marie Curie in 1902, has a half-life of 1690 years and appears in the U-238 decay chain after three stages of alpha decay. The other major group of naturally occurring radioisotopes are those produced as the result of activation of stable atoms by cosmic radiation in the upper atmosphere. Two examples of such isotopes are carbon-12 and hydrogen-3.

Radioisotopes can be produced artificially by either splitting nuclei with high-energy ions or by adding additional nucleons, usually neutrons, to the nuclei of stable atoms. Radioisotopes produced in this manner are called activation products. These products can be generated in either a high-energy accelerator, a neutron generator, or a nuclear fission reactor. An example of an activation product that is produced in this manner is cobalt-60, the isotope commonly used in medical radiation therapy. This reaction is shown below and the intermediate compound nucleus (Co-60*) is designated with an asterisk to indicate that it is in an excited state because of the binding energy of the neutron:

$$^{59}_{27}Co + {}^{1}_{0}n \longrightarrow {}^{60}_{27}*Co \longrightarrow {}^{60}_{27}Co + \gamma_c$$

The activation products can be either alpha emitters or beta emitters, depending on the target nuclei.

The other major group of manufactured radioisotopes are fission products. These products are the undesirable byproducts or the intermediate-mass nuclei formed in the fission reaction. All the fission products produced directly from fission are highly radioactive and, on the average, undergo three stages of beta-particle decay before they form a stable isotope.

Radioactive-decay rates The radioactive-decay process is a random process and the radioactive-decay rate or activity is directly proportional to the number of radioactive atoms present at any instant. The constant of proportionality, called the decay constant λ, is inversely proportional to the half-life $T_{1/2}$ of the radio-isotope. The decay rate is equal to $-dN/dt$, where N is the total number of nuclei present at time t. The negative sign indicates that the number of nuclei is decreasing with time:

$$\text{Activity} = \text{decay rate} = \frac{-dN}{dt} = \lambda N \tag{2.19}$$

Separating the variables in Eq. (2.19), integrating, and applying the boundary condition that $N = N_0$ at some arbitrary time, $t = 0$, gives

$$\lambda t = \ln\left(\frac{N_0}{N}\right) \quad \text{or} \quad N = N_0 e^{-\lambda t} \tag{2.20}$$

According to Eq. (2.19), the activity A is directly proportional to N. Thus, the activity, as a function of the initial activity A_0, is

$$\text{Activity} = A = A_0 e^{-\lambda t} \tag{2.21}$$

The activity or number of radioactive atoms present at any time t is plotted as a function of t in Fig. 2.7. It will be noted that the exponential relationship is a straight line when plotted on semilog graph paper.

The half-life $T_{1/2}$ of a given radioisotope has been defined as the time required for the number of radioactive atoms to be reduced by a factor of two. Thus, using Eq. (2.20) and the fact that $N = N_0/2$ at $t = T_{1/2}$, the following relationship is obtained:

$$\lambda T_{1/2} = \ln(2) = 0.693 \tag{2.22}$$

Most of the tables of radioisotopes list the half-life $T_{1/2}$ rather than the decay constant λ (see Apps. I and J) so that Eq. (2.22) must be used to evaluate the decay constant.

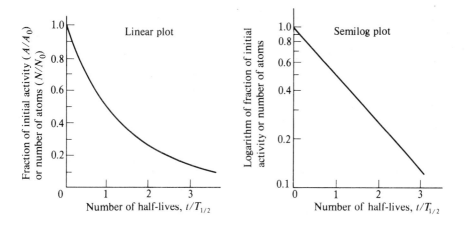

Figure 2.7 Variation of activity (A) or number of radioactive atoms (N) with time.

An interesting quantity that will prove useful in a later section on nuclear reactor kinetics is the average lifetime of a radioisotope, T_{ave}. This quantity can be found by averaging the time t over the number of atoms N from $N = N_0$ to $N = 0$:

$$T_{ave} = \frac{\int_{N_0}^0 t \, dN}{\int_{N_0}^0 dN} = \frac{1}{\lambda} = \frac{T_{1/2}}{\ln (2)} = \frac{T_{1/2}}{0.693} \tag{2.23}$$

In order to evaluate the activity of a radioisotope, one needs to determine the number of radioactive atoms present. This can be done if one knows the mass of the radioisotope or the mass of the compound containing the radioisotope. If the mass is known, the number of atoms is

$$N = \frac{m}{(MW)} (Av)nf \tag{2.24}$$

where m is the mass of the radioactive material in grams, (MW) is the molecular or atomic weight of the material in grams per gram-mole, (Av) is Avagadro's number (0.6023×10^{24} atoms or molecules per gram-mole), n is the number of atoms per molecule (one for an element), and f is the isotopic abundance or the fraction of the atoms of an element that are atoms of the isotope of interest.

There are three units of radioactivity—the becquerel, the curie, and the rutherford. The SI unit of radioactivity is the becquerel (bq) and 1 bq is the equivalent of 1 disintegration per second. The curie (Ci) is commonly used when dealing with large quantities of radioactivity and is equal to the decay rate of 1 g of pure radium-226 or 3.7×10^{10} disintegrations per second. The rutherford is a much smaller unit of radioactivity and is equal to 10^6 bq.

Example 2.5 Determine the activity, in curies, of the uranium-235 isotope in 100 kg of uranium nitride (U_3N_4) when natural uranium is used.

SOLUTION

$$m = \text{mass of } U_3N_4 = 100{,}000 \text{ g}$$

$$MW = \text{molecular weight of compound} = 3(238) + 4(14) = 770 \text{ g/g·mol}$$

$$n = 3 \text{ uranium atoms per molecule}$$

$$f = \text{isotopic abundance of U-235 in natural uranium (from App. I)}$$

$$= 0.72\% = 0.0072 \text{ U-235 atoms per uranium atom}$$

$$T_{1/2} = 7.1 \times 10^8 \text{ year (from App. I)}$$

$$N = \frac{100{,}000 \text{ g}}{770 \text{ g/g·mol}} (6.023 \times 10^{23} \text{ molecules/g·mole})$$

$$\times (3 \text{ U atoms/molecule})(0.0072 \text{ U-235 atoms/U atom})$$

$$= 1.69 \times 10^{24} \text{ atoms of U-235}$$

$$\lambda = \frac{0.693}{T_{1/2}} = \frac{0.693}{(7.1 \times 10^8 \text{ year})(8766 \text{ h/year})(3600 \text{ s/h})}$$

$$= 3.094 \times 10^{-17} \text{ disintegrations per second (U-235 atom)}$$

$$= 3.094 \times 10^{-17} \text{ bq per U-235 atom}$$

$$\text{Activity} = A = \lambda N = (3.094 \times 10^{-17} \text{ bq per U-235 atom})$$

$$\times (1.69 \times 10^{24} \text{ U-235 atoms})$$

$$= 5.228 \times 10^7 \text{ bq} = 1.413 \times 10^{-3} \text{ Ci}$$

Power from radioisotopes The power produced by a radioisotope heat source is directly proportional to the activity of the source and thus the power P at any time t is given by an expression that is similar to Eq. (2.21):

$$P = P_0 e^{-\lambda t} \tag{2.25}$$

If the radioisotope of interest is actually a daughter product or if it decays to another radioisotope, the power equation is much more complicated than that given by (2.25) because the two radioisotopes are being produced and lost at two different rates.

The power of a radioisotope heat source is equal to the activity of the isotope times the energy released per decay process. This energy is normally part electromagnetic (gamma) energy and part kinetic energy of the emitted particles. Since the gamma energy is very penetrating and may or may not be deposited in the source, it is recommended that the designer be conservative and neglect the gamma energy. The decay energy per disintegration should also include the decay energy of any product isotopes if they are radioactive with short half-lives.

When choosing a radioisotope for use as a power source, there are a number of criteria that must be considered. First, the fuel isotope should have a reasonable half-life. If the half-life is too short, the source will have very limited use and if the half-life is too long, the activity is too low to achieve a reasonable value of specific power (power per unit mass). A reasonable half-life would appear to be between 100 days and 100 years. Second, the material used to contain the radioisotope should have a reasonable value specific power, probably greater than 0.1 W/g. Third, the radioisotope and its daughter products should not emit large quantities of gamma radiation because of the biological hazard it presents. Fourth, the material containing the radioisotope should be a solid and have a high melting point, a high thermal conductivity, and it should be chemically stable and inert. Finally, the availability of the source material should be such that material and fabrication costs are not excessive.

Out of a total of around 1200 radioisotopes, only 100 or so have half-lives between 100 days and 100 years. Applying the second and third criteria to the remaining isotopes narrows the list to about 30 isotopes. When the final two criteria are imposed, eight radioisotopes are left. Four of these isotopes are alpha-emitting activation products while the other four are beta-emitting fission products. These eight radioisotopes, their properties, the materials used, the specific powers, and the power densities are listed in App. J.

Problems with radioisotope sources There are a number of problems associated with the use of radioisotope power sources. The availability of these materials is such that they are still very expensive. Plutonium-238 with a purity of 90 percent, for example, was priced at $1250 a gram in 1973, while 80 percent-pure plutonium-238 was priced at $700 a gram. Considering that plutonium metal is very dense (about 19 g/cm^3), a cubic centimeter of high-grade plutonium metal would cost around $24,000. Another problem associated with these sources is the radiation problem and also the fact that many of the fuel compounds are radiologically toxic. This means that the containment systems must be extremely leak-tight.

The power from a radioactive power source is not constant and this presents a problem in the design of any source. Since the power decays exponentially with time according to Eq. (2.25), assuming that there is no daughter-product decay, the source must be designed for the desired power at the end of service life. This means that the source produces more power than required at startup and this initial power must be "dumped" to prevent overheating.

A similar problem to that discussed above arises from the fact that there is no way to control the power output of these sources. Since there is no way to shut off the source, some provision must be incorporated in the source design to "dump" all the power if it is not needed.

Finally, a problem peculiar to only the alpha sources arises due to the buildup of helium gas in the source. The alpha particle is a helium-4 nucleus and, once it slows down, it picks up two electrons to become a helium atom. The accumulation of helium gas in the source can develop very high pressures unless an expansion volume is incorporated in the source.

Applications of radioisotope power sources Radioisotope power sources are not applicable where large quantities of power are required but they have found use in specialized applications where small, compact, reliable sources of energy are desired. These sources are used to power heart pacemakers and a number of these units have been implanted in individuals throughout the world. They are also the proposed energy source for the mechanical heart pump under development.

Radioisotope power sources have been widely used in the United States in the SNAP program. SNAP is an acronym for "Systems for Nuclear Auxilliary Power." All the even-numbered SNAP units are powered by nuclear reactors while all the odd-numbered SNAP systems are powered with radioisotopes. A typical radioisotope-powered SNAP system, using a thermoelectric-conversion system, is shown in Fig. 2.8. These systems have been used to power remote unmanned weather stations in the Arctic, Antarctic, the Apollo lunar laboratories, Coast Guard buoys, orbiting satellites, etc.

Example 2.6 Find the mass of polonium-210 metal that must be used to supply minimum thermal power of 1 kW_{th} for a period of 200 days. Also determine the initial power of the source, the power that must be initially dumped, and find the expansion volume needed in the source to assure that the internal helium-gas pressure shall not exceed 500 bar at 400°C.

SOLUTION From App. J,

$$\text{Specific power} = SP = 144 \text{ W/g}$$

$$T_{1/2} = 138.4 \text{ days}$$

$$\text{Mass fraction of Po-210 in the metal} = 80\%$$

$$\text{Final thermal power of source} = P = 1 \text{ kW}_{th}$$

$$\text{Initial thermal power of source} = P_0 = Pe^{\lambda t} = Pe^{(\ln 2)t/T_{1/2}}$$

$$= (1.0)e^{(0.693)(200)/(138.4)}$$

$$= 2.723 \text{ kW}_{th}$$

$$\text{Initial thermal power that must be dumped} = 2.723 - 1.000 = 1.723 \text{ kW}_{th}$$

$$\text{Initial mass of Po-210 metal needed} = m_0 = \frac{P_0}{SP}$$

$$= \frac{2723 \text{ W}}{144 \text{ W/g}} = 18.9 \text{ g of metal}$$

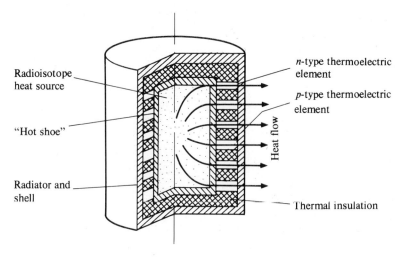

Figure 2.8 A radioisotope-powered, SNAP, thermoelectric converter.

Number of helium atoms eventually produced = number of Po-210 nuclei = N

$$N = \frac{18.9(0.8)(6.023 \times 10^{23})}{210} = 4.338 \times 10^{22} \text{ He atoms}$$

$$\text{Mass of helium} = \frac{(4.338 \times 10^{22} \text{ atoms})(4 \text{ g/g·mol})}{6.023 \times 10^{23} \text{ atoms/g·mol}}$$

$$= 0.2881 \text{ g}$$

From the ideal-gas equation,

$$PV = mRT = \frac{mR_u T}{\text{MW}}$$

$$P = 500 \text{ bar}$$

$$T = 400°C = 673 \text{ K}$$

$$\text{Required gas volume} = V = \frac{mR_u T}{(\text{MW})(P)}$$

$$V = \frac{0.0002881 \text{ kg } (0.08315 \text{ bar·m}^3/\text{kg·mol·K})(673 \text{ K})(10^6 \text{ cm}^3/\text{m}^3)}{(4 \text{ kg/kg·mol})(500 \text{ bar})}$$

$$= 8.061 \text{ cm}^3$$

2.3.3 Nuclear Fission

The fission process The fission process is very different from the radioactive-decay process in that it requires a particle interaction with the nucleus to initiate the reaction. Consequently, it is a controllable reaction or process when compared to radioactive decay. In the fission process, the heavy, fissionable isotope absorbs a low-energy neutron and forms a compound nucleus which is in an excited state because of the binding energy (7 to 8 MeV) of the absorbed neutron. Within 10^{-14} s, the excited compound nucleus either decays to the ground state with the emission of capture gamma radiation in the "radiative-capture" reaction or else the excited, compound nucleus fissions:

(Radiative-capture process)

$$^{235}_{92}U + ^{1}_{0}n \longrightarrow {}^{236}_{92}{}^{*}U \left\langle \begin{array}{l} \longrightarrow {}^{236}_{92}U + \gamma_c \\ \\ \longrightarrow FP\ 1 + FP\ 2 + (2\ \text{to}\ 3)^1 n + \gamma_f \end{array} \right.$$

(Fission process)

The radiative-capture process occurs for all isotopes but it produces only about 7 to 8 MeV of energy, and it usually converts a fissionable isotope (U-235) into a nonfissionable isotope (U-236). Except for fertile materials, which can be converted into fuel materials by this reaction, and control materials, the radiative-capture reaction is not desirable in nuclear fission reactors.

In the fission reaction, the excited compound nucleus splits into two lower-mass nuclei, called fission products, and these products are accompanied by two or three neutrons and fission-gamma radiation. The fission reaction releases a total (including decay beta and decay gamma energy from the fission products) energy of about 200 MeV per fission instead of the meager energy release from the radiative-capture reaction. The product neutrons released in the fission reaction are used to react with other fissionable atoms to sustain the fission chain reaction.

The average energy release from fission depends to some extent on the velocity of the incident neutron, the type of fuel nucleus, and the type of other material used in the reactor. In this text, it will be assumed that the average fission energy release is a constant 200 MeV per fission. This corresponds to 3.225×10^{-9} W·s per fission, or, taking the reciprocal, 3.1×10^{10} fissions per wattsecond. This means that a 3800-MW$_{\text{th}}$ reactor (the highest permitted thermal power of nuclear reactors in the United States) has a fission rate of (3800) $(10^6)(3.1 \times 10^{10})$ or 1.178×10^{20} fissions per second.

The distribution of fission energy is given in Table 2.3. It will be noted in the table that the neutrinos carry away 10 MeV or 5 percent of the fission energy which makes it unavailable. The loss of neutrino energy is compensated by the capture gamma-ray energy and the radioactive decay beta and decay gamma energy resulting from the absorption of neutrons by nonfuel nuclei. This is why the average energy from fission is somewhat dependent on the type of nonfuel material used in the reactor.

Table 2.3 Distribution of fission energy

Products of fission	Type of energy	Amount, MeV	Percentage of total	Where deposited	When deposited
Fission products	Kinetic	165	82.5	Fuel	Immediately
Fission neutrons	Kinetic	5	2.5	Moderator	Immediately
Fission gammas	Electromagnetic	7	3.5	Diffuse	Immediately
FP decay betas	Kinetic	7	3.5	Fuel	Later
Decay gammas	Electromagnetic	6	3.0	Diffuse	Later
Neutrinos†	Kinetic	10	.....	Nowhere	
Capture gammas‡	Electromagnetic	7	3.5	Diffuse	Immediately
Decay energy‡	Electromagnetic and kinetic	3	1.5	Diffuse and fuel	Later
Total usable energy		200	100.0		

† The neutrino energy is not available as usable energy.

‡ This energy is not produced directly from fission but is produced as the result of nonfission neutron absorption by the fuel and other in-core materials.

Almost all of the fission energy is deposited in the reactor core and, in fact, most of the energy is converted into thermal energy at the point where the fission occurs in the fuel. The fission products travel a very short distance and consequently their kinetic energy (over 80 percent of the fission energy) is deposited as heat in the fuel material. Most of the beta-particle energy produced by fission-product decay is also deposited in the fuel. The kinetic energy of the fission neutrons is normally deposited in light-mass material (called moderating material) which is inserted in the reactor specifically to slow down the neutrons. The fission and gamma-ray energies are also deposited in the reactor, but some of this energy is deposited in the shielding around the reactor.

Most of the binding energy from fission is released within a millisecond following fission. This includes the kinetic energy of the fission products and neutrons and the fission and capture gamma radiation. The rest of the energy, which is produced from the radioactive decay of the fission products, is delayed. This portion of the fission energy, therefore, does not go to zero immediately after a reactor shutdown. The shutdown power production produces severe emergency core-cooling problems in some modern high-powered reactors in the event of a loss of cooling accident (LOCA). In fact, these systems have an independent emergency core-cooling system (ECCS) for the removal of the decay heat following LOCA. The shutdown power of a reactor that has operated at constant power P_0 for t_0 seconds is given by the following empirical equation:

$$\frac{P_s}{P_0} = 0.1[(t_s + 10)^{-0.2} - (t_s + t_0 + 10)^{-0.2}] - 0.087[(t_s + 2 \times 10^7)^{-0.2}$$

$$- (t_s + t_0 + 2 \times 10^7)^{-0.2}] \tag{2.26}$$

P_s is the shutdown power (in the same units as P_0) and t_s is the time after reactor shutdown in seconds.

Fissionable isotopes As indicated earlier, there are a number of different heavy-mass nuclei that will undergo fission, but the availability is such that only three isotopes are commonly listed as fissionable isotopes. The three fissionable fuels are uranium-235 ($^{235}_{92}U$), uranium-233 ($^{233}_{92}U$), and plutonium-239 ($^{239}_{94}Pu$). All of these fuel isotopes are radioactive (alpha emitters) but they have very long half-lives. It will be noted that all these isotopes have odd atomic mass numbers and even atomic numbers. Evidently, the even-even atomic mass and atomic number configuration formed by neutron absorption is so unstable that the binding energy of the absorbed neutron is sufficient to cause fission. While there are many other isotopes that undergo fission following neutron absorption such as Pu-241, Pu-243, Cf-243, Cf-245, etc., the availability of these isotopes is so poor that they are not used as the initial reactor fuel. Some of these fissionable isotopes appear in irradiated nuclear fuels.

Of the three common fissionable isotopes, only uranium-235 occurs in nature. The atomic composition of natural uranium is 99.274 percent U-238, 0.72 percent U-235, and 0.006 percent U-234. Thus, the U-235 isotope comprises less than 1 percent of the natural uranium atoms. The isotopic abundance of U-235 is so low that many reactors cannot operate using natural-uranium fuel. This means that the U-235 concentration must be increased in the uranium and this fuel is said to be "enriched." An enrichment of X percent means that the fuel has a U-235 mass percentage of X percent. Any uranium that has less than the natural isotopic concentration of U-235 is called "depleted" uranium.

In order to produce enriched or depleted uranium, it is necessary to be able to separate the uranium isotopes. This is not easily accomplished because all the isotopes have the same chemical behavior. Thus, most separation techniques must depend upon physical differences in the uranium atoms.

The earliest separation method, the electromagnetic separation process, involved the ionization of uranium atoms and then the acceleration of them through a perpendicular magnetic field. The lighter uranium-235 atoms have a smaller radius of curvature. This method has now been dropped in favor of more economical and more efficient processes.

Most of the uranium separation methods currently in use employ a gaseous compound of uranium whose molecular weight difference is due solely to the atomic mass of the uranium isotopes. Uranium hexaflouride (UF_6) gas is employed as the feed material in these processes because the element flourine is composed of only one stable isotope, flourine-19.

The process that is currently employed to separate the uranium isotopes in the United States is the gaseous diffusion process. Since the average velocities of gas molecules at a given temperature depend on their molecular weights, the different isotopes will diffuse through a porous barrier at different rates, as is shown in Fig. 2.9. The amount of separation obtained in one barrier is equal to the square root of the ratio of the molecular weights of the gases, $\sqrt{352/349}$ or

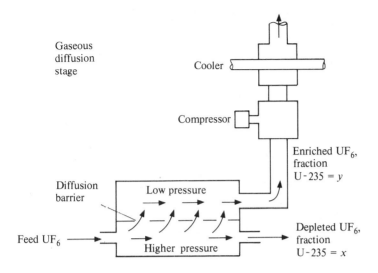

Separation factor

$$\alpha \equiv \frac{y(1-x)}{x(1-y)} = \sqrt{\frac{M_{238_{UF_6}}}{M_{235_{UF_6}}}} = \sqrt{\frac{352}{349}} = 1.00429$$

Minimum theoretical energy input for unit separative work

$$\frac{4RT_0}{(\alpha-1)^2} = 634 \text{ kW·h/kg separative work unit}$$

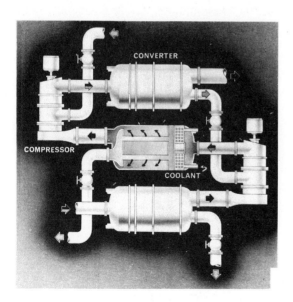

Figure 2.9 The basic gaseous diffusion isotope separator. *(Benedict, 1976.)*

1.00429. In a single diffusion stage, the uranium hexaflouride gas is pumped into a porous metallic tube and the gas diffuses through the wall. To obtain 3 percent enriched uranium from natural uranium, the diffusion-process gas must be cycled and recycled through hundreds of stages. Enriched uranium is priced according to the amount of energy or separative work units (SWU) required to produce the desired enrichment. The relationship between uranium-235 enrichment and SWU is presented in Table 2.4.

There are three government-owned gaseous diffusion plants in the United States. They are located at Paducah, Kentucky, Portsmouth, Ohio, and Oak

Table 2.4 Table of enriching services

| | Standard table of enriching services† | |
Assay, wt% U-235	Feed component (normal), kg U feed/kg U product	Separative work component, kg SWU/kg product
0.20	0	0
0.30	0.196	−0.158
0.40	0.391	−0.198
0.50	0.587	−0.173
0.60	0.783	−0.107
0.70	0.978	−0.012
0.711 (normal)	1.000	0.000
0.80	1.174	0.104
0.90	1.370	0.236
1.00	1.566	0.380
1.20	1.957	0.698
1.40	2.348	1.045
1.60	2.740	1.413
1.80	3.131	1.797
2.00	3.523	2.194
2.20	3.914	2.602
2.40	4.305	3.018
2.60	4.697	3.441
2.80	5.088	3.871
3.00	5.479	4.306
3.40	6.262	5.191
3.80	7.045	6.090
4.00	7.436	6.544
5.00	9.393	8.851
10.00	19.178	20.863
90.00	175.734	227.341
98.00	191.389	269.982

† The kilograms of feed and separative work components for assays not shown can be determined by linear interpolation between the nearest assays listed.

Ridge, Tennessee. These plants consume huge quantities of electricity and in 1962 they used a total of about 47×10^9 KW$_e$·h or about 5 percent of the total electrical energy generated in the United States that year.

There are several other uranium isotopic separation processes under evaluation and development. These include the gas-centrifuge process, the Becker-nozzle process, a gas vortex-tube process, and the laser separation process. Of these four separation processes, the greatest amount of development has been expended on the gas-centrifuge process and there are two large separation plants of this type in Europe. This method also uses uranium hexaflouride gas and is shown schematically in Fig. 2.10. The gas-centrifuge process appears to be more efficient than the gaseous diffusion process for moderate enrichments.

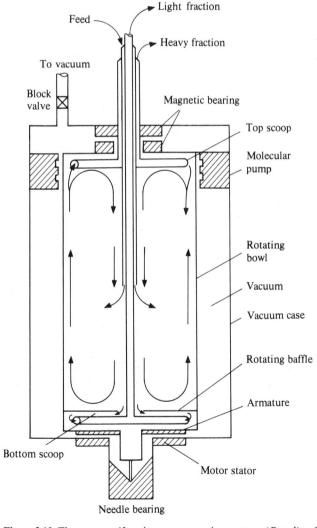

Figure 2.10 The gas-centrifuge isotope separation system. *(Benedict, 1976.)*

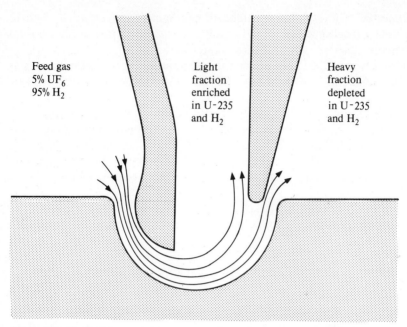

Feed gas	Light	Heavy
$5\% UF_6$	fraction	fraction
$95\% H_2$	enriched	depleted
	in U-235	in U-235
	and H_2	and H_2

Figure 2.11 The Becker-nozzle isotope separation system. *(Benedict, 1976.)*

The Becker separation process is another gas-separation process and is shown schematically in Fig. 2.11. South Africa supposedly has developed another gaseous-separation process that employs a gas vortex tube.

The laser separation process is a very promising technique that is currently under development. It involves the use of a two-stage laser to preferentially ionize the uranium-235 atoms. The ionized atoms are then collected on a charged plate or electrode as shown in Fig. 2.12. This process has the potential of separating the uranium isotopes at a fraction of the cost of other methods. It also has the potential of recovering a much higher fraction of the uranium-235 atoms in the feed material.

The other two fissionable isotopes, uranium-233 and plutonium-239, are manufactured nuclei and are produced as the result of neutron absorption by nonfuel isotopes, called fertile material. Following neutron absorption, the fertile nuclei subsequently undergo two stages of beta decay to a fuel nucleus. The fertile isotopes are thorium-232 and uranium-238 which are respectively converted to uranium-233 and plutonium-239. Since the fuel and fertile elements are different, they can be separated by chemical means.

The neutron activation and subsequent beta-decay processes for the production of uranium-233 are listed as follows:

$$^{233}_{90}\text{Th} + ^{1}_{0}n \longrightarrow ^{233*}_{90}\text{Th} \longrightarrow ^{233}_{90}\text{Th} + \gamma_c$$

$$^{233}_{90}\text{Th} \longrightarrow ^{233}_{91}\text{Pa} + ^{0}_{-1}\beta + \gamma_d \qquad (T_{1/2} = 22.4 \text{ min})$$

$$^{233}_{91}\text{Pa} \longrightarrow ^{233}_{92}\text{U} + ^{0}_{-1}\beta + \gamma_d \qquad (T_{1/2} = 27.4 \text{ days})$$

$$^{233}_{92}\text{U} \longrightarrow ^{229}_{90}\text{Th} + ^{4}_{2}\alpha + \gamma_d \qquad (T_{1/2} = 162{,}000 \text{ years})$$

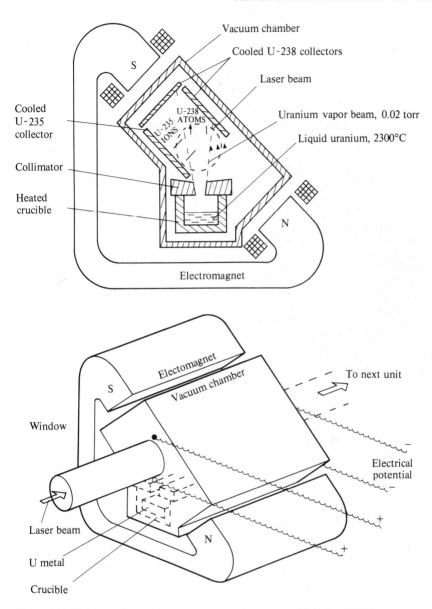

Figure 2.12 The Avco-Exxon laser isotope separation system. *(Benedict, 1976.)*

Unfortunately, an $(n, 2n)$ neutron reaction occurs during the thorium activation process that produces another uranium isotope, U-232. This isotope has a much shorter half-life than U-233 and hence has a much higher decay rate. The radiation from the U-232 nucleus and its daughter products is so high that it presents a biological hazard to the personnel handling it. Since the U-232 isotope is carried along with the U-233 isotope during the chemical separation process, fuel elements fabricated from U-233 must be fabricated behind shielding.

The neutron activation and subsequent beta-decay reactions for the production of plutonium-239 are listed as follows:

$$^{238}_{92}U + ^{1}_{0}n \longrightarrow ^{239*}_{92}U \longrightarrow ^{239}_{92}U + \gamma_c$$

$$^{239}_{92}U \longrightarrow ^{239}_{93}Np + ^{0}_{-1}\beta + \gamma_d \qquad (T_{1/2} = 23.4 \text{ min})$$

$$^{239}_{93}Np \longrightarrow ^{239}_{94}Pu + ^{0}_{-1}\beta + \gamma_d \qquad (T_{1/2} = 2.35 \text{ days})$$

$$^{239}_{94}Pu \longrightarrow ^{235}_{92}U + ^{4}_{2}\alpha + \gamma_d \qquad (T_{1/2} = 24{,}000 \text{ years})$$

The feed or fertile material for this process is the U-238 isotope which comprises over 99 percent of natural uranium. Utilization of this nonfuel isotope as a fissionable fuel increases the naturally occurring fissionable fuel reserves (U-235) by a factor of around 140. The element plutonium is a radiological poison and consequently fuel elements fabricated from Pu-239 must be fabricated in glove boxes. This makes the fabrication of Pu-239 fuel elements easier than the fabrication of U-233 elements, but not as easy as the fabrication of U-235 fuel.

Conversion of fertile material The production of the manufactured fuels is initiated with the absorption of a neutron by the fertile material. The uranium-235 fission process produces, on the average, about 2.5 neutrons per fission and only one of these neutrons is required to sustain the chain reaction. It would seem logical to use the excess fission neutrons to produce new fuel atoms in the reactor by activating the fertile isotopes. An important reactor parameter, designated here as "r," is defined as the ratio of the number of fertile atoms consumed (the number of new fuel atoms formed) to the number of original fuel atoms consumed in the fission and radiative-capture processes. If the parameter r is equal to or exceeds unity, it is called the breeding ratio and the reactor is called a breeder reactor. If the value of r is nonzero but less than unity, the reactor is called a converter reactor and r is called the conversion ratio.

The breeder reactor produces energy plus more fuel than it consumes. These reactors actually "burn" fertile material. The important operating parameter for the breeder reactor is the doubling time. The doubling time is defined as the time required for the breeder reactor to produce an amount of extra fuel equal to the reactor fuel loading. A good breeder reactor will have a short doubling time and this can be achieved by having a high breeding ratio, by operating the reactor at high power, and by having a reactor with a low fuel loading. Most proposed breeder reactors have doubling times ranging from 7 to 15 years.

While the converter reactor does not produce as much fuel as it consumes, it does convert some fertile material into fuel as long as the conversion ratio is nonzero. When the reactor is operated during one fuel cycle, N_0 original fuel atoms are consumed and $N_0 r$ new fuel atoms are produced during that fuel cycle. If these $N_0 r$ new fuel atoms are reinserted into the reactor and burned during the second fuel cycle, they will produce $N_0 r^2$ new fuel atoms during that cycle. If the process is continued, at the end of the nth fuel cycle, $N_0 r^n$ new fuel atoms are produced during the nth cycle as the result of burning N_0 original fuel atoms in the first cycle. Thus, the total number of new fuel atoms produced and burned, which is equal to the number of fertile atoms consumed, for N_0 original fuel atoms burned, is

$$N_0 r + N_0 r^2 + N_0 r^3 + \cdots + N_0 r^n = N_0 r(1 + r + r^2 + \cdots + r^{n-1})$$

The maximum fertile-atom conversion for a converter reactor occurs as the number of fuel cycles approaches infinity. As long as r is less than unity, which it is for a converter reactor, the infinite series has a finite limit,

$$\lim_{n \to \infty} (1 + r + r^2 + \cdots + r^{n-1}) = \frac{1}{1 - r}$$

Thus, for N_0 original fuel atoms consumed, a maximum of $N_0 r/(1 - r)$ fertile atoms are consumed as fuel. The maximum conversion of fertile material in any converter reactor is

$$\text{Maximum possible conversion of fertile material} = \frac{r}{1 - r} \qquad (2.27)$$

where r is the conversion ratio of the reactor. If $r = 0.8$, $r/(1 - r)$ is 4, which means that the reactor will eventually consume four atoms of fertile material for each atom of fuel material originally consumed.

It would appear that this reactor produces more fuel than it consumes (4 to 1) and hence is actually a breeder reactor. The breeder reactor, however, produces extra fuel at the end of each fuel cycle so that not all the new fuel is burned in this reactor. The maximum conversion of fertile material in a breeder reactor is infinity. If the breeding ratio of a reactor is exactly unity, all fertile material can be theoretically converted into fuel material and burned in this system, but it would not produce any extra fuel to build a similar reactor. In other words, the doubling time is infinity.

2.3.4 Nuclear Fusion

Fusion reactions The fusion process is essentially the antithesis of the fission process. In the fission process, heavy-mass nuclei are split into lighter-mass nuclei, releasing excess binding energy. In the fusion reaction, light-mass nuclei are combined in order to release excess binding energy. The fusion reaction is the general reaction that "fuels" the sun and it has been used on earth to release large quantities of energy in the thermonuclear or "hydrogen" bomb. Unfortunately, the technical problems leading to the controlled release of energy from the fusion reaction have not been solved, although a large research effort has been expended on this technology.

The five most probable fusion reactions, along with the energy released, the ignition energy, and ignition temperatures, are given below:

Reaction	Ignition	
	Energy	Temperature
$^2_1H + ^2_1H \longrightarrow ^3_2He + ^1_0n + 3.26$ MeV	50 keV	5.8×10^8 K
$^2_1H + ^2_1H \longrightarrow ^3_1H + ^1_1H + 4.03$ MeV	50 keV	5.8×10^8 K
$^3_1H + ^2_1H \longrightarrow ^4_2He + ^1_0n + 17.4$ MeV	10 keV	1.2×10^8 K
$^3_2He + ^2_1H \longrightarrow ^4_2He + ^1_1H + 18.3$ MeV	100 keV	1.2×10^9 K
$^6_3Li + ^1_1H \longrightarrow ^3_2He + ^4_2He + 4.0$ MeV	200 keV	2.3×10^9 K

It will be noted that those reactions that produce an alpha particle (^4_2He) usually have very high energy release. The combination of two neutrons and two protons, that occurs in an alpha particle, produces an extremely stable configuration with a high value of binding energy per nucleon. This probably accounts for the fact that one of the emitted particles in the radioactive-decay processes is the alpha particle.

Recently, some physicists have proposed another nuclear reaction as a possible source of nuclear energy. This reaction is called the "thermonuclear fission" reaction and involves a reaction between the hydrogen-1 and the boron-11 isotopes. The reaction is shown below:

$$^1_1H + ^{11}_5B \longrightarrow ^{12*}_6C \longrightarrow 3(^4_2He) + 8.68 \text{ MeV}$$

This reaction has several attractions. First, boron is one of the more common elements in the earth's crust and 80 percent of the boron atoms are boron-11 atoms. Boron is also much easier to isolate than the heavy-hydrogen isotopes. Second, and most important, the products of this reaction are three alpha particles and these particles are not radioactive. This reaction produces only energy and helium and is sometimes called the "super-clean" reaction.

This reaction is called a fission reaction because the excited compound nucleus, carbon-12*, breaks into three equal parts. The reaction is still a thermonuclear reaction because the reactants are positively charged nuclei. In fact, due to the high charge and great mass differential, the threshold energy for this reaction is much higher than the threshold energies of the other fusion reactions. Consequently, little experimental work will be done on this particular reaction until the other fusion systems have been developed.

Advantages and disadvantages of fusion The fusion reaction offers several advantages over the fission reaction when it comes to the conversion of nuclear energy. One major advantage over fission is that there are much greater known reserves of fusionable isotopes. In fact, there is essentially an unlimited supply of fuel. The common fuel isotope for most of the fusion reactions is deuterium, hydrogen-2, and this isotope exists in nature to about one part in 6700 parts of ordinary hydrogen. Considering the amount of hydrogen available in the waters of the world, it means that there is a tremendous supply of fuel.

Another advantage of the fusion reaction is that the products of the fusion process are not as highly radioactive as the products of fission. In the products of the five fusion reactions, listed earlier, only hydrogen-3 and the neutrons are radioactive and the neutrons soon decay to a hydrogen atom. The radioactivity produced as the result of neutron activation of the containment structure is more of a problem than that of the fusion products.

The last major advantage of fusion over fission arises from the fact that the fusion process is so hard to initiate and maintain. In fact, any slight perturbation of the system invariably terminates the reaction. This effect, along with the very small amount of reactants present in the system, circumvents the possibility of a tremendous power excursion due to an equipment failure.

The primary problem associated with the development of the fusion reactor arises from the fact that the reacting particles are both positively charged nuclei. This means that the reacting particles must have sufficient kinetic energy to overcome the coulomb repulsion forces. For minimum kinetic energy requirements, the two particles should have approximately equal masses and they should have high mass-to-charge ratios.

The minimum or threshold energies needed to initiate the reactions are listed earlier with the various reactions. These energies are commonly expressed in terms of temperature, even though the actual particle density is so low that temperature really does not have much meaning. At these very high kinetic energies, all of the electrons are stripped from the nuclei and the reactants are said to be in a state called a plasma. This is sometimes called the fourth state of matter. In the thermonuclear bomb, the ignition energy is obtained by first detonating a fission bomb. The deuterium-tritium reaction has the lowest threshold energy (mass/charge $= A/Z = 5/2$) and, for this reason, the first fusion reactor will probably operate on this reaction.

Fusion research As discussed earlier, the only significant terrestrial release of fusion energy has been in the thermonuclear bomb where the reaction is "triggered" by detonating a fission bomb to develop the necessary temperatures. The fusion reaction is the basic reaction that fuels the sun. In the sun, gravitational forces are so great that they produce extremely high temperatures near the core and these temperatures are sufficient to initiate and maintain a continuous fusion reaction.

There have been several methods and systems proposed for producing controlled fusion power. The major problem is associated with the containment of the plasma. There are two basic containment systems currently under development—magnetic-containment systems and inertial-containment systems. The earliest containment schemes for the production of controlled fusion power involve trapping the plasma in an intense magnetic field. A charged particle, such as an electron or a nucleus, spirals around a magnetic line of force as the particle travels along the line. Consequently, it has been proposed to contain the plasma in an intense magnetic field.

There are two basic types of magnetic-containment systems under evaluation today—mirror machines and tokamaks. In the common mirror machine, the magnetic field is shaped like a football like that shown in Fig. 2.13. It will be noted that the magnetic field is "pinched" at each end of the reactor. As a charged particle moves to one end of the machine, the surrounding magnetic field lines compress the amplitude of the spiralling particle until it actually turns around and spirals out again. Thus, the particle is essentially trapped in the magnetic field as if there were a mirror at each end of the machine. This basic process is essentially the same process that has led to the development of the van Allen radiation belts above the earth. The earth's magnetic force field has trapped protons and electrons as they spiral along a magnetic line of force back and forth from the north magnetic pole to the south magnetic pole.

The theta-pinch fusion device or thetatron is composed of a linear mirror machine that is used to contain the plasma. This plasma is compressed to an extremely high temperature by the sudden imposition of an intense magnetic

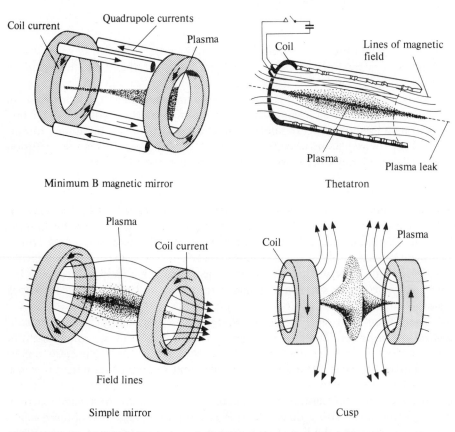

Figure 2.13 Schematic diagrams of some of the mirror fusion machines for the magnetic containment of plasma. *(Reprinted with permission of the American Nuclear Society.)*

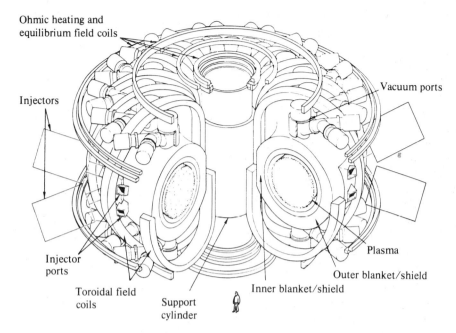

Ohmic heating and equilibrium field coils

Injectors

Vacuum ports

Injector ports

Plasma

Toroidal field coils

Support cylinder

Inner blanket/shield

Outer blanket/shield

Figure 2.14 A schematic diagram of the Argonne tokamak experimental power reactor (TEPR). *(Reprinted with permission of the American Nuclear Society.)*

field on the system. This is accomplished by discharging a huge capacitor bank through a conductor surrounding the plasma, as shown in Fig. 2.13.

The tokamak fusion device has a magnetic field shaped like a large torus. The outside of the containment vessel is wrapped with superconducting magnetic windings to achieve the desired magnetic field. Since all magnetic-containment systems propose the use of superconducting magnets, they run the gamut with regard to temperature extremes. In a superconducting magnet, the windings must be maintained around 10 K with liquid helium, while the plasma must be maintained at millions of degrees. Of all the magnetic containment systems, the tokamak is one of the most promising and successful. A conjectural arrangement for a tokamak fusion reactor is shown in Fig. 2.14.

A relatively recent entry into the development of fusion power has been the proposal to use inertial containment instead of magnetic containment. In the inertial containment system, small solid fuel pellets are imploded by the simultaneous exposure to a burst of high-energy pulsed laser and/or ion beams. This device releases a small explosion of fusion energy as the solid particles are injected into the reactor and fired. The sudden tremendous energy deposition on the outer surface of the small spherical fuel pellet is such that extremely high pressures and temperatures are generated at the center of the pellet as it implodes. These pellets will probably be fabricated of either heavy-water-tritium ice or plastic. A diagram of a proposed pulsed-laser fusion reactor and its associated power cycle is shown in Fig. 2.15.

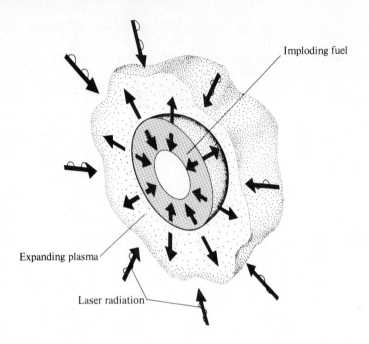

Imploding fuel

Expanding plasma

Laser radiation

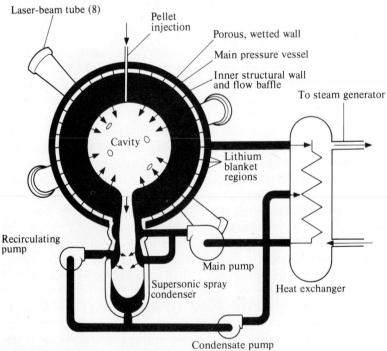

Laser-beam tube (8)

Pellet injection

Porous, wetted wall

Main pressure vessel

Inner structural wall and flow baffle

To steam generator

Cavity

Lithium blanket regions

Recirculating pump

Main pump

Supersonic spray condenser

Heat exchanger

Condensate pump

Figure 2.15 The laser fusion process and possible reactor. (*From Popular Science, vol. 209, no. 6, p. 69, December, 1976.*)

The neutrons produced in the deuterium-tritium fusion reaction can cause severe problems due to neutron activation, but most fusion systems propose using some of these neutrons to generate more fuel (hydrogen-3). In the system shown in Fig. 2.15, the neutrons react with the lithium coolant to produce tritium (hydrogen-3) and additional energy in the following reaction:

$$\frac{1}{0}n + \frac{6}{3}Li \longrightarrow \frac{4}{2}He + \frac{3}{1}H + 4.8 \text{ MeV}$$

Another scheme, called the fission-fusion hybrid reactor, calls for the use of the fusion neutrons to react with fertile material, U-238 or Th-232, to convert it into fission fuel, Pu-239 or U-233, for consumption in fission reactors. The hybrid plan effectively increases the net energy from fusion by an order of magnitude, but "purists" dislike the idea of linking the fusion reactor to the fission reaction.

In spite of the large research effort to perfect the fusion reactor, the finished product is not close at hand. To achieve a controlled fusion reaction, three things are required—a high kinetic energy (temperature), a high particle density N, and a long confinement time t. It is estimated that an exothermic fusion reaction can be attained if the ignition temperature can be attained along with an Nt product of 10^{21} ion·s/m^3. The product of Nt of 10^{21} ion·s/m^3 is called the Lawson criterion. Many of the experimental fusion machines can achieve two of the three requirements, but not the third.

2.4 SOLAR ENERGY

2.4.1 Introduction

The most abundant continuing energy source available to the human race is solar energy—specifically, the electromagnetic energy emitted by the sun. While solar energy is not being used as a primary source of fuel energy at the present time, a large research and development effort is under way to develop economical systems to harness solar energy as a major source of fuel energy, particularly for the heating and cooling of buildings.

Solar energy is very attractive because it is nonpolluting, nondepletable, reliable, and free. The two principal disadvantages of solar energy are that it is very dilute and it is not constant. The low solar energy flux dictates the use of large surface-area collectors and systems to collect and concentrate the energy. While these collection systems are relatively expensive, another possibly greater problem arises from the fact that terrestrial systems cannot expect to receive a continuous supply of solar energy. This means that some sort of energy-storage system or another conversion system is required to supply energy at night and during prolonged cloudy weather. The storage system or alternative conversion system adds significantly to the expense of the overall solar unit.

Solar energy can be converted directly into other energy forms in three separate processes—the heliochemical process, the helioelectrical process, and the heliothermal process. The principal heliochemical reaction is the photosynthesis process. As was discussed earlier, this process is the source of all fossil fuels. The principal helioelectrical process is the production of electricity by solar cells. These systems are discussed in a later section. The heliothermal process is the absorption of the solar radiation and the conversion of this energy into thermal energy. This is the only solar-conversion process that has a conversion efficiency of 100 percent and the balance of the chapter will be devoted to this process.

The amount of solar radiation on a surface is called the solar insolation. The total solar insolation on a given surface is composed of a direct (beam) component and a diffuse (scattered) component as well as short-wavelength reflected radiation from other terrestrial surfaces. The direct insolation on a surface normal to the sun's rays depends on the time of year, the time of day, and the latitude of the surface as well as the atmospheric conditions. The calculational method presented in this chapter closely parallels the procedure outlined in the ASHRAE "Handbook of Fundamentals."

2.4.2 Solar Energy Calculations

Sun times The earth travels around the sun on an elliptical orbit that is nearly circular. At the nearest point, on December 21, the earth is about 1.45×10^{11} m (89.83 million mi) from the sun, while at the furtherest point, on June 22, the earth is about 1.54×10^{11} m (95.9 million mi) from the sun. The mean sun time is the local sun time if the earth moved around the sun at a constant velocity. The elliptical orbit of the earth means that it does not move at a constant velocity and that at various times of the year the sun appears to be earlier or later than the mean sun time. The difference between the true sun time, called the apparent solar time (AST), and the mean sun time is called the "equation of time."

The equation of time is not an equation but is simply a correction factor that depends upon the time of year. These correction values range from $+16.3$ min in November to -14.4 min in February. Monthly values of the equation of time are presented in Table 2.5.

The mean sun time can be calculated directly from the local longitude. Since the earth revolves 360° in 24 h, one degree of earth rotation corresponds to [24(60)/360] or 4 min. There is an imaginary longitude line running through the approximate center of each time zone (on 15-degree longitudinal increments) called the standard meridian of the time zone. At this longitude, the mean sun time and the local standard time are identical. East or west of the standard meridian, the mean sun time is respectively later or earlier (4 min per degree of rotation) than the local standard time:

Mean sun time = MST = local standard time $\pm$ [degrees east ($+$) or

$$\text{west } (-) \text{ of the standard meridian] (4 min)} \quad (2.28)$$

Table 2.5 Parameters for solar calculations (on the 21st day of each month)

Month	January	February	March	April	May	June	July	August	September	October	November	December
Day of the year	21	52	80	111	141	173	202	233	265	294	325	355
Declination, degrees	−19.9	−10.6	0.0	+11.9	+20.3	+23.45	+20.5	+12.1	0.0	−10.7	−19.9	−23.45
Equation of time, min	−11.2	−13.9	−7.5	+1.1	+3.3	−1.4	−6.2	−2.4	+7.5	+15.4	+13.8	+1.6
Solar noon		Late		Early			Late			Early		
A, Btu/h·ft^2†	390	385	376	360	350	345	344	351	365	378	387	391
B, 1/m	0.142	0.144	0.156	0.180	0.196	0.205	0.201	0.177	0.160	0.149	0.149	0.142
C, dimensionless	0.058	0.060	0.071	0.097	0.121	0.134	0.136	0.122	0.092	0.073	0.063	0.057

† A is the apparent solar irradiation at air mass zero for each month.
B is the atmospheric extinction coefficient.
C is the ratio of the diffuse to direct normal irradiation on a horizontal surface.
Reprinted with permission from the 1974 Applications Volume, ASHRAE, "Handbook and Product Directory."

Since many time zones operate on daylight-saving time during the summer,

Mean sun time = MST = (local daylight time − 1:00) ± [degrees east (+)

$$\text{or west } (-) \text{ of the standard meridian}] \text{ (4 min)} \quad (2.29)$$

The standard meridians to the time zones in the United States are located at longitude 75° (EST), at longitude 90° (CST), at longitude 105° (MST), and at longitude 120° (PST).

Once the mean sun time has been evaluated, the apparent solar time (AST) can be determined by simply adding the equation of time to the mean sun time (MST):

$$\text{Apparent solar time} = \text{AST} = \text{mean sun time} + \text{equation of time} \quad (2.30)$$

The apparent solar time is used to evaluate some of the solar angles used in solar energy calculations.

Solar angles The axis of rotation of the earth (the polar axis) is always inclined at an angle of 23.45° from a normal to the ecliptic plane. The ecliptic plane is the plane of travel of the earth as it traverses the sun. The rotational axis of the earth is essentially unidirectional with respect to a fixed star and the inclination of the polar axis causes the earth to appear to "wobble" with respect to the sun as it moves around the sun. This phenomenon is depicted in Fig. 2.16. This "wobble" produces the seasonal variations and produces an important solar angle called the angle of declination δ.

The declination angle is defined as the angle between the sun's rays and the normal to the polar axis in the plane of the sun's rays. For the northern latitudes, δ ranges from 0° at the vernal equinox (March 21), to +23.45° at the summer solstice (June 22), to 0° at the autumnal equinox (September 23), to −23.45° at the winter solstice (December 22). The monthly values of the angle of declination are presented in Table 2.5.

There are several other angles that are essential in solar energy calculations. At a given location of latitude L, the sun's position can be defined in terms of the altitude angle β_1 and the azimuth angle α_1. The altitude angle β_1 is the angle between the sun's rays and the horizontal to the earth. The azimuth angle α_1 is the angle between the horizontal projection of the sun's rays and the due-south line going in a clockwise direction. These angles can be calculated using the latitude angle L, the declination angle δ, and the hour angle H, as follows:

$$\sin \beta_1 = \cos L \cos \delta \cos H + \sin L \sin \delta \quad (2.31)$$

$$\sin \alpha_1 = \frac{\cos \delta \sin H}{\cos \beta_1} \quad (2.32)$$

The hour angle H, like the azimuth angle of the sun's rays, is positive after noon and negative before noon. It can be calculated from the following equation:

$$H = 0.25 \text{ [number of minutes before } (-) \text{ or after } (+) \text{ noon, AST]} \quad (2.33)$$

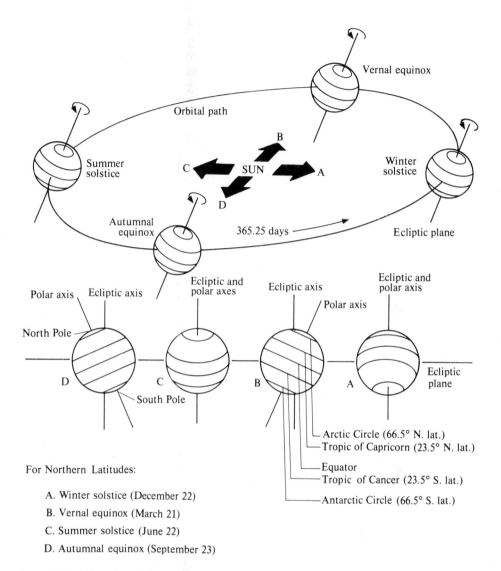

For Northern Latitudes:

A. Winter solstice (December 22)
B. Vernal equinox (March 21)
C. Summer solstice (June 22)
D. Autumnal equinox (September 23)

Figure 2.16 Orientation of the earth with respect to the sun's rays.

The values of α_1 and β_1 are tabulated in App. K along with the direct solar insolation values at the different latitudes.

In order to determine the angle between the sun's rays and the normal to the surface θ, the orientation of the surface must be established. The azimuth angle α_2 of the surface is the angle between the horizontal projection of the normal to the surface and the due-south line measured in a clockwise direction. The tilt angle β_2 of the surface is the angle between the surface and the horizontal.

Once α_1, α_2, β_1, and β_2 have been evaluated, the value of the angle θ can be evaluated from the following equation:

$$\cos \theta = \sin \beta_1 \cos \beta_2 + \cos \beta_1 \sin \beta_2 \cos (\alpha_1 - \alpha_2) \tag{2.34}$$

For a horizontal surface, $\beta_2 = 0$, the $\sin \beta_2$ is zero, the $\cos \beta_2$ is unity, and Eq. (2.34) reduces to $(\cos \theta = \sin \beta_1)$ or $(\theta = 90° - \beta_1)$. For a vertical surface, $\beta_2 = 90°$, the $\sin \beta_2$ is unity, and the $\cos \beta_2$ is zero. Equation (2.34) reduces to $[\cos \theta = \cos \beta_1 \cos (\alpha_1 - \alpha_2)]$. If the $\cos (\alpha_1 - \alpha_2)$ is negative, it means that the sun does not shine directly on the vertical wall.

Solar insolation values The amount of direct solar radiation or insolation falling on a given surface is equal to the product of the direct radiation or insolation falling on a surface normal to the sun's rays, I_{DN}, and the $\cos \theta$. The value of the direct normal insolation is a function of the thickness of the atmosphere traversed by the radiation as well as the amount of water vapor in the air and the amount of atmospheric pollution present. The atmospheric path length is usually expressed in terms of the air mass m. The parameter m is defined as the ratio of the atmospheric mass in the actual path of direct solar radiation at a given location to that present if the sun were directly overhead $(\beta_1 = 90°)$ at sea level. Outside the earth's atmosphere, $m = 0$ and at other locations, for all practical purposes, $m = 1/\sin \beta_1$.

Many people have investigated the effect of atmospheric transmittance, particularly with respect to the effect of moisture, ozone, and dust particles. The intensity of the direct normal irradiation I_{DN}, in W/m^2, at the earth's surface on a clear day may be estimated from the following equation:

$$I_{DN} = Ae^{-(B/\sin \beta_1)} \tag{2.35}$$

where A is the apparent extraterrestrial insolation (at $m = 0$) and B is the atmospheric extinction coefficient. The value of B depends on the time of year and the amount of water vapor present in the atmosphere. Monthly values of these parameters are listed in Table 2.5.

The local value of atmospheric moisture and the local elevation may be markedly different from that of the average atmosphere at sea level. The ratio of the actual clear-day direct insolation at a particular location to the value for the standard atmosphere at the same location and date is called the clearness number. The clearness number for various locations in the United States is shown in Fig. 2.17 along with the average daily solar insolation. The lines of equal insolation are called isopleths.

The total solar energy flux $I_{T\theta}$ on a terrestrial surface of any orientation and tilt with incident angle θ is equal to the sum of the direct solar component $I_{DN} \cos \theta$, the diffuse component of solar irradiation I_{DS} coming from the sky, and the reflected short-wave radiation from surrounding terrestrial surfaces I_R:

$$I_{T\theta} = I_{DN} \cos \theta + I_{DS} + I_R \tag{2.36}$$

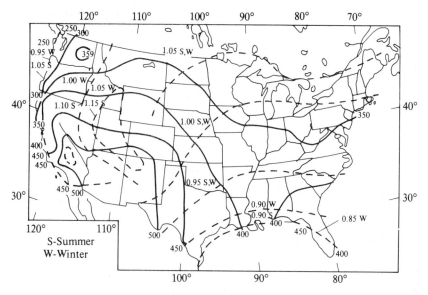

Figure 2.17 Clearness numbers for winter and summer in the United States (dashed lines) and approximate average annual insolation (solid lines), in langleys per day. *(Reprinted with permission from the 1974 Applications Volume, ASHRAE "Handbook and Product Directory.")*

The reflected component I_R depends upon the geometry and reflection characteristics of the surrounding surfaces. The amount of reflected radiation can be calculated using conventional radiation methods. Except for special applications, I_R is not normally a major component in most solar energy calculations.

The diffuse solar component I_{DS} is difficult to calculate because of the non-directional nature of the component, but for those surfaces that are exposed to direct solar radiation it is a function of the direct solar component. The diffuse solar insolation can be estimated from the following expression:

$$I_{DS} = CI_{ND}F_{ss} \qquad (2.37)$$

where C is the ratio of the diffuse to direct solar radiation falling on a horizontal surface and F_{ss} is the angle factor between the surface and the sky. The value of C is tabulated along with A and B in Table 2.4, and F_{ss} can be approximated by

$$F_{ss} = \frac{1 + \cos \beta_2}{2} \qquad (2.38)$$

The direct, normal insolation I_{DN} can be estimated from Eq. (2.35) or from the solar tables presented in App. K, as a function of the latitude, the time of year, and the apparent solar time.

The amount of solar energy absorbed by a given surface is equal to the product of the incident solar irradiation $I_{T\theta}$ and the absorptivity or emissivity of the surface for solar radiation ε_{SU}. The emissivity of a surface may be considerably different for solar radiation and ordinary thermal radiation because of

Table 2.6 Absorptivities of various surfaces for thermal radiation

Materials	Normal radiation	Solar radiation
Building materials		
Brick, red, tile, concrete	0.85–0.95	0.65–0.77
Brick, yellow and buff	0.85–0.95	0.50–0.70
Plaster	0.90–0.95	0.50–0.70
Asphalt pavement	0.90–0.95	0.82–0.88
Roofing materials		
Asphalt	0.90–0.95	0.86–0.90
Galvanized iron, dirty	0.25–0.35	0.87–0.91
Galvanized iron, new	0.20–0.30	0.64–0.68
Roofing paper	0.85–0.95	0.85–0.90
Slate	0.90–0.98	0.85–0.95
Paints		
Aluminum (bright), gilt	0.25–0.65	0.30–0.50
Black, flat	0.95–0.98	0.85–0.95
Dark (red, brown, green, etc.)	0.75–0.95	0.65–0.83
White, flat	0.95–0.99	0.12–0.25
Metals		
Aluminum, nickel, chromium (polished)	0.02–0.10	0.10–0.40
Copper, brass, monel metal (polished)	0.02–0.15	0.30–0.50
Dull aluminum, brass, copper, and polished iron	0.20–0.50	0.40–0.65
Iron oxide	0.60–0.65	0.70–0.80

the short-wavelength nature of the solar energy. The solar emissivities of some common surfaces are listed in Table 2.6. An example of the difference between solar and ordinary emissivities is flat white paint. For normal radiation, the emissivity of flat white paint ranges from 0.95 to 0.99 while the value for solar radiation ranges from 0.12 to 0.25.

Example 2.7 The southwest roof of a building is made of dirty galvanized iron and is inclined at an angle of 70° from the vertical. It is located at 38° North latitude and at a longitude two degrees west of the standard meridian for the time zone. Evaluate the combined absorbed direct and diffuse solar energy flux at 1:00 P.M., local daylight savings time on July 21.

SOLUTION Given: $\beta_2 = 20°$, $\alpha_2 = 45°$, local time = 1:00 P.M., DST, $L = 38°$
From Table 2.5: $\delta = 20.5°$, equation of time = -6.2 min,
$A = 344$ Btu/h·ft^2 = 1085 W/m^2, $B = 0.201$, $C = 0.136$
From Table 2.6: $\varepsilon_{SU} = 0.89$
Mean sun time = 1:00 − 1:00 + 0:04(−2) = −0:08 = 11:52 A.M.
Absolute solar time = MST + equation of time = 11:52 + (−6.2)

$$= 11:45.8 \text{ A.M.}$$

$$H = 0.25(11{:}45.8 - 12{:}00) = 0.25(-14.2) = -3.55°$$

$$\sin \beta_1 = \cos L \cos \delta \cos H + \sin L \sin \delta$$

$$= \cos (38) \cos (20.5) \cos (-3.55) + \sin (38) \sin (20.5) = 0.9523$$

$$\beta_1 = 72.23°$$

$$\sin \alpha_1 = \frac{\cos \delta \sin H}{\cos \beta_1} = \frac{\cos (20.5) \sin (-3.55)}{\cos (72.23)} = -0.1900$$

$$\alpha_1 = -10.95°$$

$$\cos \theta = \sin \beta_1 \cos \beta_2 + \cos \beta_1 \sin \beta_2 \cos (\alpha_1 - \alpha_2)$$

$$= \sin (72.23) \cos (20) + \cos (72.23) \sin (20) \cos (-10.95 - 45)$$

$$= 0.9533$$

$$I_{DN} = Ae^{-(B/\sin \beta_1)} = 1085e^{-(0.201/\sin 72.23)} = 878.54 \ W/m^2$$

$$F_{ss} = \frac{1 + \cos \beta_2}{2} = \frac{1 + \cos 20}{2} = 0.9698$$

$$I_{T\theta} = I_{DN} \cos \theta + CI_{DN} F_{ss} = I_{DN}(\cos \theta + CF_{ss})$$

$$= 878.54[0.9533 + 0.136(0.9698)] = 953.39 \ W/m^2$$

Absorbed solar energy flux $= \varepsilon_{SU} I_{T\theta} = 0.89(953.39) = 848.51 \ W/m^2$

The calculation methods presented in this section permit the estimation of the incident solar radiation on a surface on a clear sunny day at any time of the year. The number of clear days available at a given location can be estimated from data compiled by the U.S. Weather Bureau, now called the National Oceanographic and Atmospheric Administration (NOAA). Maps of the average monthly and average annual percentage of possible sunshine along with other atmospheric and weather data are presented in the "Climatic Atlas of the United States."

2.4.3 Solar Collection Systems

Solar energy collection systems can normally be divided into three general categories. This includes those systems that produce low-temperature (less than 150°C) thermal energy for the heating and cooling of buildings; the solar-cell conversion systems that produce electricity directly from the electromagnetic energy of the sun; and, finally, those systems that produce high-temperature thermal energy for the generation of electrical energy.

The low-temperature solar heating-cooling systems normally employ a glass- or plastic-covered flat-plate collector. The glass or plastic effectively "traps" the short-wavelength solar radiation. The heat generated in the collector is usually removed by either air or by an ethylene glycol-water solution. Most flat-plate collection systems are similar in appearance and similar in operation to the schematics shown in Fig. 2.18.

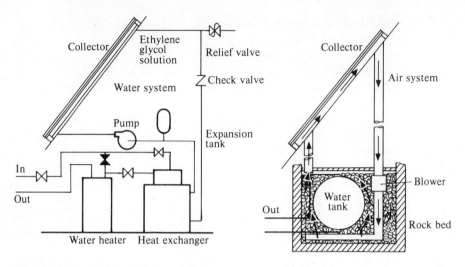

Figure 2.18 Typical flat-plate solar collection systems.

Solar cells have found widespread applications in the space program because of the high power-to-weight ratio. So far, however, high cost has precluded the use of these cells for terrestrial production of electricity. The solar-cell conversion systems are discussed in Chap. 8.

With the exception of solar cells, the production of electrical energy from solar energy requires the concentration of solar energy to achieve high temperatures. In order to attain temperatures above 150°C, the incident solar radiation must be concentrated. The degree of concentration required to obtain a given temperature can be estimated from Fig. 2.19. Solar energy can be concentrated with the use of either parabolic mirrors, with an array of focused mirrors, or with the use of concentrating lens.

One of the largest solar collection systems presently in operation is the solar furnace at Odeillo, France. This system uses a large parabolic mirror to generate very high temperatures.

The U.S. Energy Research and Development Administration (ERDA) has awarded a total of $8,000,000 to four industrial teams for the preliminary design of a 10-MW$_e$ pilot power plant. All of these proposals involve the use of a large array of mirrors to focus the solar energy on a collection tower where water is converted into high-pressure steam for the generation of electricity. This concept is called the "solar-power tower." The McDonnell-Douglas concept, shown in Fig. 2.20, uses 2470 tracking mirrors, called "heliostats," in a 1800 by 2000-ft field to focus the sun's rays on a 312-ft-high tower.

One of the more ambitious solar energy projects has been proposed by Dr. Peter E. Glaser and involves the use of a large earth satellite in a geosynchronous orbit of 22,600 mi to receive the solar energy. This satellite (see Fig. 2.21), covered with solar cells, converts the electromagnetic energy of the sun into electrical energy. This energy is then converted into microwave energy so that it can be beamed to an earth receiving system. One of the major advantages of the solar-powered satellite system (SPSS) is that it receives an almost constant supply of solar energy except for a few days around each equinox when the satellite enters the earth's shadow. It also receives a significantly higher energy flux than a comparable terrestrial system since there is no atmosphere between the sun and the SPSS.

In a proposed 5000-MW_e system, a satellite with a surface area of about 50 km^2 would transmit the microwave radiation through a transmitting antenna 1 km in diameter to a terrestrial receiving antenna 7 km in diameter. After an initial solar-electrical conversion efficiency of 15 to 20 percent, the transmission efficiency and subsequent electrical conversion is around 55 to 75 percent. The microwave radiation transmitted to the earth station is converted into 40,000 V dc where it is fed to a conventional transmission system. With the advent of a larger second-generation space shuttle, it is estimated that such a system could be built for between 25 and 30 billion dollars in 1976. This is approximately five times the cost of a comparable nuclear-powered system in 1976, although the fuel cost of the solar plant would be nil.

The satellite power plant does not appear to be a viable contender for the near-term production of electrical energy because of high cost and because of the difficult problems associated with it. It may become more attractive in the future as energy costs continue to escalate and as the cost of the space shuttle becomes relatively lower.

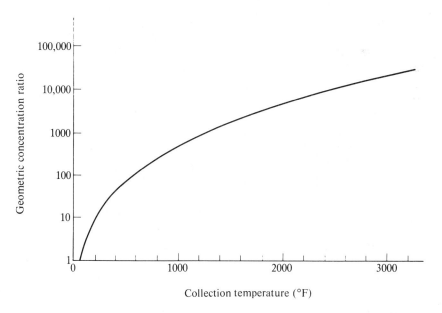

Figure 2.19 The approximate geometric concentration required for a given temperature from a high-performance solar collector.

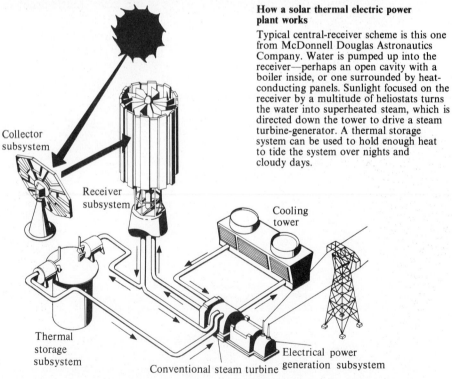

How a solar thermal electric power plant works

Typical central-receiver scheme is this one from McDonnell Douglas Astronautics Company. Water is pumped up into the receiver—perhaps an open cavity with a boiler inside, or one surrounded by heat-conducting panels. Sunlight focused on the receiver by a multitude of heliostats turns the water into superheated steam, which is directed down the tower to drive a steam turbine-generator. A thermal storage system can be used to hold enough heat to tide the system over nights and cloudy days.

Collector subsystem

Receiver subsystem

Cooling tower

Thermal storage subsystem

Electrical power generation subsystem

Conventional steam turbine

Figure 2.20 The "solar-power tower," a proposed solar electrical generation system. *(Fisher, 1975.)*

Figure 2.21 The proposed solar satellite power system for the generation of electricity. *(From Mechanical Engineering, vol. 100, no. 1, January, 1978.)*

PROBLEMS

2.1 Estimate the "as-burned" ultimate analysis for the following coals, evaluate the percentage error incurred in estimating the higher heating value with Dulong's formula, find the lower heating value (LHV), and rank the coals according to the ASTM classification system. The student should work two or three of the following parts:

(a) Luzerne County, Pa., coal with $M = 2.0$ and $A = 6.0$
(b) Coal County, Okla., coal with $M = 6.0$ and $A = 10.5$
(c) Custer County, Mont., coal with $M = 28.5$ and $A = 9.0$
(d) Stark County, N.Dak., coal with $M = 39.0$ and $A = 8.0$
(e) Wyoming County, W.Va., coal with $M = 3.0$ and $A = 4.0$
(f) Logan County, Ark., coal with $M = 3.0$ and $A = 9.0$

2.2 A sample of coal weighs 25 kg at the time it is collected. When the sample is tested, it is found that the weight has dropped to 24.6 kg as the result of moisture loss. A proximate coal analysis is then made with the following result: VM = 35.7, FC = 41.5, M = 11.6, A = 11.2, S = 4.0, and the HHV = 25,725 kJ/kg. Compute the "as-collected" proximate coal analysis. Determine the ASTM classification of the coal. Make an empirical calculation of the ultimate coal analysis using the method presented in App. D.

2.3 A 14-kg coal sample, as received, weighs 13.4 kg after air drying. When oven dried, a 2.4-g sample of the air-dried coal is reduced to a weight of 2.34 g. Find the total percentage of moisture in the as-received coal.

2.4 A 1.2-g sample of the air-dried coal of Prob. 2.3 is burned in an oxygen-bomb calorimeter. The temperature rise of 2000 g of water and the calorimeter metal is 3.62 C°, of which 0.20 C° is due to the fuse wire and acid formation. The water equivalent of the metal parts of the calorimeter is 450 g. Determine the as-received higher heating value of the coal, in kilojoules per kilogram and in British thermal units per pound mass.

2.5 A fuel oil has an API specific gravity of 32 degrees at 20°C. Compute the energy in British thermal units per gallon of oil and the higher heating value in kilojoules per kilogram. Calculate the same quantities at 60°C, if $\rho_T = \rho_{20} [1 - 0.000733(T - 20)]$, where T is in degrees Celsius.

2.6 A certain No. 2 fuel oil has a specific gravity of 36°API, while a certain No. 5 fuel oil has a specific gravity of 18°API. Compare the heating values of these oils in kilojoules per kilogram and in British thermal units per gallon.

2.7 If the cost of No. 2 fuel oil is $2.80 per million British thermal units, estimate the cost of all the available fuel oils, in cents per gallon.

2.8 If the present cost of No. 2 fuel oil is 39.9 cents per gallon, estimate the cost of all the available fuel oils in cents per gallon and in cents per million British thermal units.

2.9 A gas storage tank of 25,000 ft³ volume contains gas at a pressure of 2 bar and a temperature of 10°C. The higher heating value of the gas is 22,100 kJ/m³ at 20°C and 1 bar. Find the number of "therms" in the gas storage tank.

2.10 For Barron County, Kentucky, natural gas, find the higher heating value of the gas at 3.3 bar and 30°C, both in kilojoules per kilogram and British thermal units per cubic foot. Determine the size of the storage tank, in cubic meters, required to store 1000 therms of energy.

2.11 For Missouri natural gas, find the higher heating value of the gas in kilojoules per kilogram and in British thermal units per cubic foot if the gas is supplied at 2.5 bar and 15°C. Determine the volume of gas, in cubic feet, needed to produce 1000 therms of energy. Also, evaluate the mass fractions of each of the elements in the gas.

2.12 Repeat Prob. 2.11 for a producer gas with the following composition: 3.78% CH_4, 0.1% C_2H_6, 4.8% CO_2, 11.68% H_2, 24.4% CO, 0.6% O_2, and 54.64% N_2.

2.13 Repeat Prob. 2.11 for a carbureted water gas with the following composition: 15% CH_4, 13% C_2H_4, 34% CO, 35% H_2, 2% N_2, and 1% CO_2.

2.14 The radioisotope, Mo-99, undergoes beta decay with the emission of a 0.74-MeV gamma ray. Find the kinetic energy of the residual nucleus and the maximum kinetic energy of the beta particle. Also, determine the mass of Mo-99 that will produce an activity of 1 mCi after 100 days.

2.15 Calculate the maximum kinetic energy, in million-electronvolts, of the emitted particle and the kinetic energy of the residual nucleus resulting from (a) the alpha decay of thorium-232 if a 0.059-MeV gamma ray is emitted in the process and (b) the beta decay of strontium-90 if there is no gamma radiation emitted in the process.

2.16 Calculate the maximum kinetic energies, in million-electronvolts, of the product nuclei resulting from (a) the alpha decay of Po-210 if a 0.8-MeV gamma ray is emitted in the reaction and (b) the beta decay of Al-28 if a 1.78-MeV gamma ray is emitted in the reaction.

2.17 Radioactive C-14 (radiocarbon) is used to determine the age of materials of organic carbon. It is formed by an (n, p) reaction between nitrogen and neutrons produced by cosmic radiation. Organic substances, when living, absorb carbon dioxide from the atmosphere which contains about 0.1% $^{14}CO_2$. When dead, no more carbon dioxide is absorbed, and the proportion of C-14 in the material decreases with time as it decays. The amount of carbon in an old manuscript was determined and the fraction of radiocarbon was found to be 0.07 percent. What is its age?

2.18 Rutherford proposed that when the earth was first formed, it contained equal amounts of U-235 and U-238. There are now 137.88 atoms of U-238 for each atom of U-235. Estimate the age of the earth.

2.19 If the U-234 isotope occurring in nature is due solely to the alpha decay of U-238, calculate the isotopic abundance of U-234 compared to U-238. Assume that the rate of change of U-234 atoms with time is zero.

2.20 A sample of radioactive material is composed of two independent radioisotopes. A geiger counter, whose output is directly proportional to the total activity of the sample, gave the following results:

Time, h	0	10	20	30	40	60	90	120	180	240	360
Counts per min	5500	3883	2778	2020	1500	892	510	364	255	199	125

Plot the data on semilog graph paper and determine the half-life and the decay constant for each of the two radioisotopes. Assume that the half-life of one of the isotopes is much smaller than that of the other.

Hint: Assume that all of the activity at long times is due to the long-lived isotope. Determine the activity of the long-lived product by drawing a straight line through the end points and subtract the short-time values of that activity from the total activity to determine the activity of the short-lived radioisotope.

2.21 Repeat Prob. 2.20 for the following data:

Time, h	0	10	20	40	60	90	120	180	240	360
Counts per min	4850	2887	1761	734	380	210	152	95	61	26

2.22 It is desired to produce a Pu-238-fueled radioisotope power source which will supply a minimum of 50 W_e for a 10-year period from a thermoelectric generator with a conversion efficiency of 8 percent. Find the initial mass of radioactive material needed and the volume, in cubic centimeters, required to limit the maximum helium gas pressure to 150 bar at 300°C. There are eight stages of alpha decay in going from Pu-238 to Pb-206.

2.23 It is desired to produce a radioisotope-powered thermoelectric generator with a thermal efficiency of 7 percent, which will provide at least 50 W_e for a period of 15 years and at least 1 kW_e for a period of 180 days. The source will be constructed of PuO_2 (90% Pu-238) for the long-term requirement and Cm_2O_3-AmO_2 (using Cm-242) for the short-term requirement. Determine the minimum mass of each fuel material that should be initially loaded into the generator to meet the fuel requirements and determine the initial amount of thermal energy that must be "dumped."

2.24 Calculate the number of uranium-235 atoms that can be recovered from 1 metric ton of uranium ore. The ore contains 0.5% UO_2 by weight and 80 percent of this compound can be extracted from the ore in the refining process. Assume that 70 percent of the U-235 atoms can be effectively separated from the other uranium isotopes.

2.25 A rock returned from the surface of the moon has a ratio of Rb-87 to Sr-87 (which is produced as the result of beta decay of Rb-87) atoms of 14.45. Assuming that all of the Sr-87 was Rb-87 upon formation of the solar system, estimate the age of the solar system.

2.26 In the gaseous diffusion process, the separation factor for each stage of diffusion is equal to the square root of the ratio of the molecular weights of the UF_6 molecules. Determine the number of stages required to produce 93.6 percent enriched uranium from natural uranium and assume that the balance is U-238.

2.27 Find the energy release, in thermal megawattdays, resulting from the complete fissioning of all the atoms in (*a*) 1 g of U-235, (*b*) 1 g of U-233, and (*c*) 1 g of Pu-238. Assume that 200 MeV of energy is released in each fission reaction.

2.28 Find the energy released when one atom of U-233 fissions and produces Xe-136, another fission product, two neutrons, and two beta particles.

2.29 A U-235 nucleus absorbs a neutron and forms an excited U-236 compound nucleus. If the excited compound nucleus fissions and eventually produces Ba-138, another fission product, two neutrons, four beta particles, and gamma radiation, find the total energy released in this particular fission process. If the excited compound nucleus decays to ground state in the radiative-capture process, find the total capture gamma-ray energy released in the process.

2.30 Find the energy released when one atom of Pu-239 absorbs a neutron and the resulting excited compound nucleus:

(a) Undergoes a radiative-capture reaction.

(b) Fissions and produces the two stable fission products, Ba-135 and Pd-102.

Compare the answer in (a) with the average binding energy per nucleon in the target nucleus. Assume the incident neutron has negligible kinetic energy.

2.31 For a 3800-MW_{th} reactor operation at a load factor of 80 percent for a period of one year, find:

(a) The mass of U-235 consumed in the fission process

(b) The mass of fission products produced in the reactor of (a) if 2.5 neutrons and 200 MeV of energy are produced in the average fission

(c) The mass of deuterium and tritium consumed in the D-T reaction if the reactor is a fusion reactor

(d) The mass of boron-11 and hydrogen-1 consumed in the thermonuclear fission reaction

2.32 A given power reactor operates at a power level of 2000 MW_{th} for 200 days at which time the power level is increased to 4000 MW_{th} for another 100 days when the reactor is shut down for refueling. Calculate the reactor power level for the following times after shutdown: (a) 1 h, (b) 1 day, (c) 1 week, and (d) 1 month.

2.33 Repeat Prob. 2.32 for the case where the reactor is operated at 4000 MW_{th} for 150 days followed by a 100-day period of operation at 1500 MW_{th}, followed finally by a 50-day period of operation at 4000 MW_{th}.

2.34 Find the energy release, in thermal megawattyears, from the fusion of all the atoms of deuterium in 1 mi^3 of fresh water if it is consumed in the D-T fusion reaction.

2.35 Calculate the energy release in British thermal units per gallon for No. 2 fuel oil if all the deuterium atoms can be "burned" in the four fusion reactions involving deuterium. Assume that all the reactions proceed with equal probability and that the H-3 and He-3 nuclei needed for two of the reactions are provided from the products of the other two fusion reactions.

2.36 If the price of raw fuel is 80 cents/million Btu, find: (a) the cost of Vermillion County, Illinois, coal ($M = 9.0\%$, $A = 9.5\%$), in dollars per kilogram; (b) the cost of natural uranium oxide (U_3O_8), in dollars per kilogram, assuming that 80 percent of the U-235 atoms can be fissioned; (c) the cost of heavy water (D_2O), in dollars per kilogram, assuming that all the heavy hydrogen can be utilized in one of the four fusion reactions; and (d) the cost of boron, in dollars per kilogram, if all the boron-11 atoms can be consumed in the "thermonuclear fission" reaction and the cost of hydrogen-1 is negligible. As in Prob. 2.35, assume that all the fusion reactions involving H-2 "go" with equal probability.

2.37 Find the apparent solar times (AST) and the hour angles for the following times and locations:

(a) 10 A.M., EST, 72° West longitude on October 21

(b) 12 noon, CST, 88° West longitude on January 21

(c) 2 P.M., MST, 110° West longitude on November 21

(d) 4 P.M., PDT, 119° West longitude on July 21

2.38 Calculate the sun's altitude and azimuth angles at 40° North latitude at 3:00 P.M., AST, on November 21. Compare your values with the values in App. K.

2.39 Calculate the angle of incidence θ for a surface that faces south-southeast and is located at 38° North latitude and 92° West longitude at 11:30 A.M., CST, on November 21, if

(a) The surface is horizontal

(b) The surface is inclined at an angle of 38° with the horizontal

(c) The surface is a vertical wall

2.40 The southeast roof of a building is covered with roofing paper and is inclined at 40° from the horizontal. It is located at 30° North latitude and at 82° West longitude. Evaluate the combined absorbed solar and diffuse solar energy flux at 9:00 A.M., EST, on January 21 if the sky is clear.

2.41 The west wall of a building is constructed of red brick and is located at 45° North latitude and at 116° West longitude. Evaluate the combined absorbed solar and diffuse solar energy flux at 6:30 P.M., PDT, on May 21 if the weather is clear.

REFERENCES

ASHRAE: "Handbook of Fundamentals," American Society of Heating, Refrigeration and Air Conditioning Engineers, New York, 1974.

Benedict, Manson: Enrichment: A Critical Status Report, *Transactions of the American Nuclear Society*, Nov. 14–19, 1976, vol. 25, pp. 44–56, Washington, D.C.

Boltz, R. E., and G. L. Tuve (eds.): "Handbook of Tables for Applied Engineering Science," The Chemical Rubber Company, Cleveland, Ohio, 1970, pp. 348–349.

"Climatic Atlas of the United States," U.S. Government Printing Office, Washington, June, 1968.

De Lorenzi, Otto (ed.): "Combustion Engineering," Combustion Engineering-Superheater, Inc., New York, 1952.

Dillio, C. C., and E. P. Nye: "Thermal Engineering," International Textbook Company, Scranton, Pa., 1963.

El-Wakil, M. M.: "Nuclear Heat Transport," International Textbook Company, Scranton, Pa., 1971.

Fisher, A.: Solar Power Tower, *Popular Science*, pp. 88–90, October, 1975.

Foster, A. R., and R. L. Wright: "Basic Nuclear Engineering," Allyn and Bacon, Inc., Boston, Mass., 1976.

Post, R. F.: Nuclear Fusion, in "Annual Review of Energy," vol. I, Annual Reviews, Inc., Palo Alto, Calif., 1976.

"Steam/Its Generation and Use," 38th ed., Babcock and Wilcox Company, New York, 1972.

THREE

PRODUCTION OF THERMAL ENERGY

3.1 INTRODUCTION

Thermal energy is a basic form of energy in that all other energy forms can be completely converted into thermal energy. In fact, unless the energy is stored as some other form, it will eventually be degraded into thermal energy. The verb "degraded" is used here because the conversion of thermal energy into the other forms of energy is limited to something less than 100 percent.

3.2 CONVERSION OF MECHANICAL ENERGY

The conversion of mechanical energy into thermal energy can be summarized by one word—*friction*. In many processes, friction is considered to be an undesirable phenomenon and every effort is made to reduce or eliminate it. This is particularly true in most thermodynamic or lubrication processes. In any thermodynamic process, friction converts mechanical energy into thermal energy and makes the process irreversible.

Friction is not always bad, however. If it were not for the friction between the soles of your shoes and the floor, it would be impossible to walk. If it were not for the friction between the brake bands or disks and the drums of the automobile wheel, it would not be possible to convert the kinetic energy of the automobile into thermal energy and stop the car.

3.3 CONVERSION OF ELECTRICAL ENERGY

Electrical energy can be completely converted into thermal energy in the *joule-heating process*. This is the IE or the I^2R loss encountered when an electrical current of I A is passed through a resistance of R Ω as the result of a potential difference of E V. The resulting power loss or conversion rate is in watts.

In most electrical circuits, the joule-heating power loss is an undesirable loss, but it is one that occurs in every conductor except the superconductors that have zero electrical resistance. In some systems, such as electric ranges and furnaces, this process is useful in the production of thermal energy and there is no theoretical upper temperature limit for this conversion process as long as the conductor remains intact. In this conversion process, the energy is deposited in the volume of the conductor.

Another electrical power loss associated with alternating-current systems is the power-factor loss. When electrical energy is passed through a capacitor or an induction coil, part of the energy is stored in the electric and magnetic fields associated with each impedance. When the current reverses, the established electric and magnetic fields collapse, producing a pulse of electrical energy. If the capacitive and magnetic impedances are matched, the energy stored in the electric field is sufficient to charge the magnetic field and, conversely, there is no net power loss. If the impedances are not matched, however, the excess energy is converted into thermal energy and the power company must supply more electrical power to the customer than he is actually using.

3.4 CONVERSION OF ELECTROMAGNETIC ENERGY

The conversion of electromagnetic energy into thermal energy is accomplished in some sort of absorption process. For high-energy electromagnetic radiation, such as gamma radiation and x-rays, the absorption process is a volumetric phenomenon. For most materials, the absorption of thermal radiation is a surface-absorption process. Some materials are transparent to some wavelengths of thermal radiation and opaque to others. Glass is transparent to the ultraviolet and visible portions of the thermal spectrum but is opaque to the infrared radiation emitted by most surfaces. This leads to the so-called "greenhouse effect" and is useful in trapping solar energy.

3.5 CONVERSION OF CHEMICAL ENERGY

3.5.1 Introduction

At this instant in the history of humanity, the primary source of fuel energy is the production of thermal energy from chemical energy. The most important exothermic chemical reaction in the production of this energy is the combustion reaction. This reaction is an oxidation reaction in which the three combustible elements found in some fossil fuels, i.e., carbon, hydrogen, and sulfur, are converted respectively into carbon dioxide (CO_2), water vapor (H_2O), and sulfur dioxide (SO_2).

This section on the conversion of chemical energy deals primarily with the mechanics of the combustion reaction as well as some of the principal systems and components needed to accomplish the conversion process. A more detailed description of the actual combustion systems is presented in Chap. 4.

3.5.2 Combustion Reactions

Carbon is one of the most important combustible elements and is an essential part of any hydrocarbon compound. The oxidation of carbon is slower and more difficult than that of either hydrogen or sulfur. Although carbon has a lower ignition temperature (407°C or 765°F) than hydrogen, carbon is a high-temperature solid and as such burns relatively slowly. Consequently, in any theoretical combustion process, it will be assumed that both the sulfur and the hydrogen burn completely before the carbon burns. Furthermore, it will be assumed that all the carbon will be oxidized to carbon monoxide (CO) before any carbon is converted to carbon dioxide. This chemical reaction is

$$2C + O_2 \longrightarrow 2CO + 2Q_{C-CO} \qquad Q_{C-CO} = 110,380 \text{ kJ/(kg·mol C)}$$

In this reaction, 2 mol of carbon (24.02 kg) react with 1 mol of oxygen (32 kg) to produce 2 mol of carbon monoxide (56.02 kg).

If there is sufficient oxygen available, the carbon monoxide will then be oxidized to carbon dioxide with the release of additional energy:

$$2CO + O_2 \longrightarrow 2CO_2 + 2Q_{CO-CO_2} \qquad Q_{CO-CO_2} = 283,180 \text{ kJ/(kg·mol CO)}$$

Thus, the 2 mol of carbon monoxide (56.02 kg) react with 1 mol of oxygen (32 kg) to produce 2 mol of carbon dioxide (88.02 kg). Therefore, 64/24.02 or 2.66 kg of oxygen are required to completely burn 1 kg of carbon. This ratio is quite useful in evaluating the oxygen requirements for hydrocarbon fuels:

$$\boxed{2.66 \text{ kg } O_2/\text{kg C} \qquad \text{or} \qquad 2.66 \text{ lbm } O_2/\text{lbm C}}$$

The higher and lower heating value of carbon is 32,778 kJ/kg C or 14,093 Btu/lbm C.

Hydrogen has the highest ignition temperature (582°C or 1080°F) of the three combustible elements, but since it is a gas the kinetics are such that the combustion of hydrogen proceeds very rapidly. Consequently, if there is sufficient air, the hydrogen will burn completely to water, probably before the carbon is oxidized to carbon monoxide:

$$2H_2 + O_2 \longrightarrow 2H_2O + 2Q_H \qquad Q_H = 286,470 \text{ kJ/(kg·mol } H_2)$$

Two moles of hydrogen (4.032 kg) react with 1 mol of oxygen (32 kg) to produce 2 mol of water (36.032 kg). Thus, the mass of oxygen required to completely burn a unit mass of hydrogen is 32/4.032 or 7.94:

$$7.94 \text{ kg O}_2/\text{kg H}_2 \quad \text{or} \quad 7.94 \text{ lbm O}_2/\text{lbm H}_2$$

The higher heating value of hydrogen is 142,097 kJ/kg or 61,095 Btu/lbm H_2 and the lower heating value is 120,067 kJ/kg or 51,623 Btu/lbm H_2.

Sulfur has an ignition temperature of 243°C or 470°F, which is the lowest ignition temperature of any of the three combustible elements. While the oxidation of sulfur does release chemical energy in the following reaction, the combustion product, sulfur dioxide (SO_2), is a major atmospheric pollutant:

$$S + O_2 \longrightarrow SO_2 + Q_S \quad Q_S = 296{,}774 \text{ kJ/(kg·mol S)}$$

One mole of sulfur (32.06 kg) plus 1 mol of oxygen (32 kg) produces 1 mol of sulfur dioxide (64.06 kg). Thus, 32/32.06 or 0.998 kg of oxygen is required to burn 1 kg of sulfur:

$$0.998 \text{ kg O}_2/\text{kg S} \quad \text{or} \quad 0.998 \text{ lbm O}_2/\text{lbm S}$$

The higher and lower heating value of sulfur is 9257 kJ/kg S or 3980 Btu/lbm S.

While this completes the combustion chemistry for the three elements, a brief discussion of the source of oxygen is warranted. Almost all combustion processes rely on air as the source of oxygen. Air is composed of approximately 21 percent oxygen, by volume and by mole, and the remaining 79 percent consists primarily of nitrogen with a small amount of argon, carbon dioxide, and other gases. As far as combustion calculations are concerned, it will be assumed that air consists of 21 percent oxygen and 79 percent nitrogen on a volumetric or molar basis. These values translate to 23.2 percent oxygen and 76.8 percent nitrogen on a gravimetric or mass basis. The molecular weight of air is 28.97 kg/kg·mol or 28.97 lbm/lbm·mol.

If the combustion temperature is very high, some endothermic reactions, called dissociation reactions, take place. Some reactions of this type are shown below:

$$O_2 \longrightarrow 2O$$
$$N_2 \longrightarrow 2N$$
$$2CO_2 \longrightarrow 2CO + O_2$$

In most of the combustion processes considered in this text, the temperatures are low enough that the loss due to dissociation is negligible.

3.5.3 Theoretical Air-Fuel Ratio

The theoretical or stoichiometric air-fuel ratio gives the minimum air requirements for complete combustion of a fuel. It may be expressed in terms of mass of air per mass of fuel, in terms of moles of air per mole of fuel, or in terms of volume of air per volume of fuel. All values should normally be reported on the basis of the "as-burned" fuel analysis.

The dry, gravimetric, theoretical air-fuel ratio of a coal is determined from the as-burned ultimate analysis of the coal. This ratio is evaluated by performing an oxygen mass balance on the combustible reactants, as follows:

$$(\text{C mass fraction})(2.66) = \underline{\hspace{1cm}} \text{ kg of } O_2 \text{ to burn C in 1 kg fuel}$$

$$(H_2 \text{ mass fraction})(7.94) = \underline{\hspace{1cm}} \text{ kg of } O_2 \text{ to burn } H_2 \text{ in 1 kg fuel}$$

$$(\text{S mass fraction})(0.998) = \underline{\hspace{1cm}} \text{ kg of } O_2 \text{ to burn S in 1 kg fuel}$$

$$\text{Total} = \underline{\hspace{1cm}} \text{ kg of } O_2 \text{ to burn the combustible elements in 1 kg of fuel}$$

$$(O_2 \text{ mass fraction})(-1.0) = -\underline{\hspace{1cm}} \text{ kg of } O_2 \text{ in the fuel per 1 kg of fuel}$$

$$\text{Difference} = \underline{\hspace{1cm}} \text{ kg of } O_2 \text{ required from the air per 1 kg of fuel}$$

Dry, gravimetric (mass), theoretical air-fuel ratio $= \left(\dfrac{A}{F}\right)_{th,\,m,\,d}$

$$\left(\frac{A}{F}\right)_{th,\,m,\,d} = \frac{\text{kg } O_2 \text{ needed from the air per kg of fuel}}{0.232} \tag{3.1}$$

All the mass fractions in the above accounting system are from the as-burned ultimate fuel analysis. The factor of 0.232 in the denominator of Eq. (3.1) is the mass fraction of oxygen in the air and the ratio obtained from (3.1) can be expressed in units of kilograms of air per kilogram of fuel or in units of pounds mass of air per pound mass of fuel. The theoretical mass air-fuel ratios for most coals range between 8 and 12. This rather complex calculational procedure for determining the theoretical air-fuel ratio can be condensed into one simple equation, as follows:

$$\boxed{\left(\frac{A}{F}\right)_{th,\,m,\,d} = \frac{2.66\text{C} + 7.94H_2 + 0.998\text{S} - O_2}{0.232}} \tag{3.2}$$

Example 3.1 Calculate the theoretical air-fuel ratio for Campbell County, Tenn., coal if the as-burned moisture and ash fractions are 4 and 5 percent, respectively.

SOLUTION From App. C, the dry, ash-free ultimate analysis is: 83.1% C, 5.5% H_2, 7.4% O_2, 2.1% N_2, 1.9% S, and HHV = 34,608 kJ/kg. To convert to an as-burned basis, the multiplier is

$$(1 - M - A) = 1.00 - 0.04 - 0.05 = 1.00 - 0.09 = 0.91$$

The as-burned ultimate analysis is: 4.00% M, 5.00% A, 75.62% C, 5.01% H_2, 6.73% O_2, 1.91% N_2, 1.73% S, and $HHV = 31{,}493$ kJ/kg.

Theoretical, dry, gravimetric air-fuel ratio $= \left(\dfrac{A}{F}\right)_{\text{th, m, d}}$

$$\left(\frac{A}{F}\right)_{\text{th, m, d}} = \frac{(2.66)(0.7562) + (7.94)(0.0501) + (0.998)(0.0173) - (0.0673)}{0.232}$$

$$= 10.17 \text{ kg air/kg fuel}$$

In determining the theoretical air-fuel ratios for gaseous and liquid fuels, it is simpler to work with molar quantities than with the mass fractions of the fuel elements. Let the quantity Z be defined as the number of atoms of a given element in a mole of fuel. Z is actually the summation of the products of the mole fraction of the fuel compound and the number of moles of the particular element in the compound. The subscripts, c, h, s, o, and n, on Z, designate the summation of the products for carbon, hydrogen, sulfur, oxygen, and nitrogen, respectively. Thus, for the liquid fuel, isooctane, C_8H_{18}, $Z_c = 8$ and $Z_h = 18$. For a fuel gas composed of the following compounds: 50% CH_4, 40% C_2H_6, 5% H_2S, and 5% O_2, $Z_c = (0.5)(1) + (0.4)(2) = 1.3$, $Z_h = (0.5)(4) + (0.4) \times (6) + (0.05)(2) = 4.5$, $Z_s = (0.05)(1) = 0.05$, and $Z_o = (0.05)(2) = 0.1$.

Once the Z values have been determined, the oxygen mole balance, similar to the mass balance presented earlier, can be carried out:

$$Z_c(1) = \underline{\hspace{2cm}} \text{ moles of } O_2 \text{ needed to burn C in 1 mol of fuel}$$

$$Z_h(0.25) = \underline{\hspace{2cm}} \text{ moles of } O_2 \text{ needed to burn } H_2 \text{ in 1 mol of fuel}$$

$$Z_s(1) = \underline{\hspace{2cm}} \text{ moles of } O_2 \text{ needed to burn S in 1 mol of fuel}$$

Total $= \underline{\hspace{2cm}}$ moles of O_2 to burn combustible elements in 1 mol of fuel

$-Z_o(0.5) = - \underline{\hspace{2cm}}$ moles of O_2 in the fuel per mole of fuel

Difference $= \underline{\hspace{2cm}}$ moles of O_2 needed from the air for complete combustion of 1 mol of fuel

Dry, molar, theoretical air-fuel ratio $= \left(\dfrac{A}{F}\right)_{\text{th, mol, d}}$

$$\left(\frac{A}{F}\right)_{\text{th, mol, d}} = \frac{\text{moles of } O_2 \text{ needed from air per mole of fuel}}{0.21} \tag{3.3}$$

As before, this rather complicated calculational procedure can be condensed into one simple equation:

$$\boxed{\left(\frac{A}{F}\right)_{\text{th, mol, d}} = \frac{Z_c + 0.25Z_h + Z_s - 0.5Z_o}{0.21}} \tag{3.4}$$

For gaseous fuels, the molar and volumetric air-fuel ratios are the same and may be expressed in units of moles of air per mole of fuel or in units of cubic feet of air per cubic foot of fuel.

The dry, molar, theoretical air-fuel ratio obtained from Eq. (3.4) can be converted into a dry, gravimetric, theoretical air-fuel ratio by multiplying and dividing the molar ratio by the molecular weights of the air and the fuel, respectively:

$$\left(\frac{A}{F}\right)_{th, m, d} = \frac{28.97(A/F)_{th, mol, d}}{(molecular\ weight)_{fuel}} \tag{3.5}$$

Example 3.2 Calculate the molar air-fuel ratio when burning a LPG composed of 40% C_3H_8 and 60% C_4H_{10}. Also find the mass air-fuel ratio for the gas.

SOLUTION

$Z_c = 0.4(3) + 0.6(4) = 1.2 + 2.4 = 3.6$ moles of carbon atoms per mole

$Z_h = 0.4(8) + 0.6(10) = 3.2 + 6.0 = 9.2$ moles of hydrogen atoms per mole

$Z_s = Z_o = 0$

Theoretical, molar, dry air-fuel ratio $= \left(\dfrac{A}{F}\right)_{th, mol, d}$

$$\left(\frac{A}{F}\right)_{th, mol, d} = \frac{3.6 + 0.25(9.2) + 0 - 0}{0.21}$$

$$= 28.10\ mol\ air/mol\ gas$$

$$\left(\frac{A}{F}\right)_{th, mol, d} = \left(\frac{A}{F}\right)_{th, vol, d} = 28.10\ ft^3\ air/ft^3\ gas$$

Molecular weight (MW) of gas $= 0.4(44.094) + 0.6(58.12)$

$$= 52.51\ kg/kg \cdot mol$$

Theoretical, mass, dry air-fuel ratio $= \left(\dfrac{A}{F}\right)_{th, m, d}$

$$\left(\frac{A}{F}\right)_{th, m, d} = \frac{(28.97\ kg\ air/kg \cdot mol\ air)(28.10\ kg \cdot mol\ air/kg \cdot mol\ fuel)}{52.51\ kg\ fuel/kg \cdot mol\ fuel}$$

$$= 15.50\ kg\ air/kg\ fuel$$

3.5.4 Actual Combustion Process

The five requirements for good combustion are $MATT\rho$. $MATT\rho$ stands for: proper mixing (M) of reactants; sufficient air (A); sufficient temperature (T); sufficient time (T) for the reaction to occur; and there must be sufficient density

(ρ) to propagate the flame. Since perfect mixing is never attained in the actual combustion process, good combustion can only be assured by supplying excess air for the process. Depending on the amount of excess air supplied and the degree of mixing, the exhaust gas includes the products of complete combustion, carbon dioxide, water, and sulfur dioxide; some incomplete-combustion products including some unburned fuel, carbon monoxide, hydroxyls, and aldehydes; and nitrogen and nitrogen compounds such as nitric oxide (NO) and nitrogen dioxide (NO_2). All of these products, except water and nitrogen, are considered to be atmospheric pollutants.

There are two ways of expressing the amount of air supplied for a given combustion process—the dilution coefficient and/or the percentage of excess air. The dilution coefficient is defined as the ratio of the actual to the theoretical air-fuel ratios:

$$\text{Dilution coefficient} = \text{DC} = \frac{\text{actual } A/F \text{ ratio}}{\text{theoretical } A/F \text{ ratio}} \tag{3.6}$$

The percentage of excess air is defined by the following equations:

$$\text{Percentage of excess air} = \begin{cases} \dfrac{(\text{actual } A/F \text{ ratio}) - (\text{theoretical } A/F \text{ ratio})}{(0.01)(\text{theoretical } A/F \text{ ratio})} & (3.7) \\[2em] 100 \, (\text{dilution coefficient} - 1.0) & (3.8) \end{cases}$$

The actual air-fuel ratio for a given combustion process is normally estimated from an experimental measurement of the gaseous components of the exhaust gas. There are several ways of experimentally analyzing the concentration of gas compounds in a mixture of gases. These systems include the gas chromatograph and the orsat apparatus, among others. The gas chromatographs are very sensitive systems that can be used to detect different gas compounds, but these units are complicated and difficult to use. The orsat apparatus, on the other hand, is a relatively simple and compact portable gas analyzer that is designed to measure the concentration of some of the gaseous compounds found in the combustion products.

A typical orsat gas analyzer is shown in Fig. 3.1 and this apparatus is used to determine the volumetric or molar fractions of carbon monoxide, carbon dioxide, and oxygen in the dry exhaust gas. A 100 cm^3 exhaust-gas sample is taken at room temperature in the burette by using the leveling water bottle to collect and transfer the gas sample. Since the sample is collected at room temperature over water, it is usually assumed that any water vapor in the exhaust gas is condensed and any sulfur dioxide in the exhaust will react with the water in the flue gas or collecting bottle. Consequently, it is assumed that the sample of exhaust gas is composed of carbon dioxide, oxygen, carbon monoxide, and nitrogen.

Once the gas sample has been obtained, it is then sequentially passed through three chemical reactors in the device. The first chemical reactor contains an aqueous solution of potassium hydroxide (KOH) that preferentially removes

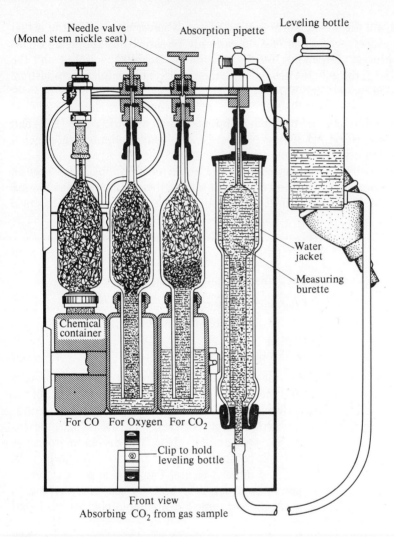

Needle valve
(Monel stem nickle seat)

Absorption pipette

Leveling bottle

Water jacket

Measuring burette

Chemical container

For CO   For Oxygen   For CO$_2$

Clip to hold leveling bottle

Front view
Absorbing CO$_2$ from gas sample

Figure 3.1 The orsat flue-gas analyzer.

any carbon dioxide in the gas sample. The second chemical reactor contains a solution of pyrogalic acid in potassium hydroxide and water and this solution preferentially removes any oxygen from the sample. The final chemical reactor contains a solution of cuprous chloride in ammonia and this absorbs any carbon monoxide present in the gas. By carefully measuring the decrease in sample volume as the gas is passed through each reactor in series and dividing each decrease by the original gas volume (100 cm^3), the volume or mole fractions of carbon dioxide, oxygen, and carbon monoxide in the dry exhaust gas are obtained. Any gas remaining (usually around 80 percent) after the sample has passed through all three reactors is assumed to be nitrogen.

The ultimate and orsat analyses are required and usually sufficient to determine the actual air-fuel ratio when burning a gaseous fuel or a liquid fuel. One important additional analysis is needed to evaluate the actual air-fuel ratio when burning a solid fuel such as coal. This analysis is the refuse analysis. The refuse analysis is simply an experimental determination of the higher heating value (HHV) of the refuse. This analysis can be reported in units of the amount of energy per unit mass of refuse (kilojoules per kilogram or British thermal units per pound-mass) or as the mass fraction or percentage of combustible in the refuse. Any combustible material appearing in the refuse is assumed to be unburned carbon. If the refuse analysis is given in terms of the higher heating value, the percentage of combustible can be found by dividing the refuse heating value by the heating value of pure carbon:

$$\text{Percentage of combustible} = \frac{100 \, (\text{HHV})_{\text{refuse}}}{(\text{HHV})_{\text{carbon}}} \qquad (3.9)$$

$$= \frac{100 \, (\text{HHV, kJ/kg refuse})}{32{,}778}$$

$$= \frac{100 \, (\text{HHV, Btu/lbm refuse})}{14{,}093}$$

Unless additional information is available, it is normally assumed that all the ash in the coal appears in the refuse pit and further that the refuse is composed of only ash and unburned carbon. If C_r is the fuel mass fraction of unburned carbon in the refuse, then R, the mass of refuse collected per unit mass of coal burned, is equal to the sum of C_r and A, where A is the ash mass fraction from the ultimate analysis. The percentage of combustible is equal to the ratio of C_r/R times 100. Thus, $(1 - C_r/R)$ is equal to $(A/R)_r$, the mass fraction of ash in the refuse. If most of the ash in the coal appears in the refuse, the refuse mass fraction can be obtained from the following equation:

$$R = \frac{\text{mass of refuse}}{\text{mass of coal}} = \frac{\text{mass fraction of ash in the coal, } A}{(A/R)_r} \qquad (3.10)$$

Once the refuse fraction R has been determined, the mass of unburned carbon in the refuse per mass of fuel consumed C_r can be found from

$$C_r = \frac{R \, (\% \text{ combustible in refuse})}{100} = \frac{C_r}{R} R \qquad (3.11)$$

$$C_r = R - A \qquad (3.12)$$

The balance of the carbon in the coal is assumed to be burned and therefore produces either carbon monoxide or carbon dioxide. C_b is defined as the mass of carbon actually burned per unit mass of fuel and is obtained from the following equation:

$$C_b = C - C_r \qquad (3.13)$$

where C is the carbon mass fraction from the as-burned ultimate fuel analysis.

Once the refuse analysis, the orsat analysis of the flue gas, and the ultimate analysis of the coal are known, the actual air-fuel ratio can be evaluated. Of all the elements in the fuel and air $(C, H_2, S, O_2, \text{ and } N_2)$, only the carbon and nitrogen can be totally accounted for in the refuse and orsat analyses. The number of moles of carbon monoxide plus the moles of carbon dioxide in the exhaust per unit mass of coal is equal to $C_b/12.01$ and the number of moles of nitrogen in the exhaust per unit mass of coal is equal to $(N_f + N_a)/28.016$, where N_f is the nitrogen mass fraction in the fuel (from the ultimate analysis) and N_a is the mass of nitrogen from the air per unit mass of coal. The ratio of the percentage of nitrogen to the percentages of carbon monoxide plus carbon dioxide from the orsat analysis is the ratio of the moles of nitrogen from the fuel and air to the moles of carbon burned in the combustion process. Using this ratio:

$$\text{Actual, mass, dry air-fuel ratio} = \left(\frac{A}{F}\right)_{act, m, d} = \frac{N_a}{0.768}$$

$$\left(\frac{A}{F}\right)_{act, m, d} = \left[\frac{(\%N_2)(28.016)}{(\%CO + \%CO_2)(12.01)}C_b - N_f\right]\Big/0.768 \qquad (3.14)$$

where the percentages of carbon monoxide, carbon dioxide, and nitrogen are obtained from the orsat analysis, C_b is obtained from the refuse and ultimate analyses, and N_f is obtained from the as-burned ultimate analysis of the coal.

A similar equation can be developed for those fuels in which the analysis is given in terms of mole fractions, again using a carbon-nitrogen balance. The calculational procedure is actually simpler because in the combustion of a gaseous or liquid fuel all the carbon in the fuel is normally converted into a gaseous product (carbon monoxide or carbon dioxide). For these fuels, the molar air-fuel ratio can be obtained from the following equation:

$$\text{Actual, molar, dry air-fuel ratio} = \left(\frac{A}{F}\right)_{act, mol, d}$$

$$\left(\frac{A}{F}\right)_{act, mol, d} = \left(\frac{\%N_2}{\%CO + \%CO_2}Z_c - 0.5Z_n\right)\Big/0.79 \qquad (3.15)$$

where Z_c are the moles of carbon atoms per mole of fuel and Z_n are the moles of nitrogen atoms per mole of fuel. Z_c and Z_n may or may not exceed unity. The actual mass air-fuel ratio can be obtained by multiplying and dividing the molar air-fuel ratio by the molecular weights of the air and fuel, respectively, as expressed in Eq. (3.5).

Theoretically, oxygen and carbon monoxide cannot appear simultaneously in the exhaust gas but they commonly both appear in the actual combustion process because of poor mixing. A commonly used rule of thumb states that the approximate percentage of excess air is five times the percentage of oxygen in the orsat analysis, if the percentage of carbon monoxide is small.

There are some circumstances in which incomplete combustion of the fuel is desired. Incomplete combustion was commonly practiced in steel mills and the resulting exhaust gas was used as a reducing atmosphere over the molten metal. This exhaust gas, called "blast-furnace gas," was then burned as a fuel gas. Burning a fuel with insufficient air gives the maximum power from a limited volume as in the internal-combustion engine and also produces higher specific power from rockets. Reducing the combustion air also reduces the formation of nitrogen oxide pollutants, although it does increase the pollution due to carbon monoxide and unburned fuel.

Example 3.3 A certain power plant burns Clay County, Mo., coal with $M = 12\%$ and $A = 10\%$, and an analysis of the refuse pit shows that the higher heating value of the refuse is 581 kJ/kg. An orsat analysis of the flue gas gives 14.91% CO_2, 3.67% O_2, and 0.15% CO. Find the dilution coefficient and the percentage of excess air.

SOLUTION Ultimate coal analysis from App. C: 79.2% C, 5.5% H_2, 8.4% O_2, 1.3% N_2, 5.6% S, and HHV = 33,422 kJ/kg.

Multiplier for as-burned analysis $= (1 - M - A)$
$$= 1.0 - 0.12 - 0.10 = 0.78$$

As-burned ultimate analysis: 12% M, 10% A, 61.78% C, 4.29% H_2, 6.55% O_2, 1.01% N_2, 4.37% S, and HHV = 26,069 kJ/kg.

%N_2 in exhaust gas $= 100.0 - (\%CO_2 + \%O_2 + \%CO)$
$$= 100 - (14.91 + 3.67 + 0.15) = 81.27\%$$

From the refuse analysis:

$$\frac{C_r}{R} = 581/32{,}778 = 0.0177 \text{ kg C in refuse/kg refuse}$$

$$\left(\frac{A}{R}\right)_r = 1.0 - \frac{C_r}{R} = 1.0 - 0.0177 = 0.9823 \text{ kg ash/kg refuse}$$

$$R = \frac{A}{(A/R)_r} = \frac{0.10}{0.9823} = 0.1018 \text{ kg refuse/kg coal}$$

$$C_r = R - A = 0.1018 - 0.10 = 0.0018 \text{ kg unburned C/kg coal}$$

$$C_b = C - C_r = 0.6178 - 0.0018 = 0.6160 \text{ kg burned C/kg coal}$$

Theoretical air-fuel ratio:

$$\left(\frac{A}{F}\right)_{th,\ m,\ d} = \frac{2.66(0.6178) + 7.94(0.0429) + 0.998(0.0437) - 0.0655}{0.232}$$

$$= 8.457 \text{ kg air/kg coal}$$

Actual air-fuel ratio:

$$\left(\frac{A}{F}\right)_{\text{act, m, d}} = \left[\frac{\%N_2}{\%CO + \%CO_2}(2.332)C_b - N_f\right]\bigg/0.768$$

$$= \frac{(81.27)(2.332)(0.6160)/(0.15 + 14.91) - 0.0101}{0.768}$$

$$= 10.081 \text{ kg air/kg coal}$$

$$\text{Dilution coefficient} = DC = \frac{(A/F)_{\text{act}}}{(A/F)_{\text{th}}} = \frac{10.081}{8.457} = 1.1920$$

$$\text{Percentage of excess air} = 100(DC - 1) = 19.2\%$$

Example 3.4 If Linn County, Mo., coal with $M = 13\%$ and $A = 11\%$ is burned with 12% excess air and 0.5% CO appears in the orsat analysis, calculate the balance of the orsat analysis. Assume that there is no unburned carbon in the refuse pit.

SOLUTION Ultimate coal analysis from App. C: 78.6% C, 5.6% H_2, 9.2% O_2, 1.3% N_2, 5.2% S, and HHV = 33,318 kJ/kg.
Multiplier for as-burned analysis $= (1 - M - A) = 1.0 - 0.13 - 0.11 = 0.76$
As-burned ultimate analysis: 13% M, 11% A, 59.74% C, 4.26% H_2, 7.07% O_2, 0.99% N_2, 3.95% S, and HHV = 25,332 kJ/kg.

$$\left(\frac{A}{F}\right)_{\text{th, m, d}} = \frac{2.66(0.5974) + 7.94(0.0426) + 0.998(0.0395) - 0.0707}{0.232}$$

$$= 8.173 \text{ kg air/kg coal}$$

$$\left(\frac{A}{F}\right)_{\text{act, m, d}} = (DC)\left(\frac{A}{F}\right)_{\text{th, m, d}} = (1.12)(8.173) = 9.153 \text{ kg air/kg coal}$$

Moles of exhaust products per kilogram of coal burned:

Moles CO/kg coal $= X$

$$\text{Moles } CO_2/\text{kg coal} = \frac{C_b}{12.01} - X = \frac{0.5974}{12.01} - X = 0.049742 - X$$

$$\text{Moles } O_2/\text{kg coal} = \frac{(DC - 1)(A/F)_{\text{th, m, d}}(0.232)}{32} + \frac{X}{2}$$

$$= \frac{(0.12)(8.173)(0.232)}{32} + \frac{X}{2} = 0.007111 + 0.5X$$

$$\text{Moles } N_2/\text{kg coal} = \frac{(A/F)_{\text{act, m, d}}(0.768) + N_f}{28.016}$$

$$= \frac{(9.153)(0.768) + 0.0099}{28.016} = 0.251264$$

$$\text{Moles dry exhaust/kg coal} = X + (0.049742 - X)$$

$$+ \left(0.007111 + \frac{X}{2}\right) + 0.251264$$

$$= 0.308116 + \frac{X}{2}$$

From orsat analysis of CO:

$$\text{Moles of CO/mole of dry exhaust} = 0.005 = \frac{X}{0.308116 + 0.5X}$$

$$0.9975X = 0.005(0.308116)$$

$$X = 0.001544 \text{ mol CO/kg coal}$$

$$\text{Moles dry exhaust/kg coal} = 0.308116 + \frac{X}{2} = 0.308888$$

Orsat analysis:

$\%\text{CO} = 0.001544/0.308888 = 0.005$	$= 0.5\%$
$\%\text{CO}_2 = (0.049742 - 0.001544)/0.308888 = 0.156$	$= 15.6\%$
$\%\text{O}_2 = (0.007111 + 0.000772)/0.308888 = 0.0255$	$= 2.55\%$
$\%\text{N}_2 = 0.251264/0.308888 = 0.8134$	$= 81.34\%$

$$\text{Total} = 99.99\%$$

In this example, one might consider how the equations would be effected if C_r were not zero. C_b would equal $(0.5974 - C_r)$ and there would be an additional term in the oxygen mole equation equal to $C_r/12.01$.

Example 3.5 A natural gas with the following molar analysis is burned in a furnace: $CO_2 = 0.5\%$, $CO = 5.0\%$, $CH_4 = 87.0\%$, $C_2H_4 = 3.0\%$, and $N_2 = 4.5\%$. An orsat analysis gives the following results: $9.39\%\ CO_2$, $3.88\%\ O_2$, and $0.83\%\ CO$. Calculate the percentage of excess air and the actual air-fuel ratio in kilograms of air per kilogram of gas.

SOLUTION From gas analysis:

$$Z_c = 0.005(1) + 0.05(1) + 0.87(1) + 0.03(2) = 0.985$$

$$Z_h = 0.87(4) + 0.03(4) = 3.6$$

$$Z_o = 0.005(2) + 0.05(1) = 0.06$$

$$Z_n = 0.045(2) = 0.09$$

$$\left(\frac{A}{F}\right)_{\text{th, mol, d}} = \frac{Z_c + 0.25\,Z_h + Z_s - 0.5\,Z_o}{0.21}$$

$$= \frac{0.985 + 0.25(3.6) + 0 - 0.5(0.06)}{0.21} = 8.833 \text{ mol air/mol gas}$$

$$\left(\frac{A}{F}\right)_{\text{th, vol, d}} = \left(\frac{A}{F}\right)_{\text{th, mol, d}} = 8.833 \text{ ft}^3 \text{ air/ft}^3 \text{ gas}$$

Molecular weight (MW) of gas $= 0.005(44.01) + 0.05(28.01)$
$$+ 0.87(16.04) + 0.03(28.05)$$
$$+ 0.045(28.016)$$
$$= 17.677 \text{ kg gas/kg} \cdot \text{mol gas}$$

From orsat analysis:

$$\%\text{N}_2 = 100 - (\%\text{CO}_2 + \%\text{O}_2 + \%\text{CO})$$
$$= 100 - (9.39 + 3.88 + 0.83) = 85.9\%$$

$$\left(\frac{A}{F}\right)_{\text{act, mol, d}} = \left(\frac{\%\text{N}_2}{\%\text{CO} + \%\text{CO}_2}\,Z_c - 0.5 Z_n\right)\bigg/0.79$$

$$= \frac{(85.9)(0.985)/(0.83 + 9.39) - (0.090)/2}{0.79}$$

$$= 10.423 \text{ mol air/mol gas}$$

$$\text{Dilution coefficient} = \frac{(A/F)_{\text{act}}}{(A/F)_{\text{th}}} = \frac{10.423}{8.833} = 1.180$$

Percentage of excess air $= (\text{DC} - 1)100 = 18.0\%$

$$\text{Actual, mass air-fuel ratio} = \left(\frac{A}{F}\right)_{\text{act, m, d}} = \frac{28.97(A/F)_{\text{act, mol, d}}}{(\text{MW})_{\text{gas}}}$$

$$= \frac{(28.97)(10.423)}{17.677} = 17.082 \text{ kg air/kg gas}$$

Example 3.6 A certain LPG has a composition of 70% C_3H_8 and 30% C_4H_{10}. Calculate the theoretical orsat analysis and the energies released per kilogram of fuel gas when it is burned with (a) a dilution coefficient of 0.9 and (b) a dilution coefficient of 0.6.

SOLUTION From the gas analysis:

$$Z_c = 0.7(3) + 0.3(4) = 3.3$$
$$Z_h = 0.7(8) + 0.3(10) = 8.6$$

Molecular weight of fuel $= 0.7(44.09) + 0.3(58.12) = 48.299$ kg/kg·mol

$$\text{HHV}_v = 0.7(\text{HHV})_{C_3H_8} + 0.3(\text{HHV})_{C_4H_{10}}$$

$$= 0.7(95,103) + 0.3(123,725)$$

$$= 103,690 \text{ kJ/m}^3 \qquad \text{at 1 atm, 20°C}$$

$$\text{Specific volume of gas} = \frac{RT}{P}$$

$$= \frac{(0.08315 \text{ bar·m}^3/\text{kg·mol·K})(293 \text{ K})}{(1 \text{ atm})(1.0133 \text{ bar/atm})(48.299 \text{ kg/kg·mol})}$$

$$= 0.4978 \text{ m}^3/\text{kg}$$

$$\text{HHV}_m = 103,690(0.4978) = 51,617 \text{ kJ/kg}$$

$$\left(\frac{A}{F}\right)_{\text{th, mol, d}} = \frac{3.3 + 0.25(8.6)}{(0.21)} = 25.952 \text{ mol air/mol gas}$$

(*a*) At 90% theoretical air:

$$\left(\frac{A}{F}\right)_{\text{act, mol, d}} = (\text{DC})\left(\frac{A}{F}\right)_{\text{th, mol, d}} = 0.9(25.952) = 23.357 \text{ mol air/mol gas}$$

$$Z_o = 23.357(0.21)(2) = 9.81 \text{ atoms/mol gas}$$

$$Z_o \text{ needed to burn } H_2 \text{ to } H_2O = 0.5Z_h = 0.5(8.6) = 4.3$$

$$Z_o \text{ needed to burn C to CO} = Z_c = 3.3$$

$$Z_o \text{ needed to burn C to CO}_2 = 2Z_c = 6.6$$

$$Z_o \text{ available to burn C} = 9.81 - 4.3 = 5.51 \qquad \text{(not enough)}$$

$$\text{Moles CO/mole fuel} = 6.6 - 5.51 = 1.09$$

$$\text{Moles CO}_2/\text{mole fuel} = 5.51 - 3.3 = 2.21$$

$$\text{Moles N}_2/\text{mole fuel} = 23.357(0.79) = 18.45$$

$$\text{Moles dry exhaust gas/mole fuel} = 18.45 + 2.21 + 1.09 = 21.75$$

Orsat analysis:

$$\%CO = 1.09/21.75 \qquad = \quad 5.01\%$$

$$\%CO_2 = 2.21/21.75 \qquad = \quad 10.16\%$$

$$\%O_2 = 0.00/21.75 \qquad = \quad 0.00\%$$

$$\%N_2 = 18.45/21.75 \qquad = \quad 84.83\%$$

$$\text{Total} = 100.00\%$$

$$\text{Energy released} = \text{HHV}_{\text{fuel}} - \frac{(\text{moles CO/mole fuel})(28.01)(\text{HHV})_{\text{CO}}}{(\text{MW})_{\text{gas}}}$$

$$= 51{,}617 - \frac{(1.09)(28.01)(10{,}110)}{48.299}$$

$$= 45{,}226 \text{ kJ/kg fuel}$$

(*b*)   At 60% theoretical air:

$$\left(\frac{A}{F}\right)_{\text{act, mol, d}} = 0.6(25.952) = 15.571 \text{ mol air/mol gas}$$

$$Z_o = 15.571(0.21)(2) = 6.540 \text{ atoms/mol gas}$$

$$Z_o \text{ available to burn carbon} = 6.54 - 4.3 = 2.24$$

This is not enough oxygen to burn the carbon to carbon monoxide so the theoretical products from the combustion of carbon will be carbon monoxide and carbon (soot).

Moles CO/mole fuel = 2.24

Moles C/mole fuel = 3.3 − 2.24

$$= 1.06 \quad \text{(carbon will not appear in exhaust)}$$

Moles N_2/mole fuel = 15.571(0.79) = 12.30

Moles dry exhaust gas/mole fuel = 12.30 + 2.24 = 14.54

Orsat analysis:

$$\%\text{CO} = 2.24/14.54 \quad = 15.41\%$$
$$\%\text{CO}_2 = 0.00/14.54 \quad = 0.00\%$$
$$\%\text{O}_2 = 0.00/14.54 \quad = 0.00\%$$
$$\%\text{N}_2 = 12.30/14.54 \quad = 84.59\%$$

$$\text{Total} = 100.00\%$$

$$\text{Energy released} = \text{HHV}_{\text{fuel}} - \frac{(\text{moles CO/mole fuel})(\text{HHV})_{\text{CO}}(28.01)}{(\text{MW})_{\text{gas}}}$$

$$- \frac{(\text{moles C/mole fuel})(\text{HHV})_{\text{C}}(12.01)}{(\text{MW})_{\text{gas}}}$$

$$= 51{,}617 - \frac{2.24(28.01)(10{,}110)}{48.299} - 1.06\frac{(32{,}778)(12.01)}{48.299}$$

$$= 29{,}844 \text{ kJ/kg gas}$$

3.5.5 Combustion Mechanics

The actual combustion process of fossil fuels proceeds in two ways. If a gaseous hydrocarbon fuel (including vaporized liquids) are mixed and heated before the actual ignition takes place, the oxygen has a chance to react with the hydrocarbon molecules in a process called hydroxylation. The compounds formed by this interaction are called hydroxylated compounds, which are unstable and are rapidly converted to aldehydes, e.g., formaldehyde (CH_2O). The aldehydes are subsequently burned to carbon dioxide and water. The resulting flame is a blue or nonluminous flame. This type of combustion is normally used in the Bunsen burner in the laboratory and in the conventional gas stove where localized heating is desired.

The other basic mode of combustion involves the introduction of fuel and air at the burner with no premixing of the reactants. This results in a very short mixing time and in a rapid heating of the fuel and air. Because of this rapid heating, the hydrocarbon compounds are "cracked" into lighter compounds and eventually into the basic elements—carbon and hydrogen. As the result of this thermal decomposition, most of the combustion occurs between elemental carbon and hydrogen. While elemental hydrogen burns with an almost invisible flame, elemental carbon burns with a characteristic yellow flame. Consequently, this mode of combustion produces a luminous flame or "yellow-flame" combustion. This type of combustion also predominates in the combustion of solid and most liquid hydrocarbon fuels. Actually, contrary to popular opinion, yellow-flame burning is usually desired in a large power boiler because it increases the radiative heat transfer from the flame to the boiler tubes, reducing the combustion temperature.

There are three basic physical methods for the combustion of fossil fuels. These include the burning bed, the traveling-flame front, and the gaseous torch. The burning-bed system is commonly employed in the combustion of solid fuels, particularly coal. The burning bed is used in the stoker furnace where the coal is burned in a stationary bed and is also employed in the fluidized-bed combustion system where the coal particles are burned in a floating bed suspended by the burning air.

The traveling-flame combustion system or mode is applicable where the reactants (fuel and oxidant) are completely mixed. When ignition occurs, the flame front progressively moves through the mixture. This process can occur very rapidly as in a spark-ignition, internal-combustion engine or more slowly as in a solid-fueled rocket.

The gaseous-torch combustion system is commonly used in large power installations and, in essence, the fuel and air are mixed and burned at the burner. Although the fuel-air mixture is burned like a gas at the burner, and, as such, is applicable to the combustion of gaseous fuels, it is also applicable to finely atomized liquid fuels and finely powdered solids, like pulverized coal. When heavy fuel oil is burned in this type of burner, it must be heated and atomized at the burner. Before a coal can be burned in this system, it must be pulverized to a texture finer than face powder. This coal is then burned in a pulverized-coal furnace.

One other combustion system is actually a combination of several of these combustion modes. This is the cyclone furnace. The cyclone furnace burns crushed coal in a flaming vortex. All of these combustion systems are discussed in greater detail, along with the associated auxiliary equipment, like fans, pumps, scrubbers, etc., in the next chapter.

3.6 CONVERSION OF NUCLEAR ENERGY

3.6.1 Introduction

As was discussed in Chaps. 1 and 2, there are three general nuclear reactions that release nuclear energy into some other energy form, usually thermal energy. These three reactions are radioactive decay, fission, and fusion. All of these reactions are discussed at length in Chap. 2, but the only reaction that presently produces sizable amounts of nuclear energy is the fission process, and the device in which this process is carried out is the nuclear fission reactor.

Radioisotope power systems are normally low-power systems and the controlled fusion reactor has yet to be developed. Consequently, the balance of this chapter is solely concerned with the nuclear fission reactor, its terminology, its principle of operation, and its makeup. The detailed description, design, and the kinetics of reactor operation will be presented in Chap. 5.

3.6.2 Nuclear Reactor Terminology

The terminology associated with the nuclear reactor and the nuclear industry is, to say the least, unusual and in some instances extremely unfortunate. Some of the terminology, which in the early stages of development seemed colorful, catchy, and cute, has subsequently returned to haunt the industry because it sounds so ominous to the average layperson who is unfamiliar with the true meaning.

The nuclear reactor is a device in which a controlled fission chain reaction is maintained in order to produce neutrons and/or energy. While there are other products from the nuclear fission reaction, e.g., the two intermediate-mass fission products and beta particles, the neutrons and energy are the most important products and are the reasons why the reactors are constructed and operated.

Every operating nuclear reactor has a critical mass (one of the ominous-sounding terms). The critical mass is simply the minimum mass of fissionable material that will just sustain a fission-chain reaction. Values of critical mass range from 200 g to more than 500 kg. The actual fuel loading for any reactor must exceed the critical mass to compensate for fuel burnup and other effects encountered during operation of the reactor.

The multiplication factor k for a reactor is defined as the ratio of the number of neutrons produced in one generation divided by the number of neutrons produced in the preceding generation. Since the fission process is initiated by neutrons, the fission rate or power is directly proportional to the neutron level in the reactor. Thus, if the multiplication factor is greater than unity, the neutron population is increasing with time, the reactor power is increasing with time, and the reactor is said to be *supercritical*. If k is exactly unity, the neutron population and the reactor power are constant and the reactor is said to be *critical*. If k is less than unity, the neutron population and the reactor power are decreasing with time and the reactor is said to be *subcritical*.

Another term which is closely related to the multiplication factor is the reactivity of the reactor, designated by the symbol ρ. In fact, the reactivity is actually defined in terms of the multiplication factor k as follows:

$$\rho = \frac{k - 1}{k} \tag{3.16}$$

Thus, for a supercritical reactor, ρ is positive; for a critical reactor, ρ is zero; and for a subcritical reactor, ρ is negative. The system reactivity is useful in the treatment of reactor kinetics and will be used extensively in Chap. 5.

Since the neutron population and the fission reaction rate is directly proportional to the thermal power of the reactor, it is important to have some quantity or term that designates the neutron level in the reactor. The term that is used is the neutron flux ϕ. The neutron flux is the number of neutrons passing through a unit area in a unit time and has units of neutrons per square meter per second (neutrons/m^2·s). The neutron flux is a scalar quantity and ranges in magnitude from 10^{15} neutrons/m^2·s for a low-powered reactor to less than 10^{20} neutrons/m^2·s for some small high-powered reactors. The neutron flux is equal to the product of the neutron density n (neutrons/m^3) and the neutron velocity v (m/s):

$$\phi = vn \tag{3.17}$$

In Eq. (3.17), the neutron density and velocity are, respectively, functions of the spatial location (core volume V) and the kinetic energy (E) of the neutrons. Consequently, an average neutron flux $\bar{\phi}$ represents an average over the volume of interest as well as over the energy range of interest. If E_L and E_H are respectively the lower and the higher energy boundaries of the neutron group under consideration, the average neutron flux is obtained from the following equation:

$$\bar{\phi} = \frac{\displaystyle\int_{E_L}^{E_H} \int_{\text{vol}} \phi(E, V)\, dV\, dE}{\displaystyle\int_{E_L}^{E_H} \int_{\text{vol}} dV\, dE} \tag{3.18}$$

It is common practice to divide the total range of neutron kinetic energy into a finite number of energy groups. The more energetic fission neutrons can have a kinetic energy as high as 10 MeV and in most nuclear reactors many of the neutrons have kinetic energies less than 0.0380 eV, which is the approximate neutron energy when it is in energy equilibrium with the surrounding material at 20°C (68°F). For most purposes, the reactor neutrons can be divided into three general groups—fast neutrons, intermediate neutrons, and thermal neutrons.

Thermal neutrons include all neutrons having a kinetic energy below 0.1 eV; intermediate neutrons include any neutrons that have kinetic energies falling between 0.1 eV and 0.1 MeV; fast neutrons are those neutrons with a kinetic energy in excess of 0.1 MeV. Thermal neutrons at 20°C have a most probable velocity of 2200 m/s, while the average velocity of the neutrons produced from fission is 2×10^7 m/s, which corresponds to a kinetic energy of about 2 MeV. All nuclear fission reactors have a fast flux and most of the reactors currently in operation also have intermediate and thermal neutron fluxes.

In detailed reactor-physics calculations, it is common practice to use many energy groups. There will be an average flux for each group of neutrons. Only three groups of neutrons will be considered in this text.

3.6.3 Neutron Reactions and Reaction Rates

Since the reactor thermal power is directly proportional to the fission rate in the reactor, the power design engineer must know how to evaluate the total and local fission rate in the reactor. Although the fission rate is not the only neutron-induced reaction that takes place in the reactor, it is the one of primary interest in the conversion of nuclear fission energy into thermal energy.

The reaction rate for a particular process in the reactor core is equal to the product of the average neutron flux $\bar{\phi}$ in the reactor core, the total number of nuclei N reacting with the neutrons to produce the reaction, and the microscopic neutron cross section σ for the given reaction:

$$\text{Neutron reaction rate} = N\sigma\bar{\phi} \tag{3.19}$$

The microscopic cross section σ is essentially a target area associated with each nucleus for producing a given reaction as the result of reacting with some kind of incident particle (the neutron in a fission reactor). The use of the microscopic cross section is not limited to just neutron reactions as it can be applied to other particles (such as protons and deuterons) that are used to bombard nuclei in accelerators, etc. The microscopic cross section has units of reactions-square meter per nuclei-incident particle.

Since only neutron-induced reactions are of primary importance in the fission reactor, this book will be concerned only with microscopic cross sections where the incident particles are neutrons. These neutron cross sections have units of reactions-square meter per nucleus-neutron. A subscript on the microscopic neutron cross sections will be used to designate the type of neutron reaction that is induced. Thus, σ_c is the microscopic capture (radiative-capture) cross section for a particular isotope and has units of captures-square meter per nucleus-neutron. The neutron-capture rate for a particular isotope in the reactor is equal to $N\sigma_c\bar{\phi}$. The microscopic fission cross section σ_f has units of fissions-square meter per fuel atom-neutron and the fission rate in the reactor core is equal to the product, $N_{fuel}\sigma_f\bar{\phi}$.

The microscopic neutron cross section σ is a function of the type of reaction, the type of target nucleus, and the kinetic energy of the incident neutrons. The microscopic neutron cross sections are determined experimentally and are normally reported in units of barns or reaction-barns per nucleus-neutron. One barn is equivalent to 10^{-28} m^2. A typical plot of the neutron cross section as a function of neutron energy is shown in Fig. 3.2. A comprehensive tabulation of cross-section data is reported in a Brookhaven National Laboratory report, BNL 325, and its supplements.

Sometimes it is convenient to use a neutron cross section for a given material rather than the microscopic cross section which is good only for a given nucleus. The material cross section is called the macroscopic cross section Σ, and this quantity is equal to the product of the microscopic cross section σ and the isotopic density (in atoms per unit volume, N/V):

$$\Sigma = \frac{\sigma N}{V} \tag{3.20}$$

The macroscopic neutron cross section has units of reactions per neutron-meter and also employs the same subscript that is used on the microscopic cross section to identify the type of neutron-induced reaction. The reciprocal of the macroscopic neutron cross section is called the neutron mean free path λ, where λ is equal to the average distance that a neutron travels before it undergoes a given type of reaction (as designated by the subscript on the cross section):

$$\lambda = \frac{1}{\Sigma} = \frac{V}{N\sigma} \tag{3.21}$$

Substituting Eq. (3.20) into Eq. (3.19) gives the following equation for the neutron reaction rate:

$$\text{Neutron reaction rate} = \sigma N\bar{\phi} = \Sigma V\bar{\phi} \tag{3.22}$$

Since there are 3.1×10^{16} fissions/MW$_{th}$·s, the fission rate in a given reactor operating at a power level of P MW$_{th}$ is

$$\text{Fission rate, fissions/s} = 3.1 \times 10^{16}\, P = \bar{\sigma}_f N_{fuel}\bar{\phi} = \bar{\Sigma}_f V_c\bar{\phi} \tag{3.23}$$

where N_{fuel} is the number of fuel atoms in the volume of the reactor core V_c, and $\bar{\sigma}_f$ and $\bar{\Sigma}_f$ are the average microscopic and macroscopic fission cross sections for the neutrons causing fission. Equation (3.23) is valuable in determining the number of fuel atoms, and hence the fuel loading for a given power and average neutron flux, or for evaluating the average neutron flux for a given reactor power and fuel loading.

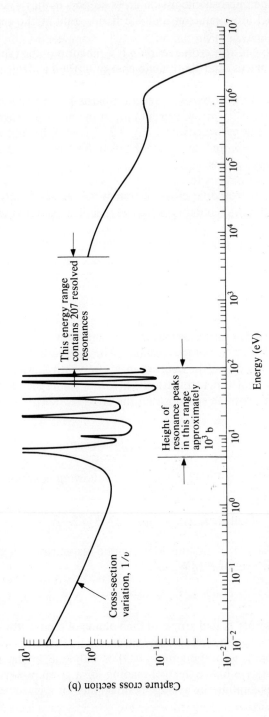

Figure 3.2 The microscopic neutron-absorption cross section for U-238. (*From "Steam/Its Generation and Use," 1972.*)

Example 3.7 (a) Determine the reactor fuel loading in kilograms of U-235 in a 1200-MW_e power reactor operating at a thermal efficiency of 33 percent, an average thermal-neutron flux of 5×10^{17} neutrons/m²·s in the core, and an average fission cross section of 420 barns. (b) If the reactor is fueled with 2.3 percent enriched UO_2, find the fuel loading of UO_2 in this system.

SOLUTION

(a)   Reactor thermal power = $P = \dfrac{1200 \text{ MW}_e}{0.33 \text{ MW}_e/\text{MW}_{th}} = 3636.36 \text{ MW}_{th}$

Fission rate = $(3.1 \times 10^{16} \text{ fissions/MW}_{th}\cdot\text{s})(3636.36 \text{ MW}_{th})$

$= 1.127 \times 10^{20} \text{ fissions/s} = \bar{\sigma}_f N_{\text{fuel}} \bar{\phi}$

$= (420 \text{ fission barns/fuel atom neutron})$

$\times (10^{-28} \text{m}^2/\text{barn}) N_{\text{fuel}} (5 \times 10^{17} \text{ neutron/m}^2\cdot\text{s})$

Number of fuel atoms in core = $N_{\text{fuel}} = \dfrac{1.127 \times 10^{20}}{(420 \times 10^{-28})(5 \times 10^{17})}$

$= 5.368 \times 10^{27} \text{ atoms of U-235}$

Mass of U-235 = $\dfrac{(5.368 \times 10^{27} \text{ atoms})(235 \text{ kg/kg}\cdot\text{mol})}{6.023 \times 10^{26} \text{ atoms/kg}\cdot\text{mol}}$

$= 2094.4 \text{ kg of U-235} = 2.0944 \text{ metric tons of U-235}$

(b)   Molecular weight of $UO_2 = 238 + 2(16) = 270 \text{ kg/kg}\cdot\text{mol}$

The U-238 atomic weight is used here because most of the uranium is U-238.

Total mass of uranium = $\dfrac{2.0944 \text{ tons of U-235}}{0.023 \text{ tons of U-235/ton of U}}$

$= 91.062 \text{ metric tons of U}$

Total mass of $UO_2 = \dfrac{91.062 \text{ tons of U}}{238 \text{ tons of U/270 tons of } UO_2}$

$= 103.306 \text{ metric tons of } UO_2$

$= 113.873 \text{ short tons of } UO_2$

3.6.4 Neutron Cross Sections

Neutron cross sections and reactions can be divided into two general categories. These two classes are the absorption cross sections and reactions and the scattering cross sections and reactions. An absorption reaction (with cross sections of σ_a and Σ_a) is any neutron-induced reaction in which the target nucleus is converted into a different isotope or isotopes. Thus, the target nucleus is destroyed in an absorption reaction. Examples of absorption reactions are the radiative-capture reaction (σ_c and Σ_c), the fission reaction (σ_f and Σ_f), the $(n, 2n)$ reaction (σ_{2n} and Σ_{2n}), the (n, α) reaction (σ_α and Σ_α), and the (n, p) reaction (σ_p and Σ_p). A scattering reaction is a neutron-induced reaction in which the target nucleus is not destroyed. Scattering cross sections are designated σ_s and Σ_s. There are two types of scattering reactions—the elastic scattering reaction (σ_{es} and Σ_{es}) and the inelastic scattering reaction (σ_{is} and Σ_{is}).

In the determination of the average neutron cross sections, $\bar{\sigma}$ and $\bar{\Sigma}$, for thermal neutrons, the following approximate rules will be applied. The scattering cross section σ_s, which for low-energy neutrons is composed of only the elastic scattering cross section, is essentially independent of the kinetic energy of the thermal neutrons. Thus, the average scattering cross section for thermal neutrons $\bar{\sigma}_{s, \text{th}}$ becomes

$$\bar{\sigma}_{s, \text{th}} = \sigma_{s, 0.0253 \text{ eV}} \tag{3.24}$$

where $\sigma_{s, 0.0253 \text{ eV}}$ is the microscopic neutron-scattering cross section for neutrons with a kinetic energy of 0.0253 eV.

The absorption neutron cross section (usually only the radiative-capture cross section for most materials) usually varies as the reciprocal of the velocity of the thermal neutrons ($\sigma_a = K_1/v = K_2/E^{1/2} = K_3/T^{1/2}$), see Fig. 3.2. Integrating this $1/v$ cross-section dependence over a Maxwell-Boltzmann, thermal-neutron distribution, gives the following relationships for the average absorption cross section, $\bar{\sigma}_a$:

$$\bar{\sigma}_{a, \text{th}} = \begin{cases} (\sigma_{a, 0.0253 \text{ eV}})(2200/v_{\text{th}})/1.128 & (3.25) \\[2mm] (\sigma_{a, 0.0253 \text{ eV}})(0.0253/E_{\text{th}})^{1/2}/1.128 & (3.26) \\[2mm] (\sigma_{a, 0.0253 \text{ eV}})(528/T_R)^{1/2}/1.128 = (\sigma_{a, 0.0253 \text{ eV}})(293/T_K)^{1/2}/1.128 & (3.27) \end{cases}$$

Most thermal-neutron cross-section data are presented at the most probable neutron velocity of neutrons in thermal equilibrium with a medium at 20°C (68°F). The energy of these neutrons is 0.0253 eV ($v_{\text{th}} = 2200$ m/s, $T_R = 528$°R, and $T_K = 293$ K). The factor of 1.128 in the absorption cross-section equations comes from the integration of the $1/v$ cross section over the thermal-neutron (Maxwell-Boltzmann) distribution.

In the intermediate-energy range, some nuclei exhibit very high cross-section peaks. These peaks are called resonances because they are similar to the amplitude plots obtained at the critical speeds of rotating machinery. The uranium-238 microscopic cross section, presented in Fig. 3.2, exhibits a strong resonance structure. These resonance peaks may correspond to different scattering and absorption reactions and are associated with nuclear energy levels in the target nucleus. Thus, heavy-mass nuclei normally exhibit strong resonance cross-section structure because of the many nuclear energy levels.

The presence of resonance cross sections in the intermediate neutron energy range makes it very difficult to determine the average neutron cross section in this energy range. These resonances also present an operational problem in some reactors due to a phenomenon called doppler broadening. Doppler broadening and the energy dependence of neutron cross sections are discussed in greater detail in Chap. 5.

Example 3.8 If the 2200-m/s value of the fission cross section for U-235 is 581 barns, find the average fission cross section for thermal neutrons, $\bar{\sigma}_{f,\,th}$, when the temperature of the material used to slow down the neutrons (moderating material) is 580°F.

SOLUTION The fission cross section is one type of an absorption cross section:

$$T_R = 460 + 580 = 1040°\text{R}$$

$$\bar{\sigma}_{f,\,th} = \frac{(\sigma_{f,\,0.0253\,\text{eV}})(528/T_R)^{1/2}}{1.128}$$

$$= \frac{(581)(528/1040)^{1/2}}{1.128} = 367.0 \text{ barns}$$

Example 3.9 Water has a density of 1000 kg/m³. Calculate the average lifetime of thermal neutrons at 60°C, the average absorption mean free path, the average scattering mean free path, and the average distance that a thermal neutron travels before some type of reaction occurs. The elemental cross sections for hydrogen and oxygen for 0.0253 eV neutrons are as follows:

	σ_a	σ_s	σ_{total}
Hydrogen	0.3320 barns	38.0 barns	38.332 barns
Oxygen	0.0002 barns	4.2 barns	4.2002 barns

SOLUTION
    Multiplier for average absorption cross sections

$$= \frac{(293/T_K)^{1/2}}{1.128}$$

$$= \frac{(293/333)^{1/2}}{1.128} = 0.8316$$

Average absorption cross sections:

For hydrogen: $\bar{\sigma}_a = 0.332(0.8316) = 0.2761$ barns

For oxygen: $\bar{\sigma}_a = 0.0002(0.8316) = 0.00017$ barns

The macroscopic cross section for water $= \Sigma_{H_2O} = \Sigma_{2H} + \Sigma_O$

$$\Sigma_{H_2O} = \Sigma_{2H} + \Sigma_O = (\rho/MW)(Av)(2\bar{\sigma}_H + \bar{\sigma}_O)$$

Average macroscopic absorption cross section for water $= \bar{\Sigma}_{a, H_2O}$

$$\bar{\Sigma}_{a, H_2O} = \frac{1000}{18.01}(0.6023 \times 10^{28})[2(0.2761) + 0.00017] \times 10^{-28}$$

$$= 18.47 \text{ per meter}$$

Average macroscopic scattering cross section for water $= \bar{\Sigma}_{s, H_2O}$

$$\bar{\Sigma}_{s, H_2O} = \frac{1000}{18.01}(0.6023 \times 10^{28})[2(38.0) + 4.2] \times 10^{-28}$$

$$= 2682.09 \text{ per meter}$$

Average total macroscopic cross section $= \bar{\Sigma}_{t, H_2O} = \bar{\Sigma}_{a, H_2O} + \bar{\Sigma}_{s, H_2O}$

$$\bar{\Sigma}_{t, H_2O} = 18.47 + 2682.09 = 2700.56 \text{ per meter}$$

Absorption mean free path $= \lambda_a = \dfrac{1}{\Sigma_a} = \dfrac{1}{18.47} = 0.0541 \text{ m} = 5.41 \text{ cm}$

Scattering mean free path $= \lambda_s = \dfrac{1}{\Sigma_s} = \dfrac{1}{2682.09} = 0.0003728 \text{ m}$

$$= 0.03728 \text{ cm}$$

Average distance traveled before some type of reaction occurs $= \dfrac{1}{\Sigma_t}$

$$\lambda_t = \frac{1}{\Sigma_t} = \frac{1}{2700.56} = 0.0003703 \text{ m} = 0.03703 \text{ cm}$$

Velocity of thermal neutrons $= v_{th} = 2694.4\left(\dfrac{T_K}{293}\right)^{1/2} = 2694.4\left(\dfrac{333}{293}\right)^{1/2}$

$$= 2872.5 \text{ m/s}$$

Average lifetime of thermal neutrons in an infinite water medium $= L_{th}$

$$L_{th} = \frac{\text{average distance traveled by a thermal neutron}}{\text{neutron velocity}}$$

$$= \frac{\text{average absorption mean free path}}{\text{velocity of thermal neutrons}}$$

$$= \frac{0.0541}{2872.5} = 0.00001884 \text{ s}$$

The microscopic neutron cross-section data and fission parameters for the three fissionable isotopes are listed in Table 3.1. These data are listed for both thermal neutrons (0.0253 eV) and for fast neutrons. The fission parameter v in this table is the average number of neutrons produced per fission. The other parameter listed in the table, η, is the average number of neutrons produced per neutron absorbed by the fuel. In order to sustain a fission chain reaction, η must exceed unity. In order to "breed" in a reactor, η must exceed 2—one neutron to continue the fission chain reaction and one neutron to be absorbed by a fertile nucleus to produce a new fuel atom.

The cross-section values and fission parameters listed in Table 3.1 point out the primary advantages of operating fission reactors with either fissions induced by low-energy neutrons (thermal reactors) or fissions induced by high-energy neutrons (fast reactors). Although the value of η exceeds 2 for all the fuels and for neutrons of all energy, the thermal-neutron values for uranium-235 and plutonium-239 are so close to 2 that it is improbable that they can be used as fuel in a thermal breeder reactor. This is due to the fact that some of the fission neutrons are absorbed by nonfuel and nonfertile material and some leak from the reactor core.

From the standpoint of breeding, uranium-233 is the best thermal-reactor fuel and it may be possible to build a thermal breeder reactor operating on the thorium-uranium-233 fuel cycle. The value of η for fast neutrons is much greater

Table 3.1 Neutron cross sections and fission parameters for the common fuel isotopes

For thermal (0.0253 eV) neutrons

Fuel isotope	Fission cross section σ_f, barns	Capture cross section σ_c, barns	Absorption cross section σ_a, barns	Scattering cross section σ_s, barns	v[†]	η[‡]
$^{233}_{92}\text{U}$	525	53	578	.....	2.51	2.28
$^{235}_{92}\text{U}$	582	101	683	10.0	2.47	2.10
$^{239}_{94}\text{Pu}$	742	286	1028	9.6	2.89	2.09

For fast (KE > 1.5 MeV) neutrons

Fuel isotope	Fission cross section σ_f, barns	Capture cross section σ_c, barns	Absorption cross section σ_a, barns	Scattering cross section σ_s, barns	v[†]	η[‡]
$^{233}_{92}\text{U}$	1.85	0.034	1.884	4.5	2.65	2.60
$^{235}_{92}\text{U}$	1.28	0.078	1.358	4.5	2.65	2.48
$^{239}_{94}\text{Pu}$	1.95	0.040	1.990	4.6	3.10	3.04

† v is the average number of neutrons produced per fission.
‡ η is the average number of neutrons produced per neutron absorbed by the fuel:
$\eta = v\sigma_f/\sigma_a = v\sigma_f/(\sigma_f + \sigma_c)$.

than 2 for all three fuel isotopes, although plutonium-239 is the best fuel in this regard. The principal advantage of fissions induced by high-energy neutrons is that high breeding ratios can be achieved with any fuel isotope.

The same table underscores the principal advantage of thermal-neutron-induced fissions. The microscopic fission cross sections for thermal neutrons are at least a factor of 280 times those for fast neutrons. The fission rate in any reactor is $\bar{\sigma}_f N_{\text{fuel}} \bar{\phi}$, where $\bar{\phi}$ is either the average fast flux $\bar{\phi}_f$ in a fast reactor or the average thermal flux $\bar{\phi}_{\text{th}}$ in a thermal reactor. If a fast and a thermal reactor are operated at the same power with the same average fluxes causing fission, the number of fuel atoms and hence the total fuel loadings are directly proportional to the microscopic fission cross sections. Even if the average fast flux exceeds the thermal flux by an order of magnitude, the uranium-235-fueled fast reactor will have a fuel loading equal to 581/12.8 or 45.4 times that of a thermal reactor operating at the same power. The low fuel requirements are the principal advantage of thermal reactors over fast reactors.

3.6.5 Classification of Nuclear Reactors

There are many ways to classify nuclear reactor systems. Some of the common classification systems include the classification with respect to the average kinetic energy of the neutrons causing fission, classification with respect to the purpose of the reactor, classification with respect to core geometry and composition, and classification with respect to the type of coolant used to transfer the thermal energy from the reactor core.

There was some discussion in the previous sections about thermal and fast reactors, but when classifying reactors according to the average kinetic energy of the neutrons causing fission, there are three principal categories of reactors, including thermal reactors, intermediate reactors, and fast reactors. In a fast reactor, the fission process is induced by fast neutrons with an average kinetic energy of a few tenths of a million-electronvolt. In the intermediate reactor, the average fission process is induced by neutrons with a kinetic energy between 0.1 eV and 0.1 MeV. In the thermal reactor, the average fission process is induced by neutrons with a kinetic energy below 0.1 eV.

The main advantage of fast reactors is that the fast-fission process produces a lot of neutrons per neutron absorbed by the fuel. This is true for any of the three fuel isotopes, and, as a result, yields very high values of breeding ratio. Consequently, it is possible to build a breeder reactor with any reactor fuel. Another advantage of these systems is that they can employ any structural material in the reactor core because the absorption cross sections of all materials are small for high-energy neutrons. The fast reactors are also usually small and compact, which makes these systems easier to "shield," but the small size increases the power density (thermal power to core volume ratio, in kilowatts per cubic meter) in the core and this complicates heat transfer and thermal transport in the core. A typical fast reactor is shown in Fig. 3.3.

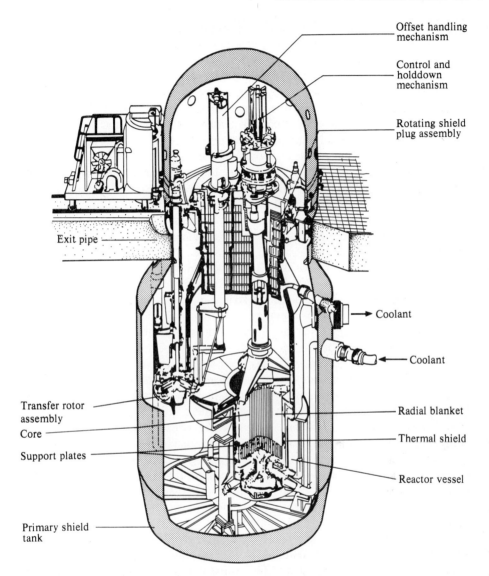

Figure 3.3 The Enrico Fermi fast-breeder reactor. *(Skrotzki and Vopat, 1960.)*

The principal disadvantage of fast reactor systems is the required high fuel loading, but there are some other disadvantages of fast reactors when they are compared to thermal-reactor systems. Fast reactors must employ enriched-uranium fuel as they cannot " go critical" with natural-uranium fuel. Another problem associated with fast neutrons is the problem associated with radiation damage of the structural materials in the reactor core. The high-energy neutrons "knock" the metallic atoms out of their normal lattice positions, causing embrittlement and swelling of the structural materials in the core. The fast neutrons also have much shorter "neutron lifetimes" than thermal neutrons and this causes some control problems under certain conditions. This problem is discussed in greater detail in the section on reactor control in Chap. 5.

The advantages of thermal-reactor systems include the fact that they require a relatively low fuel loading; they can utilize natural-uranium fuel in some systems; thermal neutrons have the longest neutron lifetimes which makes thermal reactors somewhat easier to control; and these systems have relatively low power densities since moderating material must be incorporated in the reactor core to slow down the neutrons. Some of the disadvantages associated with thermal reactors include the fact that it is almost impossible to breed with any fuel except uranium-233; the choice of structural materials is limited to only those metals and alloys with small absorption cross sections; and these thermal systems have relatively large volumes and are harder to shield. Almost all of the existing reactors in the world are thermal reactors and a typical power reactor is shown in Fig. 3.4.

Intermediate reactors have some of the advantages and disadvantages of fast and thermal reactors. It is very difficult to design an intermediate reactor system because of the resonance cross-section structure in this energy range. Only two sodium-cooled intermediate submarine reactors have been built, but these are now decommissioned. Several other intermediate reactors have been proposed but they have not been built.

There are two important products from any nuclear reactor—neutrons and energy. This leads to another major classification system that depends on how these products are utilized. This system classifies the reactors with regard to the purpose of the reactor. Under this system, there are five groups of reactors. These include research reactors, experimental reactors, production reactors, compact-power reactors, and commercial-power reactors.

Almost every large university, private, and governmental laboratory has a research reactor of some kind. These systems operate at a power ranging from a few microwatts to hundreds of megawatts. These reactors are used essentially as a source of neutrons and the thermal energy generated in these systems is commonly dissipated to the environment.

Experimental reactors are relatively low-powered (up to 10 MW_{th}) reactor prototypes that are constructed to check the feasibility and delineate operational problems of the proposed systems. This is the one reactor system that normally uses neither the neutrons nor the energy generated in the fission process. The proposed reactor concept tested in this experiment may utilize either one or both of the basic reactor products.

Production reactors are the high-powered reactors in operation at Hanford, Washington, and at Savannah River, South Carolina. These reactors were originally constructed to produce plutonium-239 fuel for the nation's weapons program. These systems are also being used to produce transuranium elements and activation products for radioisotope heat sources. Although there are a few exceptions, the energy generated in these reactors is normally dissipated to the environment.

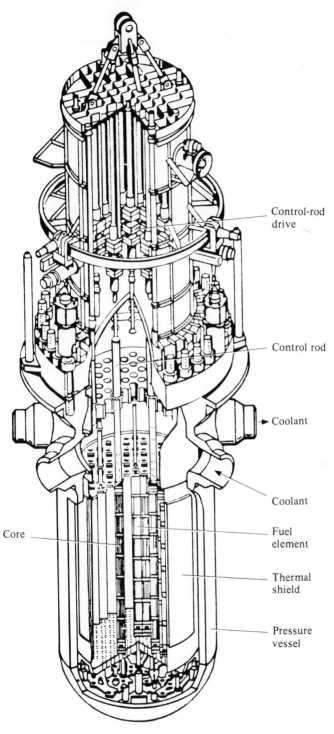

Figure 3.4 A typical pressurized-water power reactor. (*From Proceedings of the American Power Conference, 1974.*)

Control-rod drive

Control rod

Coolant

Coolant

Core

Fuel element

Thermal shield

Pressure vessel

The compact power reactors or mobile reactors are sometimes called "burner-uppers" because they are fueled with fully enriched uranium to keep the size and weight to a minimum. These systems are designed to produce only thermal energy and they are employed in the naval (submarine and surface ship) reactor program and for the army power-package reactor program. The nuclear-powered carrier *Enterprise*, for instance, is powered by eight nuclear reactors.

The commercial power reactors utilize both of the basic products from fission. The excess neutrons are used to produce new fuel (in a converter or breeder reactor) and the thermal energy is used to generate electricity. Most of the power reactors in operation in this country are called light-water reactors (LWR) because they are cooled and moderated with ordinary water.

Sometimes nuclear reactors are classified as either homogeneous or heterogeneous reactors, depending on the type of core geometry or core composition. Homogeneous reactors are those systems in which the fuel is in fluid form, including gases, liquids, and slurries. In this reactor, the fluid fuel is normally circulated from the reactor to an external heat exchanger, then to a pump and back to the reactor. In the heterogeneous reactor systems, the fuel is in solid form and the fission energy is transferred from the solid fuel assemblies (called fuel elements) to a coolant circulating through the reactor core. Most of the reactors presently in operation are heterogeneous reactors.

Homogeneous reactors offer some very attractive advantages when compared to the heterogeneous assemblies. First, this system has excellent in-core heat transfer because the fission energy is generated directly in the fuel-coolant solution. Next, these reactors have excellent control characteristics because the system tries to maintain the same average core temperature as long as the fuel concentration in the core is a constant. An increase in the fuel concentration does not increase the reactor power but simply increases the mean operating temperature in the core. The reactor power is essentially controlled by the rate at which energy is removed from the fuel solution in the external heat exchanger. If the power demand in the external heat exchanger suddenly goes to zero, the hot fuel solution entering the reactor raises the average core temperature which causes fuel to be removed from the core due to expansion. As the fuel is removed, the reactor goes subcritical and remains subcritical until the average core temperature reaches the original operating value. Because of this inherent load-following characteristic, some homogeneous reactors have no control rods.

Another major advantage of homogeneous reactors is associated with the fact that the reactor fuel can be added, removed, and reprocessed during reactor operation. This means that the reactor does not have to be shut down for refueling and that some of the fission products, particularly the high-cross-section, gaseous isotope, xenon-135, can be continuously removed during operation. As a result of this capability, the actual reactor fuel loading is less than that for a heterogeneous reactor because the heterogeneous system requires extra fuel to compensate for fuel burnup and fission-product poisoning.

Despite all of the advantages offered by homogeneous reactors, there is one major disadvantage that has prevented their widespread application. This disadvantage arises from the fact that the fuel solution also contains the highly radio-active fission products. Any leaks or component failures in the primary-reactor coolant system are extremely difficult to repair because of the presence of these fission products. Despite the problems due to maintenance and repair, some repair has been accomplished with the use of remote-handling tools in an experimental reactor system.

In some of the homogeneous reactors, the fuel solution proved to be either chemically unstable in the presence of a high-radiation environment or it proved to be highly corrosive. While the actual fuel loading in the homogeneous core is relatively small, there is a considerable portion of the total fuel inventory tied up in the external piping, heat exchanger, and pump.

Because of the problem associated with leakage from and repair of the highly radioactive primary loop, very few of the homogeneous reactors have been built. There are a few homogeneous research reactors in operation which are fueled with aqueous solutions of uranyl nitrate and uranyl sulfate. The Oak Ridge National Laboratory has constructed and tested an experimental homogeneous power reactor fueled with a molten solution of uranium and beryllium flouride salts. It is hoped that this system may permit breeding in a thermal reactor but there are no immediate plans to proceed with construction of a full-scale version of the molten-salt reactor (MSR).

The reactor fuel in a heterogeneous reactor is normally a solid and is encapsulated in a structural material, called cladding. The cladding is designed to contain all the fission products in the fuel element. This greatly facilitates maintenance and repair of the reactor system. All of the fuel material in this reactor is located in the reactor core as none is contained in the external cooling circuit.

There are a few heterogeneous reactors that have the capability of being refueled while in operation, but this requires a complicated and expensive refueling machine. This machine must be able to shield and cool the old fuel element before, during, and after extraction and it must also provide cooling to the new element during insertion. Most reactors are shut down during refueling and the fuel is loaded in batches. This means that extra fuel must be added at every reloading to compensate for fuel burnup, fission-product poisoning, etc. This excess fuel complicates the reactor control, operation, and design. The shutdown of a power reactor at periodic intervals (usually every 12 to 18 months) for refueling does not cause serious problems, however. All large power systems, fossil as well as nuclear, normally require annual shutdowns for turbine inspection and maintenance.

The heat transfer and thermal energy transport in the heterogeneous-reactor core is much more difficult than in a homogeneous reactor. The thermal energy produced in the fission process must be transferred through the fuel material by conduction, by conduction-convection through the fuel-clad interface, by conduction through cladding material, and by convection from the external surface of the cladding to the coolant.

In many of the current power reactors, it is not uncommon to find maximum fuel temperatures of 2200 to 2800°C (4000 to 5000°F) because of the low thermal conductivity of the ceramic fuel material. While the fuel temperature is very high, the cladding material usually operates at something less than 500°C (900°F). The heterogeneous fuel elements are more difficult to reprocess compared to the homogeneous reactor fuel because the cladding must be removed before the fuel can be reprocessed.

Another reactor classification system that is applicable only to heterogeneous systems groups the reactors with respect to the type of coolant used to remove the thermal energy from the core. There are five general kinds of reactor coolants, including ordinary or "light water," deuterium oxide or "heavy water," organic coolants, gases, and liquid metals.

Light-water reactors (LWR) find widespread use as both power and research reactors in the United States. This coolant has good thermal properties, it is inexpensive, and the water technology is well developed. Since water has a high hydrogen concentration, the water also commonly serves as the moderating material in a thermal reactor, to slow down the neutrons. Unfortunately, hot water is quite corrosive and it must be highly pressurized in order to operate at moderate temperatures. Another problem arises from the fact that the absorption cross section of ordinary hydrogen is high enough that these systems must employ slightly enriched-uranium fuels (normally 2 to 3 percent) in order to achieve criticality.

There are two basic types of light-water power reactors in use in this country—the pressurized-water reactor (PWR) and the boiling-water reactor (BWR). In the pressurized-water reactor, the core coolant (the primary coolant) is highly pressurized and no bulk boiling is permitted in the reactor during normal operation. The turbine steam is generated at lower pressure in a secondary water loop. This prevents contamination of the steam turbine with radioactive material from the primary coolant. A schematic diagram of a typical PWR primary cooling system is shown in Fig. 3.5. A more detailed description of the PWR is presented in Chap. 5.

The newer boiling-water reactors (BWR) have a primary cooling system in which the turbine steam is generated directly in the reactor core. This system is somewhat less complicated than the PWR primary system because it does not require the large and expensive secondary steam generator. Reactor control is accomplished with the use of jet pumps that vary the water recirculation rate in the reactor core. A schematic diagram of a typical BWR system is shown in Fig. 3.6. A more detailed description of the BWR system is given in Chap. 5.

Heavy water, which is water composed of the heavy hydrogen isotope, ^2_1H, is used to moderate and cool some power and research reactors. A number of heavy-water power reactors use heavy water for moderation and light water for the secondary system to reduce the heavy-water inventory. The Canadians have exported a number of reactors of this type in the CANDU series. A flow diagram for a CANDU reactor system is shown in Fig. 3.7.

Heavy water has many of the desirable and undesirable characteristics of light water, with one marked difference. While ordinary hydrogen has a high enough absorption cross section that it cannot be used to moderate a natural-

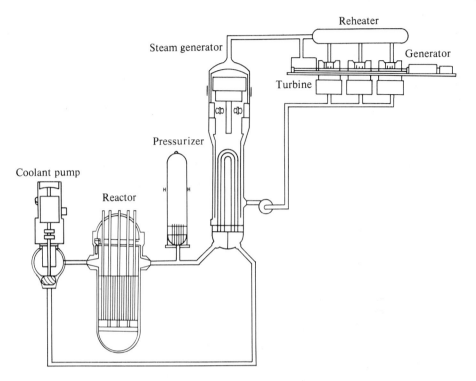

Figure 3.5 The primary and secondary cooling systems for a typical pressurized-water reactor power system. *(Courtesy of Westinghouse Electric Corporation.)*

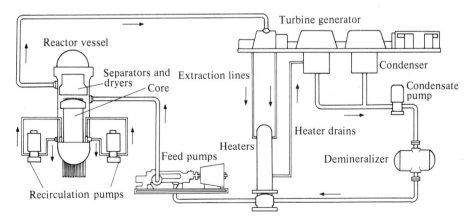

Figure 3.6 The schematic diagram of a typical boiling-water reactor power system. *(Courtesy of the General Electric Company.)*

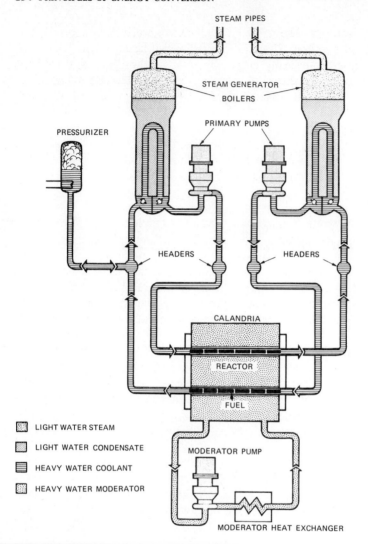

Figure 3.7 A schematic diagram of the CANDU power reactor system. *(Courtesy of Atomic Energy of Canada, Ltd.)*

uranium, thermal reactor, the heavy water has a very low absorption cross section and it can be used to moderate a natural-uranium, thermal reactor. This makes heavy-water reactor systems popular for those countries that do not have uranium-isotopic separation facilities.

Heavy hydrogen exists in nature to the extent of one part of heavy hydrogen in 6700 parts of ordinary hydrogen. Consequently, heavy water is difficult and expensive to separate from ordinary water, although it is much easier to separate than the uranium isotopes. The 1976 price of heavy water was about $55 a pound or about $450 a gallon. The major drawback of heavy-water-moderated reactors is that they are physically large compared to the light-water reactors and they require a sizable inventory of heavy water. In this country, the production reactors at the Savannah River National Laboratory, along with several research reactors including those at MIT and Georgia Tech., are cooled and moderated with heavy water.

Organic coolants, such as Dowtherm and Sanowax, have been proposed as reactor coolants and their high hydrogen concentration makes them acceptable as moderating material to slow down the neutrons. These liquids can operate at relatively high temperatures with only moderate pressurization and are much less corrosive than water. The principal problem associated with the use of organic liquids is the problem associated with the radiation damage caused by high-energy neutrons and gamma radiation. The combined radiation breaks molecular bonds, causing the fragments to recombine into longer and shorter organic molecules. The long-chained molecules have higher melting points and therefore tend to plate out on or "gunk up" the heat-transfer surfaces. Several experimental organic reactors have been built and operated, including a small power reactor at Piqua, Ohio, but all of these systems have been shut down and dismantled.

Gaseous coolants are the poorest heat-transfer and thermal-energy transport media because they have a low thermal conductivity, a low density, and a low volumetric heat capacity. This means that the gaseous coolants must be highly pressurized to be effective. For this reason, it is difficult to remove the fission-product decay power if the system undergoes a sudden depressurization. Because of their low nuclear density, gases cannot be used to moderate the neutrons in a thermal reactor.

Gases have some advantages over liquid coolants, however, including no maximum temperature limit and the fact that almost any gaseous coolant can be employed without regard to the microscopic neutron-absorption cross section because the low nuclear density produces a low macroscopic neutron-absorption cross section. It also may be possible to use gaseous coolants as the working fluid in the Brayton (gas-turbine) power cycle which takes advantage of the high maximum temperature to achieve a high thermal efficiency. Of all the gaseous coolants, the low atomic mass gases, such as helium, have the best thermal properties.

Air has been used to cool the X-10 research reactor at Oak Ridge, as well as some experimental test reactors in the early aircraft nuclear propulsion (ANP) program. Nitrogen has been used in a small Army test reactor to supply heat for a Brayton-cycle power system. Great Britain has built and exported a number of thermal power reactors cooled with carbon dioxide and moderated with graphite. These reactors are called the Calder-Hall series of power reactors.

Of all the gaseous reactor coolants, helium is probably the best. It has the best thermal properties of all the gases except hydrogen and since helium is a "noble" gas it is noncorrosive, although it is somewhat difficult to contain. Gulf

General Atomics, a division of the Gulf Oil Corporation, has developed a number of large power reactors cooled with helium and moderated with graphite. These reactors are called the high-temperature gas-cooled reactors (HTGR). A schematic diagram of the HTGR is shown in Fig. 3.8 and this reactor type is discussed in detail in Chap. 5.

Liquid metals have the best heat-transfer properties of all the proposed reactor coolants. They have very high thermal conductivities, low viscosities, and relatively high heat capacities. In addition, most liquid metals can operate at high temperatures with low system pressures. Some of the problems associated with these coolants include the problem of high radioactivity resulting from neutron activation of the coolant, the problem associated with high melting points, compatibility with containment materials, a flammability hazard, and the fact that most of these coolants are opaque to visible radiation which complicates refueling of the system.

Sodium is the most widely used liquid-metal coolant. Liquid sodium has been used to cool a series of thermal reactors moderated with graphite and has also been used to cool a number of fast reactors, including the Enrico Fermi fast-breeder reactor shown in Fig. 3.3. Sodium is inexpensive, it has excellent thermal properties, and it boils at high temperature and low pressures (870°C or 1600°F at 1 atm). Some of the disadvantages associated with liquid sodium include a relatively high melting point (97°C or 206°F), the rapid oxidation

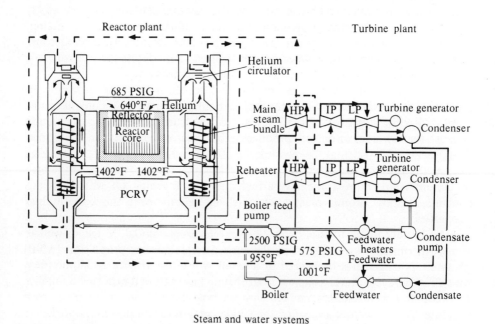

Figure 3.8 The schematic diagram for the high-temperature, gas-cooled reactor (HTGR) power system. *(Bonilla, 1957.)*

(combustion) when exposed to air, and the rapid oxidation in water which generates large quantities of hydrogen. Although sodium has a low absorption cross section, it does become highly radioactive with a half-life of 15 h. Because of this last problem, sodium-cooled power reactors employ a secondary sodium loop between the radioactive primary sodium loop and the steam system to prevent a chemical reaction between the radioactive sodium and the water in the event of a leak in the steam generator. Sodium is the proposed reactor coolant for the liquid-metal fast-breeder reactor (LMFBR) which is under evaluation and may be built at Oak Ridge, Tennessee. This reactor will also be covered in Chap. 5.

3.6.6 Reactor Physics

A detailed study of reactor physics is a major undertaking and is beyond the scope of this text. However, it is felt that the nonnuclear engineer can obtain an understanding of what is occurring in a reactor by examining an elementary neutron balance in a reactor. This can be done from qualitative considerations, without delving deeply into the field of reactor physics.

There are three ways in which neutrons are "lost" in a nuclear fission reactor. These include neutron-absorption processes or reactions, neutron leakage from the reactor core, and, finally by radioactive decay since the neutrons are radioactive. The absorption reactions for the fuel and nonfuel materials in the core include fission, radiative capture, and particle reactions, where applicable. In general, only three materials in the reactor core should have high macroscopic neutron-absorption cross sections. These are fuel, fertile, and absorber-control materials. All other materials should have low macroscopic absorption cross sections so that they do not compete with these materials for the neutrons.

The neutron-leakage rate from the reactor core is a function of the core size and, for thermal-reactor systems, the properties of the slowing-down or moderating media. Fast reactors are generally small compact systems because the fission neutrons travel a very short distance before they are absorbed while the thermal systems are normally considerably larger. The actual leakage rate in the thermal reactor depends on the type of moderating material employed as well as the reactor size. Light-water reactors are usually considerably smaller than the heavy-water and graphite-moderated systems simply because fewer scattering collisions are required to slow the neutrons to thermal energy.

The neutron-leakage rate from a given reactor can be reduced by increasing the reactor size, and it can also be decreased by surrounding the core with a material that reflects a portion of the leakage neutrons back into the core. Such a material is called a reflector and, for a thermal reactor, is usually composed of a moderating material. In a fast reactor, the reflector, sometimes called a "tamper," is composed of a heavy-mass material.

Outside the atomic nucleus, all neutrons are radioactive with a half-life of 11.7 min. Since the maximum lifetime of the average neutron in a nuclear reactor is less than 1 ms, the loss due to radioactive decay is very small. The actual fraction of neutrons lost in 0.001 s is equal to $e^{-(0.001)(\ln 2)/(11.7)(60)}$ or 0.00001 percent. Consequently, most neutron balances normally assume that the loss rate due to radioactive decay is negligible.

A possible neutron balance (not necessarily typical of any particular reactor) is given in Table 3.2. In this balance it is assumed that we start with 1000 neutrons that have an energy distribution as given in Fig. 3.9.

Several conclusions can be made from the neutron balance presented in Table 3.2. First, the reactor would be classified as a converter reactor because $(20 + 450 = 470)$ fuel (U-235) atoms are destroyed while only $(36 + 350 = 386)$ new fuel (Pu-239) atoms are produced. This means that the conversion ratio for this reactor is $386/470$ or 0.821. It should be noted that the new fuel atoms are plutonium-239 atoms and, unless the balance is made at the reactor startup, some consideration should be given to the burnup of plutonium. In a typical light-water reactor, at the end of core life, about half the fissions are produced by plutonium-239.

Another conclusion that can be made from the neutron balance is that the reactor is just critical. The fast and intermediate neutrons react with uranium-235 to produce 45 neutrons for the next generation, the fast neutrons react with uranium-238 to produce 10 new fission neutrons, and the thermal neutrons react with uranium-235 to produce 945 neutrons. Thus, $(45 + 10 + 945 = 1000)$ new fission neutrons are produced in the next generation from the 1000 fission neutrons in this generation, or $k = 1000/1000 = 1.00$.

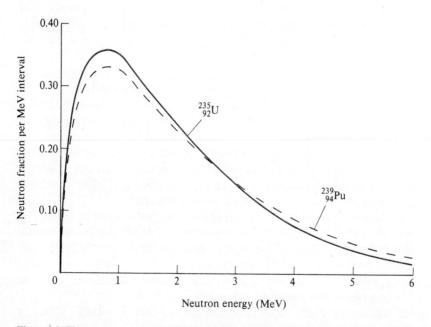

Figure 3.9 The energy spectrum of fission neutrons. *(From "Steam/Its Generation and Use," 1972.)*

Table 3.2 Possible neutron balance in a reactor

Start with 1000 fission neutrons.

    I. The slowing-down process (2 MeV to 0.1 eV)

        A. Assume that 30 of the fast and intermediate neutrons "leak" from the core during the slowing-down process.

        B. Assume that 70 fast and intermediate neutrons are absorbed by the core materials.

Burned fuel

           1. Assume that 20 neutrons are absorbed by U-235.

              a. 2 of these neutrons are lost in the radiative-capture reaction and produce U-236.

              b. 18 neutrons induce fission.

                  i. Assume the average value of v is 2.5.

New fission neutrons

                  ii. These 18 fissions produce 2.5(18) = 45 neutrons.

           2. Assume that 40 neutrons are absorbed by U-238.

              a. 4 neutrons produce fast fission of U-238.

                  i. Assume the average value of v is 2.5.

New fission neutrons

                  ii. These 4 fissions produce 2.5(4) = 10 neutrons.

New fuel atoms

              b. 36 of the neutrons are absorbed in the radiative-capture process and produce new fuel atoms (U-238 to Pu-239).

           3. Assume that 10 of the neutrons are absorbed by nonfuel and nonfertile materials in the reactor core.

              a. 2 neutrons are absorbed by cladding material.

              b. 2 neutrons are absorbed by coolant.

              c. 1 neutron is absorbed by moderator material.

              d. 1 neutron is absorbed by a fission product.

              e. 1 neutron is absorbed by the structural material.

              f. 3 neutrons are absorbed by the control-rod material.

    II. 900 of the original 1000 neutrons reach thermal energy.

        A. Assume that 20 of the thermal neutrons "leak" from the core.

        B. The balance of the thermal neutrons are absorbed by the in-core materials.

Burned fuel

           1. Assume that 450 thermal neutrons are absorbed by U-235.

              a. 72 of these neutrons (about 16 percent) are lost in the radiative-capture reaction and produce U-236.

              b. 378 of the thermal neutrons cause fission of the U-235.

                  i. Assume the average value of v is 2.5.

New fission neutrons

                  ii. These fissions produce 2.5(378) = 945 neutrons.

           2. Assume 350 of the thermal neutrons are absorbed by U-238.

              a. All of these reactions are radiative-capture reactions.

New fuel atoms

              b. All of these reactions will eventually produce 350 Pu-239 atoms.

           3. Assume that 100 of the thermal neutrons will be absorbed by the other materials in the core.

              a. 10 neutrons are absorbed by cladding material.

              b. 14 neutrons are absorbed by the coolant.

              c. 6 neutrons are absorbed by moderator material.

              d. 10 neutrons are absorbed by fission products.

              e. 6 neutrons are absorbed by the structural material.

              f. 54 neutrons are absorbed by the control rods.

This reactor can be made supercritical or subcritical by simply moving the control rods. If the rods are withdrawn from the core, the number of neutrons absorbed by the rods will be reduced. This means that the leakage rate and absorption rate by the other materials, including uranium-235, will be slightly increased. If the absorption rate in uranium-235 is increased, the number of fission neutrons produced will increase, k will exceed unity, the reactor is supercritical, and the power level is increasing with time. On the other hand, if the rods are inserted into the core, they will absorb proportionally more of the neutrons, reducing the number of absorptions by uranium-235. This means that there will be fewer fission neutrons produced, k will be less than unity, the reactor will be subcritical, and the power is decreasing with time.

The control of nuclear reactors is markedly different from the control of internal-combustion engines or other prime movers. If an increase in power is desired from an internal-combustion engine, one simply opens the throttle until the desired power is reached and then the throttle is left in that new position. In a nuclear reactor, however, the rods are withdrawn to increase the system reactivity and make the system supercritical. This condition is maintained until the desired power level is attained. When the new desired power level is reached, the control rods are reinserted to make the reactor just critical again. Barring temperature and other minor reactivity effects, the initial and final control-rod positions will be the same, although the reactor power level may be markedly higher.

In order to estimate the multiplication factor k for a given reactor assembly, the modified one-group critical equation, obtained from reactor physics, may be used. This equation is

$$k = \frac{\eta f p \varepsilon}{1 + L^2 B^2} e^{-B^2 \tau} \tag{3.28}$$

where
η = number of neutrons produced per neutron absorbed by the fuel
f = thermal utilization factor, which is the fraction of thermal neutrons absorbed by the fuel
p = resonance escape probability, which is the probability that a neutron will reach thermal energy without being absorbed
ε = fast fission factor, which is the ratio of total to thermal fissions
B^2 = reactor buckling, which is a function of the reactor size; increasing the reactor dimensions reduces the buckling, which has units of reciprocal meters squared
L = thermal diffusion length, which is a function of the diffusion coefficient and the thermal macroscopic absorption cross section of the core materials, in meters
τ = Fermi age, which is a function of the moderating material, in square meters
$e^{-B^2 \tau}$ = the fast nonleakage probability, which is the probability that a fast neutron will not leak from the core
$1/(1 + L^2 B^2)$ = thermal nonleakage probability and is the probability that a thermal neutron will not leak from the reactor core

Using the neutron balance presented in Table 3.2, the values for the parameters in the critical equation become

$v =$ average number of neutrons produced from fission
$\quad = 1000/(18 + 4 + 378) = 2.50$
$\eta = v(\sigma_f/\sigma_a)_{th} = 2.50(378/450) = 2.10$
$\quad f =$ (fuel absorptions/total absorptions)$_{th} = 450/880 = 0.51136$
$e^{-B^2\tau} =$ (total neutrons − fast leakage)/(total neutrons)
$\quad = (1000 - 30)/1000 = 0.970$
$p =$ fraction of fast neutrons not absorbed in the slowing down process

$$= \frac{1000 - \text{leakage} - \text{fast and intermediate absorptions}}{1000 - \text{fast leakage}}$$

$\quad = (1000 - 30 - 70)/(1000 - 30)$
$\quad = 900/970 = 0.92784$
$\varepsilon =$ (total fissions)/(thermal fissions) $= (18 + 4 + 378)/378 = 1.0582$
$1/(1 + L^2B^2) =$ (thermal neutrons − thermal leakage)/(thermal neutrons)
$\quad = (900 - 20)/900 = 0.977778$

Multiplication factor $= k = (2.1)(0.51136)(0.92784)(1.0582)(0.97)(0.977778)$
$\quad = 1.00000$

3.6.7 Composition of a Nuclear Power Reactor

Reactor core The reactor core is that part of the nuclear reactor which contains the fuel and fertile materials and in which almost all of the fission energy is released. The core volume not only includes the fuel and fertile materials but it also includes the coolant, instrumentation, control elements, cladding, moderator (for thermal reactor), etc. The core of homogeneous reactors consists of the fluid fuel and either a solid or a liquid-solution moderator. The fuel of a homogeneous power reactor may be composed of molten plutonium, molten bismuth-uranium alloy, or a solution of molten flouride salts. Aqueous solutions of uranyl nitrate or uranyl sulfate have been used in some experimental power and research reactors.

Heterogeneous power reactors have a core composed of solid fuel elements that are cooled by a liquid or gas. The solid fissionable elements are metals at room temperature but are not stable during long periods of irradiation and thermal cycling. Only limited success has been achieved in the use of metallic fuel alloys as most of the alloys are also unstable.

Almost all of the present power reactors employ oxides of the fuel isotopes. The oxide compounds, uranium dioxide and plutonium dioxide, are compatible with the water coolant in LWR's. They are stable in the high radiation fields and they can undergo a high percentage of fuel burnup. About the only problem associated with these fuel compounds is that they have very low values of thermal conductivity which produces very high operating temperatures in the fuel.

Fortunately, these materials have very high melting points and can operate at these temperatures. Some research reactors use uranium dioxide particles dispersed in an aluminum matrix. Carbides of the fuel isotopes (plutonium carbide and uranium carbide) appear to be quite promising and there is a lot of developmental work being done on these systems. The fuel carbides may prove to be the fuel material in future reactors.

In any of the heterogeneous systems, the fuel is encapsulated in a structural material, called cladding. The purpose of the cladding material is to contain the fission products in the fuel element, thereby reducing the radioactive contamination of the coolant. Cladding materials should have good physical properties, including a high thermal conductivity, a high melting point, good mechanical properties, and should be stable in a high-neutron and radiation environment. In addition, the cladding material should have a low macroscopic neutron-absorption cross section so that the cladding does not compete with the fuel for the neutrons. The cladding material should be chemically stable so that it will experience little or no corrosion in the reactor, and it should be capable of being easily dissolved or stripped from the fuel during reprocessing of the fuel element.

A number of different structural materials have been used as cladding materials. Aluminum has been used as cladding material in a number of water-cooled, research reactors, but its relatively low melting point rules it out for application in power reactors. The British used a magnesium-based alloy in their Calder-Hall reactor series cooled with carbon dioxide. Stainless steels were used in some of the early water-cooled reactors, but these alloys have fairly high neutron-absorption cross sections. The 300-series stainless steels are used as cladding and structural materials in sodium-cooled reactors because these alloys are compatible with alkali metals. Almost all of the light-water power reactors use a zirconium-based alloy as the cladding material. Most of the fuel elements are fabricated from zirconium-alloy tubes filled with oxide fuel and these tubes are mounted in a square matrix. A typical LWR fuel element is shown in Fig. 3.10. Graphite is used as the cladding material in the HTGR fuel elements. While graphite has an excellent thermal conductivity and also a very high melting point, it is fairly porous and fission products diffuse through it.

In a thermal-reactor system, some sort of moderating material is incorporated in the reactor core to slow down the fission neutrons. This material should have a high macroscopic neutron-scattering cross section, a low macroscopic absorption cross section and a low atomic mass to get the maximum kinetic energy transfer from the neutrons to the moderator atoms. In addition, the moderator must be resistant to the high radiation levels in the reactor core, it should have good physical properties, and it should be compatible with the other materials in the core. Actually only three elements are used in materials to moderate the neutrons. The moderator elements or isotopes are ordinary hydrogen (mostly $_1^1H$), heavy hydrogen or deuterium ($_1^2H$), beryllium, and carbon.

Ordinary hydrogen is often employed as the moderating material in the form of water, solid, and liquid organic compounds, and in the form of an

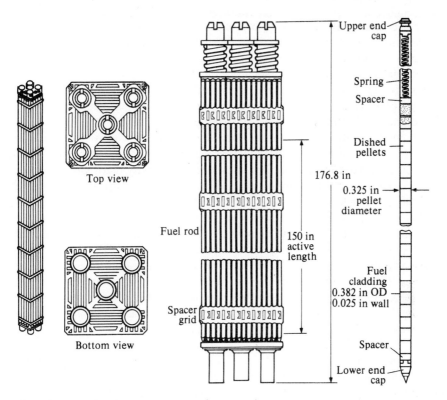

Figure 3.10 A typical fuel element for a light-water reactor (LWR). *(From Proceedings of the American Power Conference, 1974.)*

intermetallic compound, zirconium hydride. While ordinary hydrogen is commonly used in thermal reactors, the light-hydrogen isotope has a sufficiently high absorption cross section that reactors using it must employ enriched uranium. In many power and research reactors, light water is used as both the moderator and the coolant.

Heavy hydrogen is used as the moderating material in a number of thermal reactors in the form of heavy water, D_2O. The heavy-hydrogen isotope has such a low absorption cross section that reactors moderated with it can employ natural-uranium fuel. Because of the high cost associated with heavy water, some reactors use the heavy water only as a moderator and use ordinary water as the reactor coolant. As discussed earlier, Canada is a major proponent of the heavy-water-moderated, light-water-cooled power reactor.

The other two moderators are solids at operating temperatures. Beryllium is the best metallic moderator and can be used in the form of either pure metal or as beryllium oxide (BeO). Beryllium is expensive and difficult to machine and consequently is not widely used in reactors. The other solid moderating material is carbon in the form of graphite. Although carbon has the highest atomic mass of any of the moderating materials, it is relatively inexpensive and it has good high-temperature properties and excellent heat capacity. Carbon also has such a low neutron-absorption cross section that it can be used in thermal reactors fueled with natural uranium.

All heterogeneous reactors producing significant quantities of power employ coolants in the form of gases or liquids to transport thermal energy from the core. Coolants should have desirable physical properties, including a high thermal conductivity, a high specific heat, a high density, a low viscosity, and, if it is a liquid, it should have a high boiling point, a low vapor pressure, and a low melting point. In addition, the coolant should be chemically inert with the other in-core materials, it should not experience radiation damage in the intense in-core radiation field, it should have a very low macroscopic neutron-absorption cross section, and it should not become highly radioactive. Typical reactor coolants include light and heavy water, organic liquids, gases, and liquid metals.

Another component found in any reactor core is the neutron source. The neutron source provides a source of neutrons to bring the startup neutron level in the core to the point where it can be monitored by the in-core instrumentation. Thus, the reactor power and the rate of change of the neutron population can be checked at all times during the startup period. Monitoring the neutron level during the startup period prevents any inadvertant power excursions during this interval.

There are two basic types of neutron sources. Both sources use beryllium as the basic material along with either an alpha-emitting radioisotope or a radioisotope that emits an energetic gamma ray. The (α, n) source is composed of a mixture of beryllium and an alpha-emitting radioisotope, such as plutonium-239. The resulting alpha particles react with the beryllium nuclei to produce neutrons in the following reaction:

$$\ce{^9_4Be} + \ce{^4_2\alpha} \longrightarrow \ce{^{13*}_6C} \longrightarrow \ce{^{12}_6C} + \ce{^1_0}n + \gamma_\alpha$$

This source is normally withdrawn from the core region into a low-flux region at low to moderate power levels to prevent "burnup" of the alpha source by the neutrons.

The other basic neutron source is the photoneutron or (γ, n) source. This source is composed of a mixture of beryllium and a radioisotope that emits an energetic gamma ray. These gamma rays react with the beryllium nuclei in one of the following reactions:

$$\gamma + \ce{^9_4Be} \longrightarrow \ce{^{9*}_4Be} \Big\langle \begin{array}{l} \longrightarrow \ce{^8_4Be} + \ce{^1_0}n + \gamma_\gamma \\ \longrightarrow 2\ce{^4_2\alpha} + \ce{^1_0}n + \gamma_\gamma \end{array}$$

In order for the photoneutron reaction to take place, the incident gamma-ray energy must exceed the threshold energy of 1.62 MeV. The isotope that is commonly used to produce such a high-energy gamma ray is antimony-124 (Sb-124). This radioisotope is produced by bombarding stable antimony

(42.75% Sb-123) with neutrons to generate antimony-124 in the radiative-capture process. The antimony-124 radioisotope has a half-life of 60 days and emits an energetic gamma ray during the decay process. This source is normally left in the reactor during operation of the reactor in order to keep the neutron activation of the antimony at the saturated level.

All forms of instrumentation are employed in the core of a nuclear reactor, ranging from the usual flow nozzles or orifices and thermocouples to the more sophisticated neutron detectors and fission-product leak-detection systems. Of all this instrumentation, the neutron-detection systems are probably the most important for proper operation of the reactor. The detection of neutrons is difficult because they are not charged particles. In order to detect neutrons, they must first react with some other nuclei in a reaction that produces a charged particle or particles. The resulting ionization produced by the charged particle is then detected.

Most nuclear reactors use gas-filled ionization chambers to detect the ionizing radiation. These chambers are similar to that shown schematically in Fig. 3.11. As ionizing radiation in the form of gamma rays or charged particles pass through the gas they strip electrons from the gas molecules. This produces ion pairs along the path of the ionizing particle or radiation. If a voltage is applied across the center electrode and the casing, as indicated in Fig. 3.11, the resulting electric field can separate these ions before they have a chance to recombine. When these ions reach the center electrode or case, they induce a voltage drop that is amplified and detected in the triode grid.

The response of a radiation-detection chamber to a fixed quantity of ionization as a function of the voltage difference across the electrodes is shown in Fig. 3.12. As the electrode voltage is increased to around 200 V, the signal or ionization collected in this region (region II in Fig. 3.12) levels off on a plateau called the "ionization region." In this region, essentially all of the ion pairs produced by the ionizing radiation are separated by the electric field and are collected at the electrodes producing a signal that is equal to the amount of ionization produced in the chamber. As the electrode voltage is increased, the

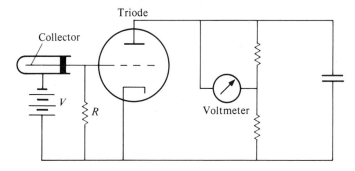

Figure 3.11 A typical radiation-detection chamber and electrical circuit. (*Glasstone and Sesonske, 1963.*)

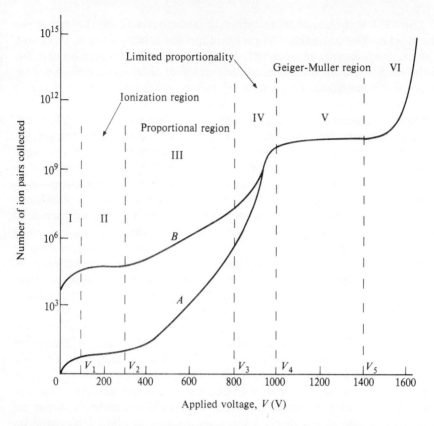

Figure 3.12 The response of a gas-filled radiation-detection chamber to primary ionizations of 10 ion pairs (curve *A*) and 50,000 ion pairs (curve *B*). (*Foster and Wright, 1973.*)

output signal also increases in an area of operation called the "proportional region" (region III in Fig. 3.12). The higher signal is generated because the separated primary ions, produced by the original radiation, are accelerated to the point where they produce secondary ion pairs, thereby giving a higher signal. As the electrode voltage is increased further, another signal plateau, called the Geiger-Muller region (region V), is attained. At this point, the gas in the tube is completely discharged because of the secondary ionization produced by the high voltage.

The Geiger-Muller region of operation is an excellent mode for the detection of any type of radiation because the production of one ion pair causes a complete discharge of the tube and generates a pulse. Consequently, geiger tubes are commonly employed in the general detection of the presence of radiation. However, since 1000 primary ions also generate one pulse, it is impossible to use these tubes to determine the amount of primary ionization produced by the ionizing particle or radiation.

In the reactor core, it is necessary to be able to discriminate between the heavy ionization caused by the charged particles generated by a neutron reaction and the ionization produced by the ever-present gamma radiation present in the reactor core. Consequently, almost all neutron detectors operate as ionization chambers where the signal or pulse is directly proportional to the amount of primary ionization produced in the chamber.

There are three basic types of in-core neutron detectors. These include the fission chamber, the compensated-ion chamber, and the uncompensated-ion chamber. The fission chamber is a gas-filled ionization chamber that has the inside surface of the chamber casing plated with a thin layer of uranium-235. Some of the neutrons entering the chamber react with the uranium-235 atoms in the fission process. One of the resulting fission products, with an average kinetic energy of 80 MeV, enters the gas region generating very heavy ionization in the gas. It is very easy to discriminate this single pulse from the ionization caused by the gamma radiation penetrating the chamber. The fission chamber is used during the initial stages of reactor startup when there may be very few neutrons and a great deal of gamma radiation. This is particularly true if the reactor had just been shut down. The fission chamber, like the (α, n) neutron source, is withdrawn into a shielded area as the reactor power is increased to prevent burnup of the fissile lining.

The compensated-ion chamber is actually a detector that is composed of two ion chambers. One chamber is filled with boron triflouride (BF_3) gas. The neutrons entering the chamber react with the boron in an (n, α) reaction and the resulting alpha particle causes considerable ionization in the chamber. The chamber with the boron triflouride gas is sensitive to both neutrons and gamma rays. The other chamber is filled with an inert gas with a low neutron-absorption cross section so that it is sensitive to only the gamma radiation. The difference in the two outputs gives a signal that is proportional only to the neutron level. The compensated-ion chamber is used from the intermediate-power to the full-power range.

The uncompensated-ion chamber is a detector that is composed of a single boron trifluoride gas chamber. Since this chamber is sensitive to both neutrons and gamma rays, it is used only during high-power operation where the ionization caused by the (n, α) reaction overshadows that caused by the gamma radiation.

All reactors have some sort of in-core control system to regulate the fission rate and hence the power level. There are two basic types of reactor control systems. One system employs the movement of fuel and/or moderator while the other system employs the movement of neutron absorbers to control the system reactivity.

Insertion of fuel and/or moderator in the reactor core increases the reactivity of the reactor but these systems are difficult to design properly. Movement of the moderator alone does not usually have a great effect on the fission rate and, consequently, it is difficult to satisfy all the reactor control requirements with a system of this type. Movement of fuel into and out of the core does have a

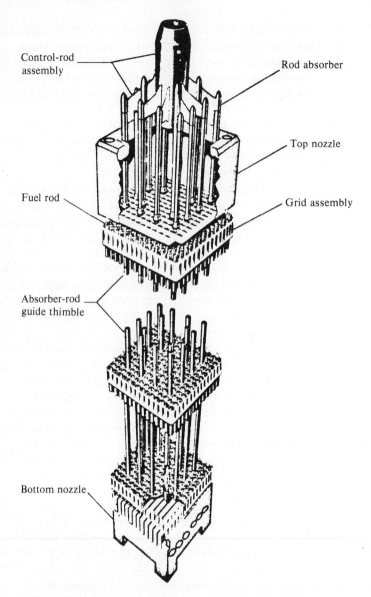

Control-rod assembly

Rod absorber

Top nozzle

Fuel rod

Grid assembly

Absorber-rod guide thimble

Bottom nozzle

Figure 3.13 Cutaway of a typical pressurized-water reactor control element assembly (CEA). (*Courtesy of Westinghouse Electric Corporation.*)

marked effect on the fission rate but a control system of this type is extremely difficult to design because of the problems associated with heat generation and removal.

The control systems of most power reactors utilize some sort of absorber control. This system controls the fission rate by inserting or withdrawing a substance that has a high macroscopic neutron-absorption cross section. This respectively decreases or increases the system reactivity and the fission rate. While there is considerable heat generated in these rods, it is much smaller than that generated in a fuel element. A number of different absorber materials have been used in nuclear reactors including hafnium, samarium, europium, gadolinium, cadmium, and indium, but the common absorber material is boron in the form of solid boron carbide (B_4C).

There are many kinds and shapes of control rods employed in various reactors. The common types of control rods used in light-water power reactors are either cruciform control rods or the "finger" control assemblies. The cruciform rods are cross-shaped rods that slide between the fuel assemblies. The older PWR and all the BWR systems use this type of control rod. The newer PWR reactors use the "finger" control element assembly in which control rods are inserted into guide tubes in several adjacent fuel elements. A typical control assembly of this type is shown in Fig. 3.13.

The control rods must be actuated from either the top or the bottom of the reactor core. Almost all of the PWR's use control-rod-drive systems that pull the control rods from the top of the reactor core with the use of electromagnets or magnetic jacks. In the event of an emergency, the magnet power is terminated and the rods fall into the core under the influence of gravity. This operation is called a reactor "scram." In a BWR, the top of the reactor is commonly filled with steam-drying equipment and consequently the cruciform rods are driven from the bottom of the core. In the event of a reactor "scram," the rods are driven into the core by the potential energy stored in a hydraulic accumulator-actuator.

The control rods essentially complete those components that are found in the reactor core except for the structural material. In some power reactors, fertile material is loaded into certain regions of the core and these regions are called "blankets." This is done to improve the conversion ratio of the reactor.

Components external to the core The core regions of almost all reactors are immediately surrounded by a neutron reflector. In thermal reactors, the reflector is composed of a moderating material, while the reflector of a fast reactor is composed of a heavy-mass material. The neutron reflector slows down and returns leakage neutrons in both of these systems, thereby reducing the critical mass and also reducing (improving) the maximum-to-average power ratio in the reactor core.

Most power reactors employ a thermal shield around the reflector. This shield attenuates the gamma and fast-neutron fluxes so that they do not cause excessive heating and radiation damage in the containment vessel and the biological shield. Thermal shields are required in both fast and thermal reactors. In a light-water reactor, the thermal shield is usually composed of alternate layers of water and steel.

The primary containment vessel is placed around the thermal shield and essentially contains the reactor coolant. This vessel, along with the rest of the primary cooling loop, is the second barrier for the containment of the fission products as the fuel-element cladding serves as the first barrier. In a LWR, the primary containment vessel is actually a large metallic pressure vessel. These vessels are usually fabricated from low-carbon steel, clad on the inside with stainless steel to minimize corrosion problems. The wall thickness of the pressure vessel normally runs from 8 to 10 in and the internal cladding is around $\frac{1}{2}$ in in thickness. These vessels range in dimension up to 21 ft in diameter and 40 to 60 ft in length, with an overall weight of more than 700 ton. They are so massive that they must normally be shipped by barge. A typical reactor pressure vessel and internals are shown in Fig. 3.14.

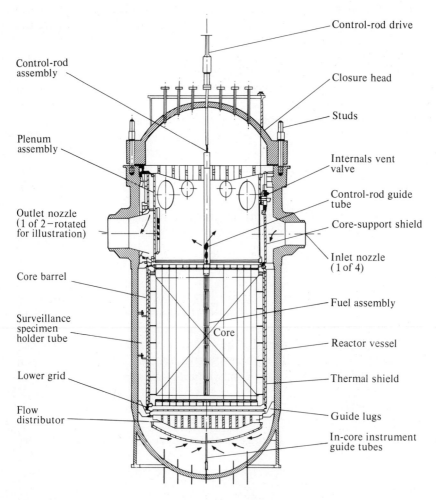

Figure 3.14 The primary pressure vessel of a pressurized-water reactor and its internals. (*From "Steam/Its Generation and Use," 1972.*)

In the high-temperature, gas-cooled reactor (HTGR), the containment vessel is constructed of prestressed concrete. This vessel is designed to contain the helium coolant and all the heat-exchange equipment is located in the shield.

The biological shield is put around the reactor vessel to protect operating personnel from the radiation emanating from the reactor core. This shield is normally constructed of concrete with a total thickness of 8 to 12 ft. Heavy aggregate is sometimes used in the concrete to reduce the shield thickness. Heavy metals, such as lead or even depleted uranium, make excellent gamma shields, but they are not employed around reactors because a light-mass material, such as the water in concrete, is needed to attenuate the neutrons.

The final major system or component associated with a power reactor is the secondary containment system. This system is designed to contain all the coolant and fission products that might be released in the event of a sudden failure of the primary coolant system. In any LWR, this means that the secondary containment must hold all the steam generated from a sudden depressurization of the primary system plus that released from a simultaneous failure of the secondary cooling system if it is a PWR. There are two basic types of secondary containment systems. One of these systems, like that shown in Fig. 3.15, uses a large volume to reduce the maximum pressure to a value that can be tolerated by the structure. These systems are the large spherical or cylindrical buildings

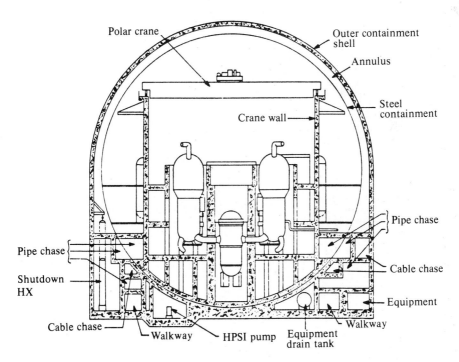

Figure 3.15 The spherical containment system for a pressurized-water reactor (PWR). *(From Proceedings of the American Power Conference, vol. 37, 1975.)*

associated with many reactor installations. These structures are designed to be leak-tight and entry to and egress from the building must take place through an air lock. This assures that the integrity of the secondary containment system is never violated. The LWR secondary containment vessels usually incorporate water-spray systems which are mounted high in the vessel to condense a lot of

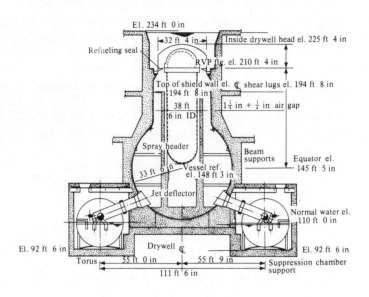

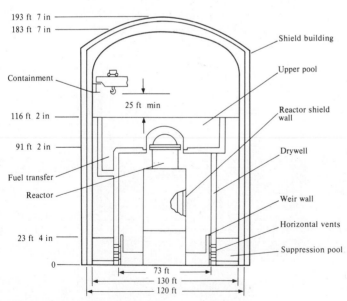

Figure 3.16 The water, vapor-suppression secondary containment system. (*From Proceedings of the American Power Conference, vol. 35, 1973.*)

the vapor generated in the accident and reduce the building pressure as soon as possible.

The other type of secondary containment system is called the vapor-suppression system. In this system, the primary pressure vessel and most of the piping is contained in a "dry well" located within the biological shield. Any steam released from the primary system is forced to pass through a water moat (the General Electric system) or through a bed of ice (the Westinghouse system). These systems are designed to "kill" or condense most of the steam generated in

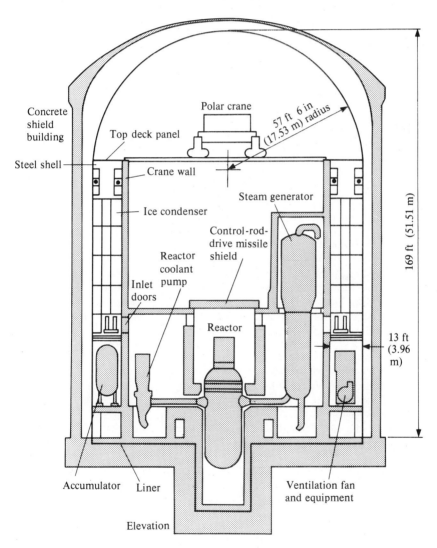

Figure 3.17 The ice, vapor-suppression secondary containment system. *(Courtesy of Westinghouse Electric Corporation.)*

a cooling-system failure, and keeps the maximum pressure in the external build-ing to a minimum. Not only does the water and ice "kill" the steam, but it also removes a lot of the fission products that might be released with the steam from a failure or meltdown of the cladding. Since the pressures generated outside the "dry well" are very low, the balance of the secondary containment system is simply a leak-tight building. Some of the vapor-suppression secondary contain-ment systems are shown in Figs. 3.16 and 3.17.

PROBLEMS

3.1 Calculate the theoretical air-fuel ratios, in kilograms of air per kilogram of fuel, for the following fuels:

(a) Schuylkill County, Pa., coal with $M = 4.5\%$ and $A = 9.7\%$

(b) Pike County, Ky., coal with $M = 2.5\%$ and $A = 3.3\%$

(c) Morton County, N. Dak., coal with $M = 37.0\%$ and $A = 4.2\%$

(d) Number 2 fuel oil with $C = 87.2\%$, $H_2 = 12.5\%$, and $S = 0.3\%$

(e) Number 5 fuel oil with $C = 87.85\%$, $H_2 = 11.25\%$, and $S = 0.9\%$

(f) Ohio-A natural gas

Use the values from the Appendices in the back of the book.

3.2 In a pulverized-coal boiler, the amount of unburned carbon must be calculated separately from the refuse analysis of the ashpit and from the flyash carried up the stack in the flue gas and removed in the precipitator. Calculate the total quantity of unburned carbon per kilogram of coal from the following data:

Ash in the fuel (from the ultimate fuel analysis) = 12.5%

Ashpit refuse = 0.1 kg refuse/kg coal

Combustible in refuse = 25%

Combustible in precipitator refuse = 40%

3.3 If Westmoreland County, Pa., coal (VM = 26.5%) with $M = 3.0\%$ and $A = 8.5\%$ is burned with 18 percent excess air, calculate the actual air-fuel ratio and the theoretical orsat analysis.

3.4 During the test of a boiler, the following data were obtained: weight of coal fired per hour, 9958 kg; weight of ashes per hour, 2920 kg; combustible in the ash, 22.84%. Ultimate analysis of the coal: $C = 60.31\%$, $H_2 = 4.06\%$, $O_2 = 6.71\%$, $N_2 = 1.13\%$, $S = 5.23\%$, and ash = 22.56%. Orsat analysis: $CO_2 = 10.69\%$, $O_2 = 8.23\%$, and $CO = 0.23\%$. Find the mass of dry flue gas per kilogram of coal fired, the percentage of excess air, the dilution coefficient, and the total weight of flue gas and weight of carbon monoxide generated in kilograms per hour.

3.5 If Greenbrier County, W. Va., coal, with $M = 3.0\%$ and $A = 8.0\%$, is burned with 15 percent excess air and the orsat analysis gives 0.7% CO, find the balance of the orsat readings, assuming that there is no combustible in the refuse.

3.6 Repeat Prob. 3.5 with 0.5% CO in the orsat analysis and a refuse analysis that shows 20 percent combustible.

3.7 A coke-oven gas has the following molar composition: $CO_2 = 1\%$, $CO = 8\%$, $CH_4 = 25\%$, $H_2 = 48\%$, and the balance is nitrogen. Calculate the actual air-fuel ratios in cubic meters of air per cubic meter of gas and in kilograms of air per kilogram of gas, and determine the theoretical orsat analysis when the gas is burned with 15 percent excess air.

3.8 The carbureted water gas, in the Appendix, is burned and the resulting orsat analysis gives: 17.19% CO_2, 3.15% O_2, and 0.96% CO. Find the actual mass air-fuel ratio and the percentage of excess air.

3.9 Palestine, Ill., natural gas has the following composition: 0.5% CO_2, 3.9% N_2, 0.5% O_2, and the balance is methane.

(a) Find the theoretical mass and volumetric air-fuel ratios.

(b) If the dry exhaust contains 0.5% CO when this gas is burned with a dilution coefficient of 1.15, determine the orsat analysis.

3.10 If 1 kg of propanol (propyl alcohol, C_3H_7OH) with a higher heating value of 33,492 kJ/kg is used as a fuel, find the theoretical orsat analysis and the energy released per unit mass when the fuel is burned with:

(a) A dilution coefficient of 0.9

(b) A dilution coefficient of 0.65

3.11 If 1 gal of ethanol (C_2H_5OH) is burned with dilution coefficients of 0.85 and 0.65, find the theoretical orsat analysis and estimate the energy released per gallon of fuel. The density of ethanol is 0.791 kg/liter.

3.12 Natural gas with the following analysis is burned in a furnace: $CO_2 = 0.4\%$, $CH_4 = 97.3\%$, and $N_2 = 2.3\%$. The orsat analysis of the flue gas shows: $O_2 = 5.6\%$, $CO_2 = 8.5\%$, and $CO = 0.1\%$. Find the percentage of excess air and the actual mass air-fuel ratio.

3.13 A producer gas with the following molar composition is burned in a furnace: $CO = 21\%$, $H_2 = 31\%$, and $N_2 = 48\%$. The orsat analysis of the flue gas reads: $CO_2 = 8.6\%$, $O_2 = 6.4\%$, and $CO = 0.2\%$. Calculate the percentage of excess air and the actual mass air-fuel ratio.

3.14 The Bruce Mansfield Power Plant of the Pennsylvania Power Company is a three-unit, coal-fired system with a total output of 2751 MW_e. Assume the system burns Elk County, Pa., coal with moisture and ash fractions of 0.03 and 0.09, respectively. The plant heat rate is 8800 Btu/kW·h and the higher heating value of the refuse is 455 Btu/lbm. The orsat analysis gives: 13.77% CO_2, 4.40% O_2, and 0.99% CO. Find:

(a) The maximum coal rate for the plant at full power, in tons per hour

(b) The percentage of excess air

3.15 What will be the temperature, in degrees centigrade, of a block of graphite containing thermal neutrons with a most probable velocity of 2600 m/s?

3.16 The University of Missouri-Columbia is one of the highest-powered research reactors in the country and operates at a maximum thermal power of 10 MW. If the average water and beryllium temperature is 60°C and the fuel loading is about 6 kg of uranium-235, find the average thermal-neutron flux in the reactor.

3.17 A 3800-MW_{th} pressurized-water power reactor operates at an average thermal flux of 5×10^{17} neutrons/m²·s and the average water temperature is 343°C (650°F). The reactor is fueled with uranium dioxide that is enriched to 2.3 percent. Find the mass of uranium dioxide in the reactor.

3.18 A boiling-water reactor power plant produces 1200 MW_e with a thermal efficiency of 32 percent. If the system is fueled with 2.2 percent enriched uranium in the form of uranium dioxide (UO_2) and operates at a saturation pressure of 85 bars with an average thermal-neutron flux of 4×10^{17} neutrons/m²·s, find the total loading of uranium-235 in the system and also determine the total loading of uranium dioxide. If the uranium dioxide has a density of 10 g/cm³, find the total fuel volume in the core, in cubic feet. Also determine the volume fraction of the fuel if the core is 12 ft long and 12 ft in diameter.

3.19 Beryllium oxide (BeO) has a density of 3.0 g/cm³. Calculate the total average distance traveled by a thermal neutron in an infinite medium, the average distance that the same neutron travels before some event (absorption or scattering) occurs, and determine the average thermal-neutron lifetime if the oxide is at an average temperature of 1040°F. The 2200-m/s neutron cross sections are:

For Be atoms:	$\sigma_a = 0.01$ b	$\sigma_s = 7.0$ b
For O atoms:	$\sigma_a = 0.0002$ b	$\sigma_s = 4.2$ b

3.20 Calculate the total neutron reaction rate, the absorption rate, the capture rate, the scattering rate, and the thermal power in watts per cubic centimeter of uranium dioxide (UO_2). The density of uranium dioxide is 10.3 g/cm³, the moderator temperature is 800°F, and the average thermal-neutron flux is 5.0×10^{17} neutrons/m²·s. The material cross sections for 2200 m/s neutrons are:

U-235:   Use the data in Table 3.1
U-238: $\sigma_a = 2.71$ b $\sigma_s = 8.3$ b $\sigma_c = \sigma_a$
Elemental oxygen: Use the data from Prob. 3.19

Assume that the uranium is natural uranium and neglect the uranium-234 isotope (assume it has the same properties as uranium-238). In addition to the quantities asked for earlier, determine the average neutron lifetime (λ_a/v_{th}) and the breeding/conversion ratio (U-238 absorptions/U-235 absorptions).

3.21 100,000 thermal neutrons are inserted into an infinite lattice of uranium nitride (U_3N_4) at a temperature of 600°F, using natural uranium. Uranium nitride has a density of 10 g/cm³. Using the cross-section data below, find:

   (a) The scattering, absorption, fission, and total macroscopic neutron cross sections
   (b) The average lifetime (λ_a/v_{th}) of these neutrons
   (c) The number of the original neutrons absorbed by each of the three uranium isotopes (in the natural uranium) and by the nitrogen atoms
   (d) The multiplication factor for the lattice, assuming the fission neutrons are produced at thermal energy
   (e) The breeding/conversion ratio for this lattice
The microscopic neutron cross-section data for 0.0253-eV neutrons are as follows:

U-234: $\sigma_a = \sigma_c = 105.0$ b $\sigma_s = 9.0$ b $\sigma_f = 0.0$
U-235:   Use the data in Table 3.1
U-238:   Use the data in Prob. 3.20
Elemental nitrogen: $\sigma_a = \sigma_c = 1.85$ b $\sigma_s = 10$ b $\sigma_f = 0.0$

REFERENCES

Bonilla, C. F. (ed.): "Nuclear Engineering," McGraw-Hill Book Company, New York, 1957.

De Lorenzi, Otto (ed.): "Combustion Engineering," Combustion Engineering Superheater, Inc., New York, 1952.

Dillo, C. C., and E. P. Nye: "Thermal Engineering," International Textbook Company, Scranton, Pa., 1963.

Foster, A. R., and R. L. Wright, Jr.: "Basic Nuclear Engineering," Allyn and Bacon, Inc., Boston, Mass., 1973.

Glasstone, S., and A. Sesonske: "Nuclear Reactor Engineering," D. Van Nostrand Company, Inc., New York, 1963.

Lish, K. C.: "Nuclear Power Plant Systems," Industrial Press, Inc., New York, 1972.

Potter, P. J.: "Steam Power Plants," The Ronald Press Company, New York, 1972.

Proceedings of the American Power Conference, 1974, vol. 36, pp. 70–134, Illinois Institute of Technology, Chicago, Ill.

Skrotzki, B. G. A., and W. A. Vopat: "Power Station Engineering and Economy," McGraw-Hill Book Company, New York, 1960.

"Steam/Its Generation and Use," 38th ed., Babcock and Wilcox Company, New York, 1972.

FOUR

FOSSIL-FUEL SYSTEMS

4.1 INTRODUCTION

This chapter presents a more detailed description and the principal of operation of fossil-fueled steam-generating systems and the associated auxiliary equipment. Specifically, this material covers the fluid machinery needed in the operation of the steam generator, including pumps and fans, the different furnace types, and, finally, the types of boilers and the operational problems associated with them.

4.2 FLUID MOVING SYSTEMS

4.2.1 Introduction

Two basic fluid moving systems are employed in almost all steam generator systems. These are the pumps needed to supply the boiler feed to the generator and the air compressors or fans needed to supply combustion air to the furnace. An important parameter for these systems is the mechanical efficiency, η_{mech}, which is a measure of the machine's ability to transmit mechanical work to the fluid flowing through the device. The relationship for the mechanical efficiency for fluid moving systems is given by

$$\eta_{\text{mech}} = \frac{\text{ideal work input}}{\text{actual work input}} \tag{4.1}$$

This relationship is the reciprocal of the equation for the mechanical efficiency of a prime mover.

Since the boiler feed pumps handle essentially incompressible water and the fans operate over a very narrow pressure range, both of these systems can be assumed to pump an incompressible fluid. For geometrically similar centrifugal machines, operating at the same efficiencies, the pressure rise ΔP across the device, the volumetric flow rate Q through the device, and the input power requirements P are related by the following equations:

$$\Delta P = K_1 \rho N^2 D^2 \tag{4.2}$$

$$Q = K_2 N D^3 \tag{4.3}$$

$$P = \frac{Q \, \Delta P}{\eta_{\text{mech}}} = K_3 \rho N^3 D^5 \tag{4.4}$$

where ρ is the fluid density, N is the angular velocity, and D is the diameter of the system. These relationships are used to develop the so-called fan or pump laws.

4.2.2 Boiler Feed Pumps

The boiler feed pump supplies high-pressure liquid to the boiler and commonly operates over a very wide range of pressures. The centrifugal pump is commonly used for this purpose and the performance of these systems is usually expressed in terms of the "specific speed" N_s of the pump. The specific speed of a given pump is defined as the revolutions per minute of a geometrically similar pump reduced in size which will produce a volumetric flow rate of 1 gal/min against a total pressure rise of 1 lb/in². The specific speed of a given pump can be determined from a known value of Q, ΔP, and the angular velocity N:

$$N_s = \frac{N Q^{1/2}}{\Delta P^{3/4}} \tag{4.5}$$

The general relationship between the shape of an impeller and the specific speed of the pump is shown in Fig. 4.1.

A certain operating condition that should be avoided in the operation of any pump is to operate it with cavitation. Cavitation occurs when the liquid pressure on the surface of the impeller falls below the vapor pressure of the fluid. This causes vapor bubbles to form on the surface of the impeller and these bubbles collapse as they move into a region of higher pressure. The sudden collapse of these bubbles causes severe impact on the impeller and this action can cause severe erosion of the impeller surface. Not only can cavitation physically damage the pump but it also lowers the mechanical efficiency and makes the pump noisy. This condition can be alleviated by increasing the fluid pressure at the pump inlet. This pressure, minus the vapor pressure of the liquid, is called the net positive suction head or NPSH. The minimum value of the NPSH to assure cavitation-free operation is specified by the pump manufacturer.

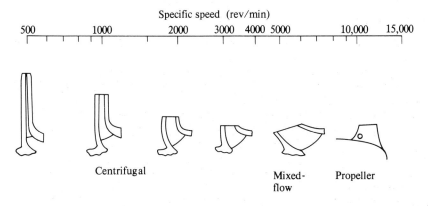

Figure 4.1 The variation of the specific speed of a pump with impeller shape. *(From Skrotzki and Vopat, 1963.)*

Some of the boiler feed pumps are very large components with input power requirements of more than 60,000 hp. A list of some high-pressure pumps is shown in Table 4.1. A cross-sectional view of a typical pump is shown in Fig. 4.2. These large boiler feed pumps are usually driven by steam turbines.

4.2.3 Combustion Air Systems

There are two general types of air compressors—positive-displacement air compressors and dynamic air compressors. In the positive-displacement compressor, the impeller or piston forcibly displaces the air volume to compress it. A common positive-displacement air compressor is the reciprocating compressor which is characterized by relatively low capacity, high head, and pulsating flow.

Table 4.1 Large boiler feed pumps

Power station	Pump speed, rev/min	Flow rate, gal/min	Discharge pressure, lb/in² gage	Pump power, hp
Glen Lyn 6	5500	3,200	2452	5,650
Muskingum River 4	5500	3,200	2503	5,650
Clinch River 3	5500	3,200	2490	6,300
Breed 1	3600	7,200	4500	21,600
Philip Sporn 5	3500	3,600	4500	10,800
Big Sandy 1	5500	4,400	2868	8,500
Tanners Creek 4	4500	8,900	4575	26,800
Cardinal 1 and 2	4560	9,500	4695	29,100
John E. Amos 1 and 2	4840	12,550	4570	37,500
Gen. James M. Gavin 1 and 2	4160	21,600	4625	63,000

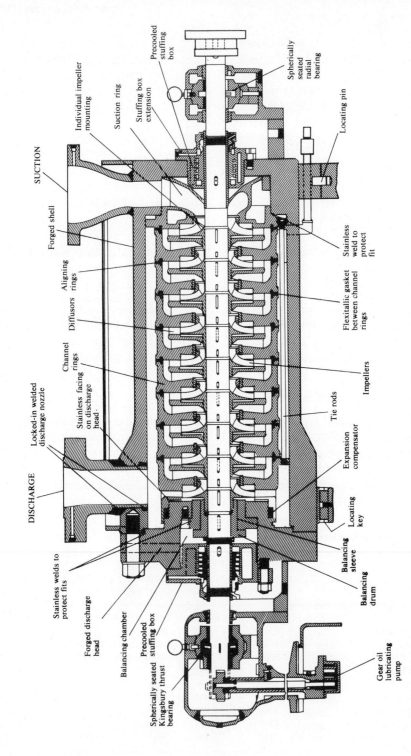

Figure 4.2 Cross section of a large boiler feed pump. (*Courtesy of Ingersoll-Rand Company.*)

Precooled stuffing box

Spherically seated radial bearing

Individual impeller mounting

Suction ring

Stuffing box extension

Locating pin

SUCTION

Forged shell

Stainless weld to protect fit

Aligning rings

Diffusors

Flexitallic gasket between channel rings

Channel rings

Stainless facing on discharge head

Impellers

Locked-in welded discharge nozzle

Tie rods

DISCHARGE

Expansion compensator

Stainless welds to protect fits

Locating key

Forged discharge head

Balancing chamber

Balancing sleeve

Precooled stuffing box

Balancing drum

Spherically seated Kingsbury thrust bearing

Gear oil lubricating pump

160

These machines are commonly used to supply high-pressure station air for control and other purposes but they are not normally used to supply combustion air. The rotary air compressors are also positive-displacement machines and include the sliding-vane air compressor, the Roots blower, the screw compressor, etc. These machines are characterized by moderate pressures, medium capacity, and continuous flow. Rotary compressors are used to supply air to supercharge engines but they are not widely used to supply combustion air to large combustion systems.

In the dynamic air compressor, the high-velocity impeller transfers momentum from the impeller to the air. The two types of dynamic machines are the axial-flow and centrifugal compressors. These machines are characterized by relatively high capacity, low head, and steady flow. The axial-flow compressors are commonly used in the gas-turbine and turbojet engines and they are becoming increasingly popular for large combustion systems. Centrifugal compressors, commonly called "fans," are widely used to supply the large volumes of combustion air for electric power generation.

Combustion air fans usually have a very high flow rate but a total pressure rise of less than 2 or 3 lb/in^2. The three types of centrifugal fans include the fans with backward-inclined blades, fans with radial blades, and fans with forward-curved blades. The different blade orientations give the fans different operating characteristics. A comparison of the operating characteristics of the fans is given in Table 4.2 and the characteristics as a function of the volumetric flow rate are shown in Fig. 4.3.

Since the pressure rise across any fan is relatively small, the air flow through the fan can be assumed to be incompressible. In this case, the volumetric flow Q, the pressure rise across the fan ΔP, and the power requirement of the fan P varies according to the so-called "fan laws." These laws are essentially the same as Eqs. (4.2), (4.3), and (4.4), except that the impeller diameter is constant.

Some of the newer fossil-fuel combustion systems are using very large axial-flow fans to supply combustion air to the furnace. The Martins Creek power station of the Pennsylvania Power and Light Company has two new 820-MW_e fossil systems that employ axial-flow fans for both the forced-draft and the induced-draft fans. Axial-flow units are used instead of centrifugal fans because

Table 4.2 Operating characteristics of centrifugal fans

Quantity	Backward inclined	Radial	Forward curved
Size of impeller	Medium	Medium	Small
Mechanical efficiency	High	Average	Average
Stability	Good	Good	Poor
Fan speed, rev/min	High	Average	Low
Impeller-tip speed	High	Medium	Low
Resistance to erosion	Medium	Good	Poor

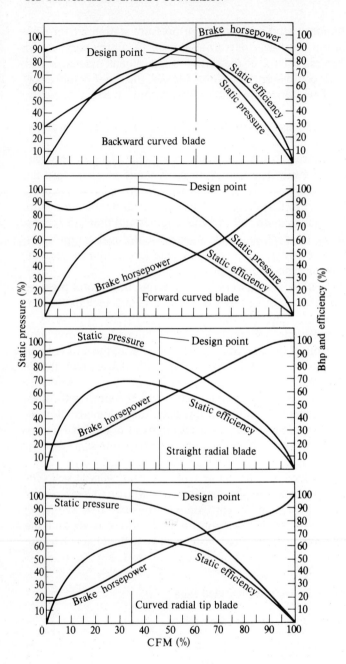

Figure 4.3 Operating characteristics of centrifugal fans. *(From Engineer's Reference Library (Fans), by Editors of Power Magazine, McGraw-Hill, Inc., 1969.)*

the gas flow rate is regulated by varying the pitch of the fan blades. This permits these units to operate at much higher efficiencies than comparable centrifugal units operating at off-load conditions. Figure 4.4 shows the single-stage, forced-draft fan and the two-stage, induced-draft fan used in the Martins Creek station. This station has two forced-draft and two induced-draft fans.

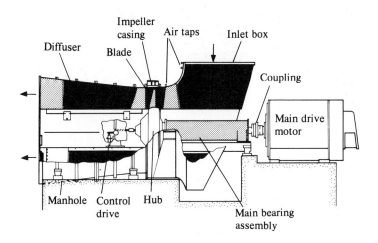

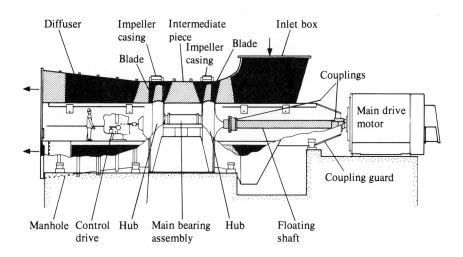

Figure 4.4 A single-stage, 5000-hp forced-draft fan and a two-stage, 12,000-hp induced-draft fan. *(From Editors of Power Magazine, 1976.)*

4.2.4 Draft Systems

The air required for combustion can be supplied by a natural-draft system, by a mechanical-draft system, or, as is usually the case, by a combination of these two systems. In the natural-draft system, the air flow is produced by a driving force established by the difference in the specific weights of the exhaust gas in the stack and the atmospheric air. The driving force is equal to the product of the difference in specific weights and the height of the stack. This driving force is balanced by the pressure drop in the draft system, the furnace, and the exhaust systems.

The air flow rate in a natural-convection system can be increased by raising the exhaust-gas temperature and/or by increasing the stack height. An increase in exhaust-gas temperature decreases the gas density (specific weight) thereby increasing the head but it also increases the system losses, which is undesirabie. Increasing the stack height not only increases the gas-flow rate but it improves the dispersion of exhaust products. Stack heights of 700 to 1200 ft are fairly common. Despite the use of very tall smoke stacks or chimneys, the natural-draft systems cannot normally supply sufficient air flow for good combustion. Consequently, supplementation is required by some sort of mechanical-draft system.

There are two basic types of mechanical-draft systems—the forced-draft (f-d) system and the induced-draft (i-d) system. In the forced-draft system, the main fan pumps combustion air into the furnace. The forced-draft fan pumps only cold, clean combustion air and the positive pressure in the combustion chamber improves the gas side-convection coefficient. Unfortunately, the positive pressure in the furnace produces a major problem because any leakage from the furnace is likely to cause the combustion products to be released in areas where personnel may be working. This is a major drawback of the forced-draft system when it is used alone.

The minimum ideal mass flow rate of the forced-draft fan is equal to the product of the actual mass air-fuel ratio and the fuel rate:

$$\text{Minimum mass flow rate for f-d fan} = (\text{fuel rate})\left(\frac{A}{F}\right)_{\text{act}, m, w} \qquad (4.6)$$

The air-fuel ratio used in (4.6) should include the moisture in the atmospheric air. The flow rate given by Eq. (4.6) should be increased by 20 to 40 percent to account for leakage. Most forced-draft fans are high-speed centrifugal fans with backward-curved blades.

Many large positive-pressure furnaces use another large fan along with the forced-draft fan, called a gas-recirculation fan. This fan pumps exhaust gas back into the combustion chamber to cool the combustion chamber and control the rate of energy deposition in the boiler. It also provides some control of the nitrogen oxide (NO_x) emissions. This fan pumps relatively hot exhaust gas which may be loaded with flyash but it handles only a small portion of the total exhaust-gas flow.

In the induced-draft mechanical-draft system, the fan draws combustion products from the combustion chamber and pumps them into the stack. This produces a negative pressure in the combustion chamber, eliminating the leakage problem. This fan also handles hot combustion gas, drastically increasing the capacity of the induced-draft fan. It also pumps some flyash in coal-fired systems and this can cause an erosion problem as well as vibration problems due to ash deposits building up on the blades. For a given combustion rate in the furnace, the mass flow rate through the induced-draft fan is around 10 percent higher than the mass flow rate through the forced-draft fan. Theoretically, for each kilogram of fuel burned, the forced-draft fan pumps only the actual air-fuel ratio, but the induced-draft fan pumps the actual air-fuel ratio plus all of the 1 kg of fuel except for the refuse fraction R:

$$\text{Mass flow rate for i-d fan} = (\text{fuel rate})\left[\left(\frac{A}{F}\right)_{\text{act, } m, w} + 1.0 - R\right] \quad (4.7)$$

As is the case for the forced-draft fan, the flow rate given by Eq. (4.7) is a minimum rate and should be increased by 20 percent or so to account for infiltration due to leakage into the furnace and ductwork.

The volumetric flow rates for the forced-draft and induced-draft fans can be obtained by dividing the mass flow rates by the inlet gas density:

$$\text{Volumetric flow rate} = \frac{\text{mass flow rate}}{\text{inlet gas density}}$$

$$= \frac{(\text{mass flow rate})(R_u T)}{(P)(\text{MW of gas})} \quad (4.8)$$

Equation (4.8) is straightforward for the forced-draft fan because the molecular weight of the pumped gas (air) is 28.97 kg/kg·mol, but it is more difficult for the exhaust gas because the molecular weight of the exhaust must be calculated. Sometimes it is assumed that the molecular weight of the exhaust is the same as that for air and this is usually a pretty good approximation.

Many combustion systems employ forced-draft and induced-draft fan systems to produce a " balanced-draft " system. In this system, the combustion chamber operates at or near atmospheric pressure. The forced-draft fan overcomes the pressure drops in the fuel handling and air-supply ductwork while the induced-draft fan and the stack overcome the pressure drops in the exhaust-gas system, including the precipitator or baghouse filter, if one is used. With the balanced-draft or negative-draft systems, minor firebox inspections and maintenance are possible while the unit is in operation. This is very difficult to accomplish in a positive-pressure furnace.

As was mentioned earlier, most forced-draft fans are high-speed centrifugal fans with backward-curved blades, but the induced-draft fans are usually low-speed centrifugal fans with forward-curved blades. The lower speed of the induced-draft fan reduces the erosion problem. The fans used to supply combustion air for a typical fossil-fueled power system are large, expensive machines. The Labadie power station of the Union Electric Company has four 600-MW$_e$ units and each unit has two 1750-hp forced-draft fans and two 4000-hp induced-draft fans. These fans are driven by large alternating-current induction motors. A

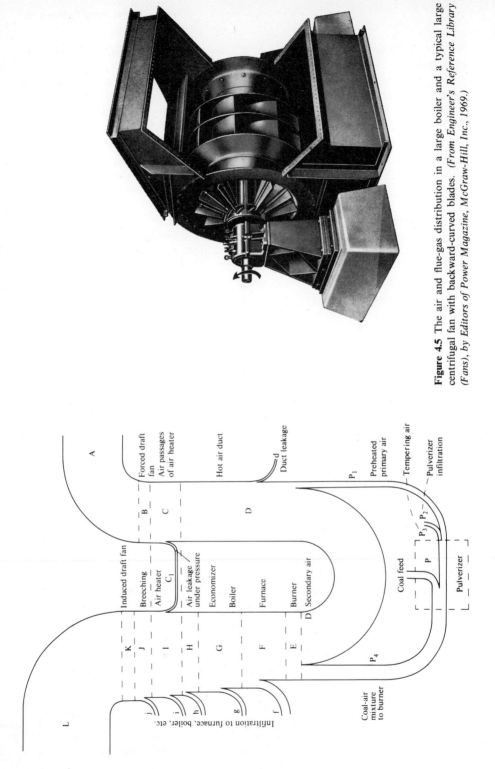

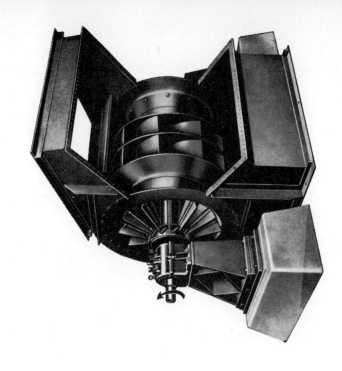

Figure 4.5 The air and flue-gas distribution in a large boiler and a typical large centrifugal fan with backward-curved blades. *(From Engineer's Reference Library (Fans), by Editors of Power Magazine, McGraw-Hill, Inc., 1969.)*

typical combustion air fan is shown in Fig. 4.5 along with a diagram showing the air and flue-gas distribution in a large coal-fired steam generator.

Example 4.1 A 600-MW_e power plant burns Lafayette County, Mo., coal with average moisture and ash fractions of 14 and 11 percent, respectively. This plant operates with a heat rate of 8863 Btu/kW·h. An analysis of the refuse pit gives a higher heating value of 2605 kJ/kg refuse. An orsat analysis of the flue gas gives 13.78% CO_2, 4.49% O_2, and 0.75% CO. Find:

(a) The thermal efficiency of the power plant
(b) The coal rate
(c) The dilution coefficient and the percentage of excess air
(d) The capacity of the f-d fan, in ft³/min and in kg/min, if atmospheric conditions are 50°C, 0.93 atm, and the relative humidity is 50 percent
(e) The capacity of the i-d fan, in ft³/min and in kg/min, if the exhaust gas is at 288°C and 0.88 atm
(f) The size of the motors required to drive the i-d and f-d fans, in horsepower, if there is a 25-in water differential across each unit and the mechanical efficiency of each fan is 86 percent
(g) The dew-point temperature of the exhaust gas leaving the i-d fan

SOLUTION

(a) Thermal efficiency of plant $= \dfrac{3413 \text{ Btu}_e/\text{kW}_e \cdot \text{h}}{\text{heat rate, Btu}_{\text{th}}/\text{kW}_e \cdot \text{h}}$

$$= \frac{3413}{8863} = 0.3851 = 38.51\%$$

(b) Coal analysis (from App. C): 78.6% C, 5.6% H_2, 9.3% O_2, 1.3% N_2, 5.2% S, and HHV = 33,160 kJ/kg

Multiplier $= (1 - M - A) = (1.0 - 0.14 - 0.11) = 0.75$

As-burned coal analysis: 14.0% M, 11% A, 58.95% C, 4.2% H_2, 6.975% O_2, 0.975% N_2, 3.9% S, and HHV = 24,870 kJ/kg

Refuse analysis:

$(\text{HHV})_r = 2605$ kJ/kg

$$\frac{C_r}{R} = \frac{(\text{HHV})_r}{(\text{HHV})_C} = \frac{2605}{32,778} = 0.0795 \text{ kg } C_r/\text{kg } R$$

$$\frac{A}{R} = 1.0 - \frac{C_r}{R} = 1.0 - 0.0795 = 0.9205 \text{ kg } A/\text{kg } R$$

$$R = \frac{A_{\text{ult}}}{A/R} = \frac{0.11}{0.9205} = 0.1195 \text{ kg } R/\text{kg coal}$$

$$C_r = R - A = 0.1195 - 0.11 = 0.0095 \text{ kg } C_r/\text{kg coal}$$

$$C_b = C_{\text{ult}} - C_r = 0.5895 - 0.0095 = 0.58 \text{ kg } C_b/\text{kg coal}$$

$$\text{Coal rate} = \frac{\text{thermal power}}{(\text{HHV})_{\text{coal}}}$$

$$= \frac{(6 \times 10^8 \text{ W}_e)(10^{-3} \text{ kJ}_e/\text{W}_e\text{·s})}{(0.3851 \text{ kJ}_e/\text{kJ}_{\text{th}})(24{,}870 \text{ kJ}_{\text{th}}/\text{kg coal})}$$

$$= 62.65 \text{ kg coal/s} = 225{,}530 \text{ kg coal/h} = 248.6 \text{ short tons/h}$$

(c) Theoretical dry air-fuel ratio $= \left(\dfrac{A}{F}\right)_{\text{th, } m, d}$

$$= \frac{2.66 \text{ C} + 7.94 \text{ H}_2 + 0.998 \text{ S} - \text{O}_2}{0.232}$$

$$= \frac{2.66(0.5895) + 7.94(0.042) + 0.998(0.039) - 0.06975}{0.232}$$

$$= 8.063 \text{ kg air/kg coal}$$

Actual dry air-fuel ratio $= \left(\dfrac{A}{F}\right)_{\text{act, } m, d}$

$\% \text{ N}_2 \text{ (from orsat analysis)} = 1.0 - 0.1378 - 0.0449 - 0.0075 = 0.8098$

$$= 80.98\%$$

$$\left(\frac{A}{F}\right)_{\text{act, } m, d} = \frac{(28.01 \text{ } C_b)(\% \text{ N}_2)/(12.01)(\% \text{ CO} + \% \text{ CO}_2) - \text{N}_{2, \text{ult}}}{0.768}$$

$$= \frac{28.01(0.58)(80.98)/(12.01)(13.78 + 0.75) - 0.00975}{0.768}$$

$$= 9.804 \text{ kg air/kg coal}$$

$$\text{Dilution coefficient} = \frac{(A/F)_{\text{act}}}{(A/F)_{\text{th}}} = \frac{9.804}{8.063} = 1.2159$$

Percentage of excess air $= 100 \text{ (dilution coefficient} - 1)$

$$= 100(1.2159 - 1) = 21.59\%$$

(d) Moisture in the combustion air:

Partial pressure of H_2O in air

$$= \text{(relative humidity)(saturation pressure at } T_{\text{in}})$$

$$p_w = 0.5(1.7888) = 0.8944 \text{ lb/in}^2$$

Specific humidity $= \omega = \dfrac{0.622 \text{ } p_w}{p_{\text{air}}} = \dfrac{0.622(0.8944)}{0.93(14.7) - 0.8944}$

$$= 0.04354 \text{ kg H}_2\text{O/kg dry air}$$

Mass flow rate through f-d fan $= \left[\left(\dfrac{A}{F}\right)_{\text{act, } m, d}(1 + \omega)\right](\text{fuel rate})$

$$= [9.804(1 + 0.04354)](62.65)(60)$$

$$= 38,458 \text{ kg/min}$$

Air density $= \rho_a = \dfrac{P}{RT} = \dfrac{0.93(14.7)(144)(28.97)}{1545(582)} = 0.06342 \text{ lbm/ft}^3$

Water vapor density $= \rho_w = \rho_a \dfrac{18.02}{28.97} = 0.03945 \text{ lbm/ft}^3$

Volumetric flow rate through f-d fan

$$= (\text{fuel rate})\left[\left(\dfrac{A}{F}\right)_{\text{act, } m, d}\left(\dfrac{1}{\rho_a} + \dfrac{\omega}{\rho_w}\right)\right]$$

$$= 62.65(60)\left[9.804\left(\dfrac{1}{0.06342} + \dfrac{0.04354}{0.03945}\right)\right]\dfrac{1}{0.4536}$$

$$= 1,370,750 \text{ ft}^3/\text{min}$$

(e) Mass flow rate through i-d fan

$$= \left[\left(\dfrac{A}{F}\right)_{\text{act, } m, d}(1 + \omega) + 1.0 - R\right](\text{fuel rate})$$

$$= [9.804(1 + 0.04354) + 1.0 - 0.1195](62.65)(60)$$

$$= 41,768 \text{ kg exhaust/min}$$

Mole fractions of exhaust gas:

Mol O_2/kg coal $= \dfrac{0.232[(A/F)_{\text{act}} - (A/F)_{\text{th}}]}{32} + \dfrac{\text{mol CO}}{2} + \dfrac{C_r}{12.01}$

Mol CO/kg coal $= \dfrac{C_b(\% \text{ CO})}{12.01(\% \text{ CO} + \% \text{ CO}_2)}$

Mol H_2O/kg coal $= \dfrac{M + (A/F)_{\text{act, } m, d}\omega + 9H_2}{18.02}$

Mol SO_2/kg coal $= \dfrac{S}{32.06}$

Mol N_2/kg coal $= \dfrac{0.768(A/F)_{\text{act, } m, d} + N_{2, \text{ult}}}{28.01}$

Values:

$$\text{Mol CO/kg coal} = \frac{0.58(0.75)}{12.01(0.75 + 13.78)} = 0.002493$$

$$\text{Mol CO}_2\text{/kg coal} = \frac{0.58(13.78)}{12.01(0.75 + 13.78)} = 0.045800$$

$$\text{Mol H}_2\text{O/kg coal} = \frac{0.14 + 9.804(0.04354) + 9(0.042)}{18.02}$$

$$= 0.052434$$

$$\text{Mol SO}_2\text{/kg coal} = \frac{0.039}{32.06} = 0.001216$$

$$\text{Mol N}_2\text{/kg coal} = \frac{0.768(9.804) + 0.00975}{28.01} = 0.269162$$

$$\text{Mol O}_2\text{/kg coal} = \frac{0.232(9.804 - 8.092)}{32} + \frac{0.002493}{2} + \frac{0.0095}{12.01}$$

$$= 0.014450$$

Total moles of flue gas/kg coal

$$= 0.002493 + 0.045800 + 0.052434 + 0.001216 + 0.269162$$

$$+ 0.014450$$

$$= 0.385555$$

$$\text{Total mass of flue gas/kg coal} = \left(\frac{A}{F}\right)_{act,\,m,\,d} (1 + \omega) + 1.0 - R$$

$$= 9.804(1 + 0.04354) + 1.0 - 0.1195$$

$$= 11.1114 \text{ kg flue gas/kg coal}$$

$$\text{Molecular weight of flue gas} = \frac{11.1114}{0.385555} = 28.819 \text{ kg/kg·mol exhaust}$$

$$\text{Density of inlet flue gas} = \frac{P}{RT} = \frac{0.88(14.7)(144)(28.819)}{1545(1010)}$$

$$= 0.03440 \text{ lbm/ft}^3$$

Volumetric flow rate in i-d fan

$$= \frac{\text{mass flow rate of gas}}{\text{gas density}}$$

$$= \frac{41,768 \text{ kg/min}}{(0.4536 \text{ kg/lbm})(0.03440 \text{ lbm/ft}^3)} = 2,676,800 \text{ ft}^3\text{/min}$$

(f) Pressure rise across both fans

$$= 25 \text{ in } H_2O = \frac{25}{12}(62.4) = 130 \text{ lb/ft}^2 = \Delta P$$

$$\text{Horsepower of f-d fan} = \frac{(\text{volumetric flow rate}) \Delta P}{33,000 \, \eta_{\text{mech}}}$$

$$= \frac{1,370,750(130)}{33,000(0.86)} = 6279 \text{ hp}$$

$$\text{Horsepower of i-d fan} = \frac{2,676,800(130)}{33,000(0.86)} = 12,260 \text{ hp}$$

(g) Mole fraction of H_2O in exhaust $= \dfrac{\text{mol } H_2O/\text{kg coal}}{\text{mol exhaust/kg coal}}$

$$= \frac{0.052434}{0.385555} = 0.01360$$

$$\text{Exhaust-gas pressure leaving fan} = 0.88(14.7) + \frac{130}{144} = 13.84 \text{ lb/in}^2 \text{ abs}$$

Partial pressure of water vapor in exhaust

$$= (\text{gas pressure})(\text{mole fraction})$$

$$= 13.84(0.1360) = 1.882 \text{ lb/in}^2 \text{ abs}$$

Dew-point temperature of exhaust gas

$$= \text{temperature at } P_{\text{sat}} = 1.882 \text{ lb/in}^2 \text{ abs}$$

$$= 124°F = 51°C \text{ (from steam tables)}$$

4.3 COMBUSTION METHODS AND SYSTEMS

4.3.1 Gas-Fired Systems

Gaseous fuels, including natural gas, are the easiest fuels to burn. The gas needs little or no preparation before combustion. It must be simply proportioned, mixed with air, and ignited. This can be accomplished in several ways.

The atmospheric gas burner is one of the more common gas burners. In these systems, the momentum of the incoming gas is used to draw the primary air into the burner in a process called aspiration. The operation of these systems is normally satisfactory with primary air-gas premix from 30 to 70 percent. Secondary air is drawn in around the burner to complete the combustion process. Two typical atmospheric burners are shown in Fig. 4.6.

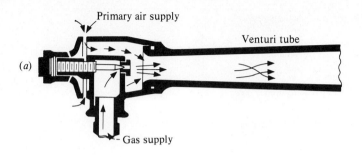

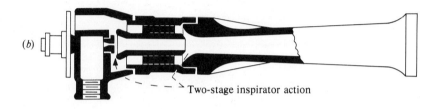

Figure 4.6 Atmospheric gas burners. (*a*) Atmospheric gas burners pull in their primary air for combustion by the action of a stream of low-pressure gas expanding through an orifice. (*b*) Two-stage burner operates on high-pressure gas; passes it through two venturi sections in series. Primary air enters shutter, at left, under induction. (*From Bender, 1964.*)

Refractory gas burners are commonly used in steam generators. The combustion air is drawn in around a burner which has multiple gas jets that discharge the gas into the airstream in such a way that the violent agitation produces good mixing. The gas-air mixture is then discharged into a short mixing tube or tunnel of refractory material. The refractory tube protects the metal burner from the high temperature. Three refractory gas burners are shown in Fig. 4.7.

Another common gas burner is the fan-mix burner. In this system, the fuel gas issues from nozzles mounted at an angle in a rotating spider and the resulting reaction force turns the spider plus an attached fan. This fan produces thorough mixing and the combustion occurs very close to the burner. A typical fan-mix burner is shown in Fig. 4.8.

4.3.2 Oil-Fired Systems

Oil is somewhat more difficult to burn than natural gas because the burner must prepare the fuel for combustion as well as proportion it, mix it with air, and burn it. This is particularly true when the unit is fired with a residual fuel oil. There are several ways to prepare the fuel oil for combustion, including vaporization or gasification of the oil by heating it within the burner, or atomization of

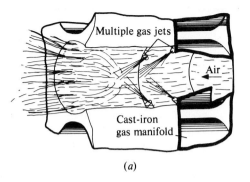

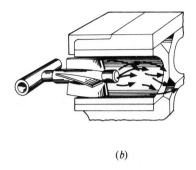

(a)   (b)

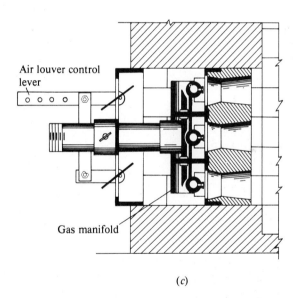

(c)

Figure 4.7 Refractory gas burners. (a) Premixing of fuel gas and air needed for combustion takes place in a mixing chamber outside the furnace proper. (b) Vanes placed in the path of incoming air to this tunnel burner act to impart swirling motion to stream. (c) Gas issues from a number of spuds connecting to vertical and horizontal manifolds. Primary air enters around the spuds. *(From Bender, 1964.)*

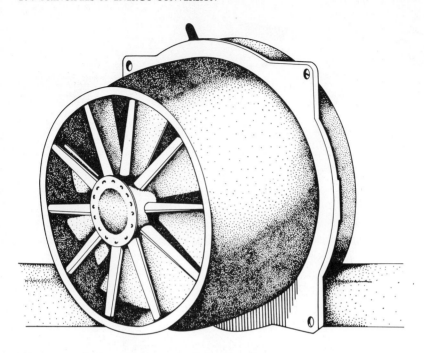

Figure 4.8 The fan-mix gas burner. *(From Bender, 1964.)*

the oil into the airstream. The vaporization technique is particularly well suited for the light fuel oils.

Atomization of the oil droplets can be accomplished with the use of high-pressure air or steam, or the oil film can be torn apart by centrifugal force. Examples of two atomizing burners using air and steam are shown in Fig. 4.9. Steam or air atomization is best suited for variable load and can cover a wide range of capacity without changing the tip or gun assembly. Mechanical atomization is best suited for steady loads and high capacities but has a fairly limited capacity range. A common rotary-cup burner is shown in Fig. 4.10. This burner employs a high-speed (3500 rev/min) horizontal rotary cup to spin the oil off the rim and into the airstream by centrifugal force. This particular mechanical burner has a very wide capacity range (16 to 1).

Any fuel-oil combustion system is commonly composed of an oil-storage tank, pumping and heating equipment, an oil-supply header, the burners, and an oil-return line. The schematic diagrams of two fuel-oil supply systems are shown in Fig. 4.11.

4.3.3 Coal Combustion Systems

Stoker furnaces The stoker furnace is one of the oldest furnace types and is still in use today. Its somewhat limited capacity does not lend this type of furnace to

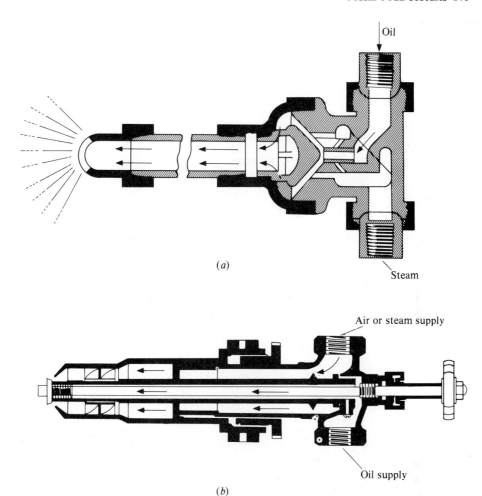

Figure 4.9 Atomizing oil burners. (a) In typical design of internal-mixing steam-atomizing burner, changing tip alters the range of capacities handled. (b) Steam-atomizing burner of external-mixing type, above, brings oil and atomizing medium, steam, together at the burner tip. Register, below, has damper vanes to regulate the air suppl⋅⋅d. (*From Skrotzki and Vopat, 1963.*)

large power applications but it is widely used where relatively limited amounts of process steam are desired. In this unit, the coal is introduced on a grate and it is finally burned in a stationary bed. The stoker furnace burns crushed coal and some of the combustion air, called primary air, is introduced below the burning bed. The primary air initiates the combustion process and also cools the grate. The balance of the combustion air, called secondary or overfire air, is introduced above the burning bed to complete the combustion process.

There are several different types of stoker furnaces. The more common types include the chain-grate and traveling-grate stokers, the vibrating-grate stoker,

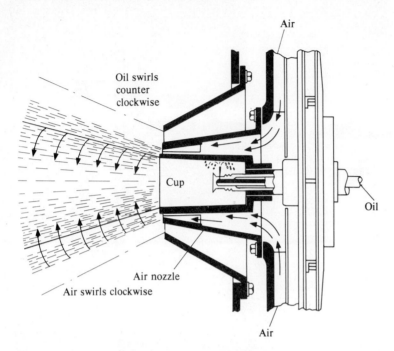

Air

Oil swirls
counter
clockwise

Cup

Oil

Air nozzle

Air swirls clockwise

Air

Figure 4.10 Rotary-cup oil burners. Cup revolving counterclockwise breaks up oil film at rim by centrifugal force and discharges into a clockwise airstream. *(From Bender, 1964.)*

the underfeed stoker, and the spreader stoker. The chain-grate or traveling-grate stoker is the simplest and least expensive of the stoker furnaces and a typical unit is shown in Fig. 4.12. The grate is composed of a continuous traveling link chain with narrow gaps between the links to permit the flow of primary air. This particular unit is not well suited to the combustion of caking coals because there is insufficient agitation of the burning bed to break up the fused coal lumps.

The vibrating-grate stoker is very similar to the chain-grate stoker, except in the operation and orientation of the grate. In this unit, the coal grate does not move continuously in one direction; instead, the grate is inclined toward the ashpit and the grate is vibrated. The raw coal flows from the storage-inlet hopper to the grate and then the particles move down the grate under the influence of the vibration and gravity, burning as they progress down the grate. The ash falls off the end of the grate into the ashpit. Since there is considerable movement of the fuel bed, this stoker can burn most caking coals.

The underfeed stoker furnace is one of the more common types of stoker furnaces. In this type of stoker, the coal is pushed into the bottom of the fuel bed in one or more feed troughs called retort trenches. The coal is advanced in the retort trench by rams, pusher rods, or screws. A single-retort, horizontal-feed underfeed stoker furnace is shown in Fig. 4.13. In the horizontal underfeed unit

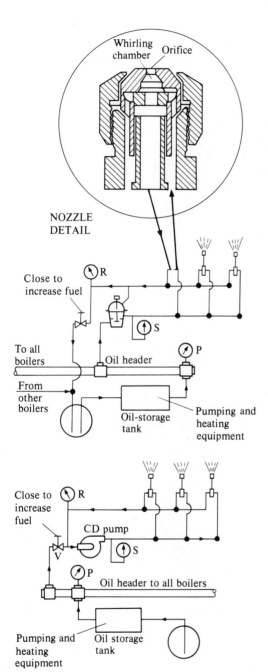

Whirling chamber   Orifice

NOZZLE DETAIL

R

Close to increase fuel

S

To all boilers

From other boilers

Oil header

P

Oil-storage tank

Pumping and heating equipment

Close to increase fuel

R

CD pump

S

V

P

Oil header to all boilers

Pumping and heating equipment

Oil storage tank

Figure 4.11 Fuel-oil supply and distribution systems. *(From Skrotzki and Vopat, 1963.)*

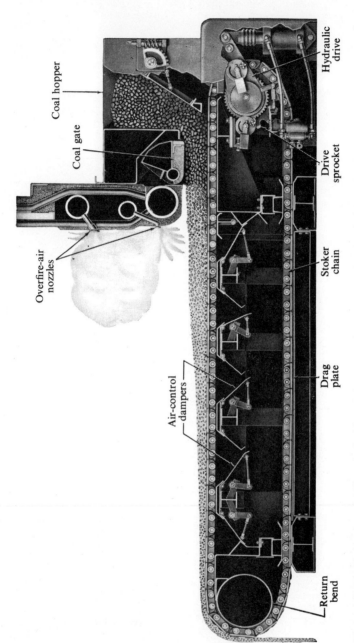

Figure 4.12 The Babcock and Wilcox jet-ignition chain-grate stoker. (*From "Steam/Its Generation and Use," 1972.*)

Coal hopper

Coal gate

Overfire-air nozzles

Air-control dampers

Hydraulic drive

Drive sprocket

Stoker chain

Drag plate

Return bend

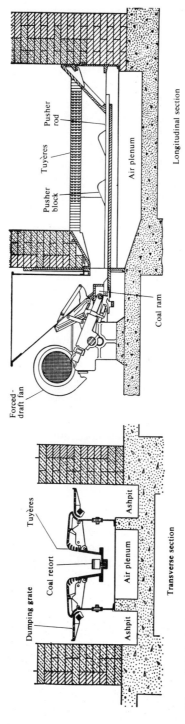

Figure 4.13 A single-retort horizontal-feed underfeed stoker furnace. (*From "Steam/Its Generation and Use," 1972.*)

179

in Fig. 4.13, the ash flows off the side of the grate. In the gravity-feed furnace, the grate is inclined at an angle of 20 to 25 degrees and the ash falls off the back of the grate. Multiple retort furnaces have grate-release rates as high as 720,000 Btu/h·ft². Because of the ramming action in the bottom of the burning bed of coal, the agitation probably makes this particular unit the best system for burning caking coals.

The newest type of stoker furnace is the spreader stoker and this unit has found widespread application because of its relative simplicity. This system has high capacity, it is low in initial cost, and it is relatively insensitive to the coal characteristics. In this stoker furnace, the raw coal is literally thrown over the burning bed by rotating paddles or vanes, or it is blown over the bed by a pneumatic spreader system powered by high-pressure air or steam. The coal then falls on a moving or traveling grate, similar to that of a chain-grate stoker, except that the grate moves back toward the coal injector and the ashpit lies directly below the coal hopper, as shown in Fig. 4.14. A typical injection system for the spreader stoker is also shown in Fig. 4.14. This furnace has a faster response than the other stoker furnaces but it also requires more combustion air and it has greater problems with flyash emissions than the other stoker units.

Pulverized-coal furnaces The pulverized-coal furnace burns finely powdered coal and air in a gaseous torch. This combustion system can produce much higher capacities than the stoker furnaces; it gives fast response since there is little unburned fuel in the combustion chamber; it reduces the amount of excess air required for combustion and this reduces the NO_x emissions; it can burn all ranks of coal from anthracitic coals to lignitic coals; and this unit permits combination firing. The term, combination firing, refers to the capability of burning coal, oil, or natural gas. Normally, only one type of fuel is burned at a time, although two different fuels can be simultaneously burned for short periods of time. The pulverized-coal furnace finds widespread application in coal-fired power plants.

Despite the numerous advantages listed above, the pulverized-coal system has a number of problems associated with it. Probably the major problems arise from the use of pulverized coal. The pulverized-coal system requires a complicated coal pulverizer with its accompanying power demand and associated high maintenance costs. Flyash erosion and pollution also complicate the operation of the unit and increase the required maintenance of the exhaust system. The pulverized-coal systems have higher initial cost and require larger volumes for the combustion process.

There are many different kinds of coal pulverizers but they all employ one or a combination of three basic pulverizing actions—crushing, impact, and attrition. Attrition is a grinding action caused by two coal particles rubbing together. The common coal pulverizer grinds the coal to a consistency finer than most face powders. The standard pulverizer specifications require that at least 98 percent of the output particles pass through a 50-mesh screen (50 wires per inch)

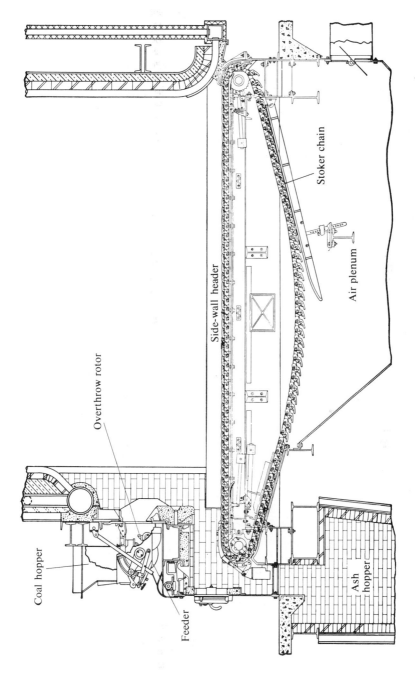

Figure 4.14 The spreader stoker furnace and the coal-injection system. (*From "Steam/Its Generation and Use," 1972.*)

Coal hopper

Overthrow rotor

Feeder

Side-wall header

Stoker chain

Air plenum

Ash hopper

181

yielding a maximum particle diameter of less than 0.3 mm. Not only does the pulverizer grind the coal to a fine powder, but the coal is usually dried in the pulverizer. The primary air is usually heated to 260 to 315°C (500 to 600°F) and is pumped by a separate fan through the pulverizer where it dries the coal and transports it to the burner.

The simplest type of coal pulverizer is the ball mill or the ball-and-tube mill. This system is composed of a large drum rotating on a horizontal axis and is partially filled with steel balls or tubes. The most widely used coal pulverizers are probably the ball-race mills, the roll-and-race mills, and the bowl mills. These mills have a series of large steel balls or rollers running against a rotating race. There are a number of other kinds of pulverized-coal mills, including the hammer-mill pulverizers and the steam-jet pulverizers, among others. Some of the common pulverizers are shown in Fig. 4.15.

There are many types of pulverized-coal burners, but the circular and cell-type burners are commonly employed with the capability of combination firing. A burner of this type is shown in Fig. 4.16. These burners usually fire horizontally and are commonly either fired from the furnace corners so that the flames are directed toward an imaginary circle (tangential firing) or the burners are mounted in the wall opposite to each other and the flames are directed toward each other (opposed firing).

In many pulverized-coal furnaces, the burners can be tilted up or down to permit some control of the energy deposition in the boiler. If the burners are tilted down, more heat is transferred to the evaporator section of the boiler adjacent to the burners and less heat is transferred to the pendant superheaters and reheaters in the top of the boiler. If the burners are tilted up, less heat is transferred to the evaporator section and more heat is transferred to the super-heat and reheat sections. See the tangentially fired system in Fig. 4.17.

The ash from a pulverized-coal furnace can be handled in one of two ways. In the "wet-bottom" or "slag-tap" furnace, the ash is handled in molten or liquid form. The combustion chamber of this unit is usually insulated and some of the burners are directed toward the furnace floor. A coal with a low ash-softening temperature is desired for this system and the slag pool must be maintained at a high enough temperature to assure that the hot slag will flow continuously over a water-cooled weir or ring into a quenching tank filled with water. Approximately 50 percent of the flyash is usually trapped in the pool of molten slag.

The other general ash system handles the ash in solid form. The coal used in this furnace should have a high ash-softening temperature (at least 1300°C, or 2400°F). The walls of the combustion chamber are water cooled and the ash particles are cooled in suspension. Some of the ash settles to the furnace floor but about 80 percent of the ash leaves with the exhaust gas, entrained as flyash. Because of the high flyash concentration in the exhaust gas, this system will probably require one or more mechanical dust collectors plus an electrostatic precipitator or baghouse filter or their equivalent.

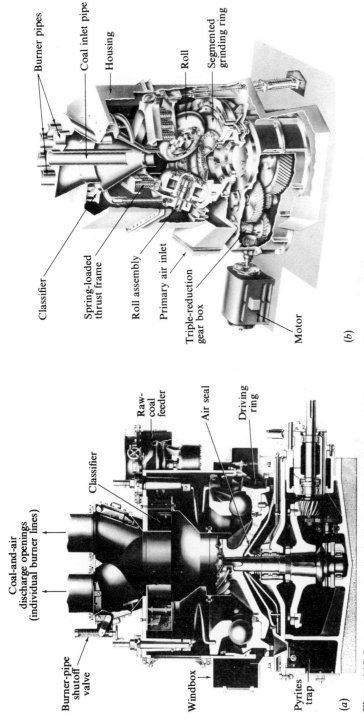

Figure 4.15 Some common coal pulverizers. (*a*) The Babcock and Wilcox type EL single-row ball-and-race pulverizer. (*b*) The Babcock and Wilcox type MPS mill. (*Courtesy of Babcock and Wilcox, from "Steam/Its Generation and Use," 1972.*)

Burner pipes

Coal inlet pipe

Housing

Roll

Segmented grinding ring

Classifier

Spring-loaded thrust frame

Roll assembly

Primary air inlet

Triple-reduction gear box

Motor

(*b*)

Raw-coal feeder

Air seal

Driving ring

Classifier

Coal-and-air discharge openings (individual burner lines)

Burner-pipe shutoff valve

Windbox

Pyrites trap

(*a*)

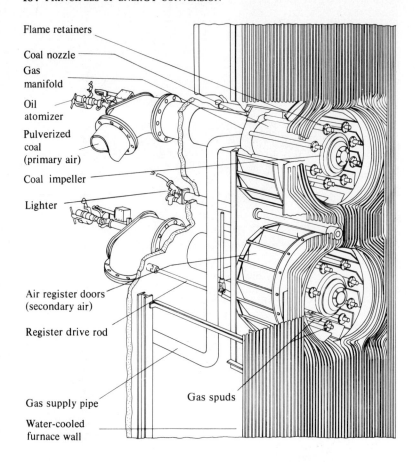

Flame retainers

Coal nozzle

Gas manifold

Oil atomizer

Pulverized coal (primary air)

Coal impeller

Lighter

Air register doors (secondary air)

Register drive rod

Gas supply pipe

Gas spuds

Water-cooled furnace wall

Figure 4.16 Pulverized-coal burners. (*From Bender, 1964; "Steam/Its Generation and Use," 1972.*)

Cyclone furnaces The cyclone furnace is a combustion system that employs a number of independent combustion chambers (as many as sixteen in a large power plant) and all of these chambers feed hot exhaust gas into a large, water-cooled boiler. A typical cyclone, like that shown in Fig. 4.18, consists of a small cylindrical chamber (about 4 ft in diameter) that feeds a large (10 to 12 ft in diameter) insulated cylinder, and this large cylinder discharges the hot gases into the heat-transfer section of the boiler.

The cyclone furnace burns crushed coal ($\frac{1}{4}$-in or smaller), not pulverized coal. The coal and primary air are fed tangentially into the small combustion chamber and the secondary air is introduced tangentially into the large insulated cylindrical chamber. Molten ash collects on the walls of the large chamber. A small quantity of air, called tertiary air, is introduced into the vortex of the cyclone to break up the vacuum.

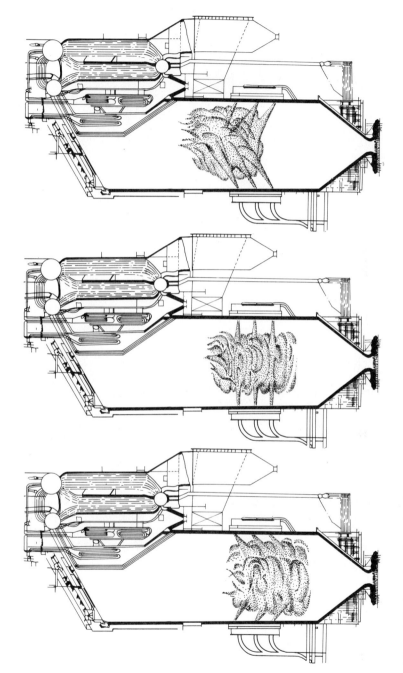

Figure 4.17 Operation of a tangentially fired, pulverized-coal furnace with tiltable burners. (*From Skrotzki and Vopat, 1963.*)

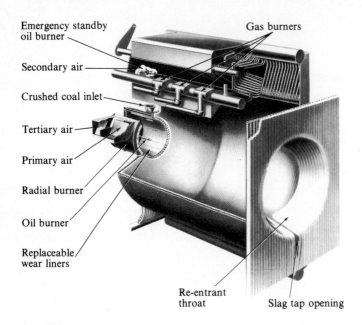

Emergency standby
oil burner

Gas burners

Secondary air

Crushed coal inlet

Tertiary air

Primary air

Radial burner

Oil burner

Replaceable
wear liners

Re-entrant
throat

Slag tap opening

Figure 4.18 The cyclone coal combustion system. *(Courtesy of Babcock and Wilcox, from " Steam/Its Generation and Use," 1972.)*

The strong vortex flow throws any large unburned chunks of coal into the molten-ash collection on the side of the cylindrical wall where the secondary air completes the combustion process. The main combustion chamber of a cyclone unit operates at a temperature of 1760°C (3200°F) and the molten ash runs out of the cyclone burner through a slag-tap opening onto the furnace floor. Once on the furnace floor, the molten ash falls through an opening, called the "monkey," into a water-filled slag tank. This furnace traps approximately 75 percent of the ash in the molten slag.

The cyclone furnace commonly operates as a positive-pressure furnace. The draft system for these units is composed of one or two large forced-draft fans plus several intermediate-sized gas-recirculation fans. The positive-pressure firebox does present some problems with respect to leakage and maintenance. Consequently, some of these units have been retrofitted with induced-draft fans so that the furnace can operate as a balanced-draft system.

The cyclone furnace became a very popular combustion system in the 1950s and 1960s, but it is currently being displaced in popularity by the pulverized-coal combustion system. There are two significant environmental problems associated with the operation of cyclone furnaces. First, these systems have difficulty burning low-sulfur coals because of slagging problems. Second, because of the high combustion temperature required in the combustion chamber, a significant amount of NO_x is formed and these compounds are important atmospheric pollutants.

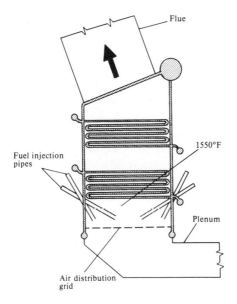

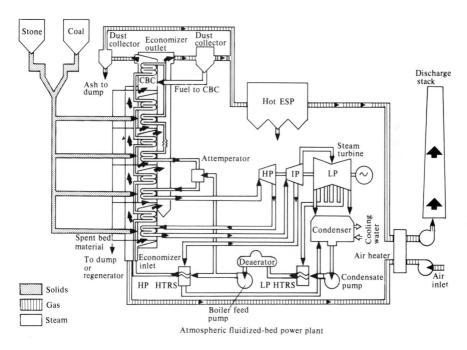

Figure 4.19 The fluidized-bed coal combustion chamber. *(From Mesko and Gamble, 1974.)*

Fluidized-bed combustion systems The fluidized-bed furnace is a radically new type of combustion system. In this unit, crushed coal, ash, and either crushed dolomite or limestone are mixed in a bed and this bed is then levitated by the combustion air entering the bottom of the furnace. The boiler evaporator tubes are directly immersed in the fluidized bed and the direct contact between the burning coal particles and the tubes produces very high heat-transfer rates, reducing the size of the unit. This also produces very low combustion temperatures.

The major advantage of the fluidized-bed system lies in the area of pollution control and abatement. The relatively low combustion temperatures, 820 to 950°C (1500 to 1600°F), produced by the placement of the evaporator tubes directly in the fluidized bed, greatly inhibits the formation of nitrogen oxides (nitric oxide and nitrogen dioxide). The addition of either dolomite (a calcium-magnesium carbonate) or limestone (calcium carbonate) produces a chemical reaction between the calcium or magnesium and the sulfur dioxide to produce calcium or magnesium sulfites and sulfates. These calcium- and magnesium-sulfur salts are solids and are trapped in the combustion chamber. Thus, the fluidized-bed combustion system, similar to that shown in Fig. 4.19, traps the sulfur in the furnace thereby permitting the combustion of high-sulfur coal.

The major disadvantage of this particular system is due to the fact that the combustion air must be supplied at a high enough pressure that it can support the fluidized bed. This means that the pressure rise across the fan or compressor must be significantly higher than that for a conventional coal-fired furnace with an accompanying increase in the required fan power. Another disadvantage (or maybe advantage?) associated with the fluidized-bed system is the fact that it is a relatively new concept and, as such, has not been as fully developed as the other types of combustion systems.

4.4 STEAM GENERATORS

4.4.1 General

The steam generator, or boiler, is a combination of systems and equipment for the purpose of converting chemical energy from fossil fuels into thermal energy and transferring the resulting thermal energy to a working fluid, usually water, for use in high-temperature processes or for partial conversion to mechanical energy in a turbine. In most modern large power plants, one boiler is used to supply steam to one steam-turbine generator unit. The boiler complex includes the air-handling equipment and ductwork, the fuel-handling system, the water-supply system, the steam drums and piping, the exhaust-gas system, and the pollution-control system.

The heat-transfer sections of a large boiler include the evaporator, the super-heater, the reheater, the air preheater, and the economizer sections. The evapora-tor, superheater, and reheater surfaces are called primary heat-transfer surfaces while the air preheater and economizer surfaces are called secondary heat-transfer surfaces. The heat-transfer sections and an energy flow diagram for a typical large pulverized-coal boiler are shown in Fig. 4.20.

4.4.2 Boiler Types and Classifications

Steam boilers can be classified in many ways but there are actually two major groups of steam generators, depending on the orientation of the water-steam and hot-gas flow paths. These two general classifications are the fire-tube boilers and the water-tube boilers. The common fire-tube boiler is essentially composed of a water-filled pressure vessel containing a number of tubes which are the passageways for the hot exhaust gas, and the heat is transferred from the hot gas to the water in the vessel. This system is the simplest and least expensive steam generator that can be constructed.

In this fire-tube system, the high-pressure water is placed on the external surface of the tubes. Since any tube has essentially half the strength for external pressure that it has for internal pressure, these systems are limited to relatively low steam pressures. The maximum drum diameter is around 8 ft and the maximum steam pressure is limited to about 17 atm (250 lb/in^2 abs), although most of the systems normally operate at a pressure of about 10 atm (150 lb/in^2 abs).

Another problem associated with the fire-tube boiler is the consequence of a major tube failure. In the event of a sudden tube failure, the high-pressure water flows into the hot combustion chamber generating large quantities of steam in the furnace. This is likely to produce a steam explosion in the furnace.

The entire heating surface in the drum of a fire-tube boiler is devoted to the production of saturated steam, although a separate superheater section can be added. There are several different kinds of fire-tube boilers but the horizontal-return-tubular boiler (hrt) is one of the more popular types. The fire-tube steam generator is commonly employed in small industrial plants and these systems can be purchased in the form of complete operational packages. A typical two-pass packaged steam generator is shown in Fig. 4.21.

The water-tube boilers are best suited for high-pressure, high-capacity steam generators. The high-pressure water and the steam flows from tube headers or drums through tubes in the furnace walls or in tube bundles mounted in the exhaust-gas duct. These tubes range from 2 to 4 in in diameter and can withstand boiler pressures up to 350 atm. The water-tube steam generators can be classified according to the shape of the tubes, the position of the drums, the number of drums, the method of circulation, the type of service (stationary or marine), or the capacity and thermal conditions of the outlet steam. Because of the large exhaust-gas ductwork in a water-tube boiler, there is much less danger of a steam explosion in the event of a catastrophic tube failure. Normally, a tube failure necessitates a boiler shutdown only because of excessive water loss.

All water-tube generators may be classified as either natural-circulation or forced-circulation boilers. Most of the older steam generators are natural-circulation systems. In these units, the fluid flow is induced by the difference in the specific weight of the water in the vertical supply tubes and the average specific weight of the water-steam mixture in the evaporator tubes where the evapora-

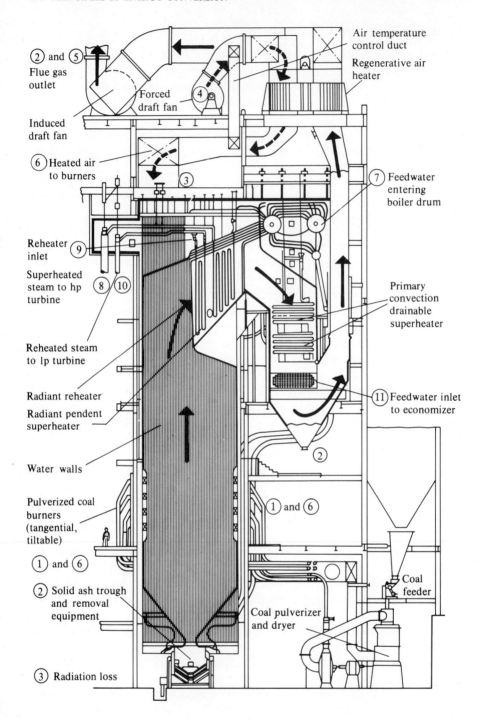

② and ⑤ Flue gas outlet

Induced draft fan

Air temperature control duct

Regenerative air heater

Forced draft fan

④

⑥ Heated air to burners

③

⑦ Feedwater entering boiler drum

Reheater inlet ⑨

Superheated steam to hp turbine ⑧ ⑩

Primary convection drainable superheater

Reheated steam to lp turbine

Radiant reheater

⑪ Feedwater inlet to economizer

Radiant pendent superheater

②

Water walls

① and ⑥

Pulverized coal burners (tangential, tiltable)

① and ⑥

Coal feeder

② Solid ash trough and removal equipment

Coal pulverizer and dryer

③ Radiation loss

Figure 4.20

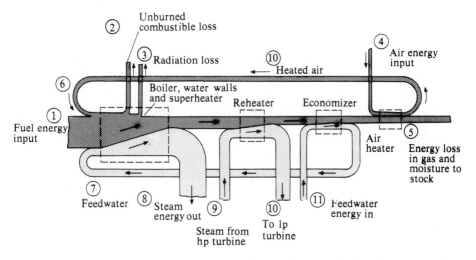

Figure 4.20 The layout and energy flow diagram for a typical pulverized-coal steam generator. *(From Bender, 1964.)*

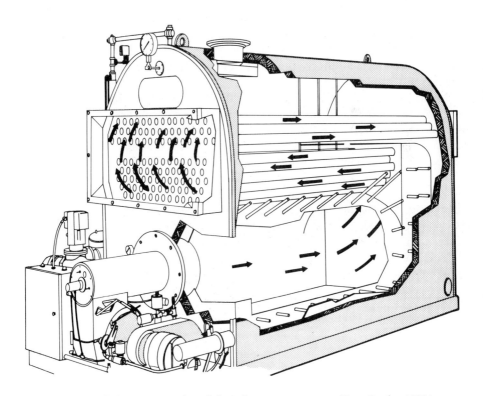

Figure 4.21 A typical two-pass, packaged, fire-tube, steam generator. *(From Bender, 1964.)*

tion occurs. The driving force for fluid flow through the evaporator section is equal to the product of the difference in the average specific weights and the height of the evaporator tubes. This driving force is balanced by the frictional pressure drop in the supply and evaporator tubes. Most of the pressure drop occurs in the evaporator tubes as the result of the two-phase flow of water and steam.

The principle of circulation in any natural-circulation boiler is shown in Fig. 4.22. The saturated water flows from the steam drum high in the boiler, through the supply or "downcomer" tubes, located in the cooler part of the boiler, to the bottom or "mud" drum. From the mud drum, the water flows back to the steam drum through the evaporator or "riser" tubes. In the steam drum, the steam and water are separated and the steam is washed and dried before it is sent to the superheater. The internals of a typical steam drum are also presented in Fig. 4.22. The lower drum is commonly called the mud drum because any water impurities naturally accumulate in this drum. Periodically, some of the water in the mud drum is vented ("blowdown") to the atmosphere and fresh water ("makeup") is added to lower the overall concentration of impurities in the water system.

As the steam pressure is increased, the density and specific gravity difference between the saturated steam and water decreases, requiring a greater height for adequate flow (up to 15 stories high). At the critical point of water ($705.4°F$ and 3206.2 lb/in^2 abs) the difference is zero and there is no natural-circulation driving force. Actually, at pressures greater than about 160 atm (2400 lb/in^2 abs), the density difference is effectively too small for natural-circulation units and these systems must employ forced circulation.

In a forced-circulation boiler, the fluid is pumped through the evaporator section of the boiler. This permits operation of the cycle at very high pressures, even above the critical pressure. High-pressure operation theoretically improves the efficiency of the basic steam cycle. The forced-circulation system eliminates the need for boiler height; it is lighter in weight; it uses smaller tubes and drums or no drums at all; and the lower total water content in the boiler reduces the danger of a steam explosion. In addition to the problems associated with the main circulation pump and potentially higher operating pressures, some of these systems also require extremely pure water—orders of magnitude purer than that required for natural-circulation systems.

There are many different kinds of forced-circulation boilers, depending upon the circulation paths in the evaporator. The most widely used forced-circulation boiler system in the United States is the universal-pressure (UP) or Benson boiler. A schematic diagram of the Benson boiler and some of the other forced-circulation systems are shown in Fig. 4.23. In the Benson boiler, the water is pumped to about 340 atm (5000 lb/in^2 abs) in the main feed pump (FP). The compressed water is then piped to the economizer section (E), through the evaporator tubes (T), through a transition section (TS), and finally through a convection superheater (CS), where it is exhausted to the turbine at a pressure of

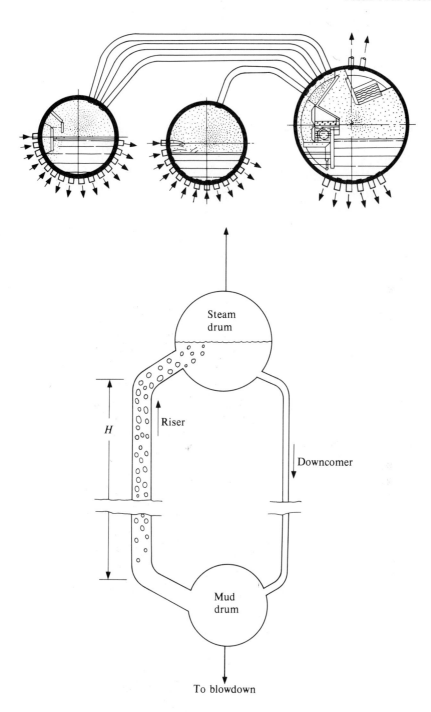

Figure 4.22 A typical steam drum and the schematic of a natural-circulation boiler. *(From Bender, 1964.)*

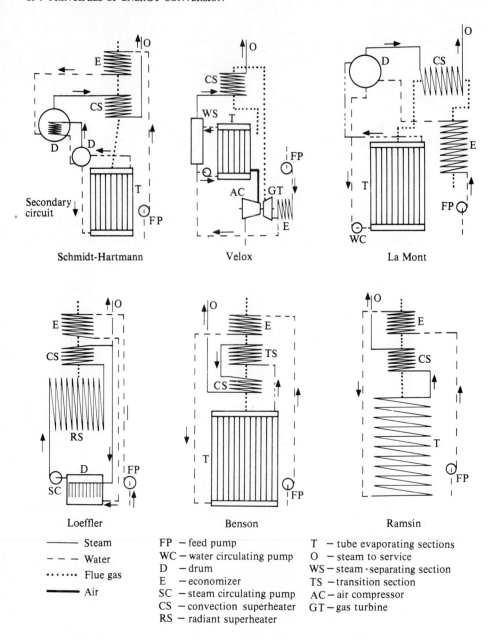

Schmidt-Hartmann

Velox

La Mont

Loeffler

Benson

Ramsin

—— Steam	FP — feed pump	T — tube evaporating sections
– – – Water	WC — water circulating pump	O — steam to service
······ Flue gas	D — drum	WS — steam-separating section
—— Air	E — economizer	TS — transition section
	SC — steam circulating pump	AC — air compressor
	CS — convection superheater	GT — gas turbine
	RS — radiant superheater	

Figure 4.23 Schematic diagrams of forced-circulation boilers. *(From Skrotzki and Vopat, 1963.)*

about 240 atm (3500 lb/in² abs). This supercritical boiler system requires extremely pure water with an impurity concentration of only a few parts per billion. Any impurities at all in the water will normally be deposited in the boiler tubing. The Ramsin boiler is essentially identical to the Benson boiler except that the evaporating section is composed of inclined "T"-bundle coils arranged in a spiral array.

The LaMont boiler, shown schematically in Fig. 4.23, closely resembles the natural-circulation boiler because it has a steam drum. Since the water-steam mixture is separated in the drum, the system must operate at a pressure below the critical pressure. The Velox boiler is an interesting unit because it combines the steam or Rankine power cycle with the gas-turbine or Brayton power cycle. This system is very different from the so-called "combined-cycle" systems that are becoming increasingly popular. In these latter systems, the gas-turbine discharge is used as the source of hot boiler combustion air. In the Velox system, the combustion air is compressed to between 2 to 3 atm in an air compressor which is driven by a gas turbine. The gas turbine then receives the hot gas after it has gone through the steam boiler where it produces steam at 14 to 80 atm (200 to 1200 lb/in² abs). This unit has a relatively high overall thermal efficiency but the steam-generator shell must be a pressure vessel which significantly increases the cost.

As mentioned earlier, some of the forced-circulation systems require ultra-pure water, but the Loeffler and the Schmidt-Hartmann boilers can operate while using water with a relatively high impurity content. Both of these units must operate at pressures below the critical pressure. In the Loeffler boiler, the water is evaporated by part of the superheated steam and the resulting saturated steam is pumped through the superheater section. In the Schmidt-Hartmann boiler, the water is boiled by the condensation of steam in a high-pressure, closed, natural-circulation secondary system.

Despite some of the advantages realized by the forced-circulation boiler systems, the only unit that has seen widespread application in the United States is the universal-pressure or Benson boiler. Forced-circulation systems are fairly common in Europe. Recently, difficulties encountered in the operation of the universal-pressure boilers in the United States have reversed the trend to these systems. Many utilities are presently returning to the use of 160-atm (2400-lb/in² abs) natural-circulation boilers.

4.4.3 Primary Boiler Heating Surfaces

The primary heat-transfer surfaces in the boiler include the evaporator section, the superheater section, and the reheat section if the system employs a reheat turbine. The evaporative surface is usually located in the hottest part of the boiler near the combustion zone because the boiling water in the tubes protects the tube material from excessive temperatures. The evaporator sections include the water walls, the water floor, and the water screens which are used to direct the flow of hot exhaust gas. Three general types of water-wall surfaces are shown in Fig. 4.24. These include bare tubes, tubes with extended surfaces or metal-block facing, and studded tubes packed with refractory insulation or tile facing. This last arrangement is used where little heat transfer is desired such as in the large cylinder of a cyclone burner.

Superheater sections are the heat-transfer surfaces in which heat is transferred to the saturated steam to increase its temperature. Superheaters are particularly important in the production of steam for steam turbines to reduce the moisture content of the steam as it expands through the turbine. Superheaters are commonly classified as either radiant superheaters, convection superheaters, or combined superheaters, depending on how they receive the thermal energy. It

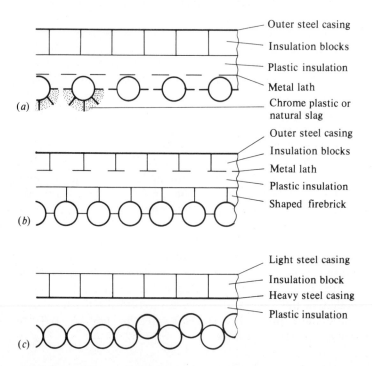

Figure 4.24 Evaporative water-wall surfaces used in steam boilers. (a) Steel lugs or longitudinal fins are often welded to nontangent wall tubes. In some designs the lugs protrude from the tube into the furnace and are covered with a chrome-base refractory or slag. To assure furnace tightness, adjacent fins, right, are often welded. (b) With moderate cooling a key requirement, this design has tubes spaced apart and wall surface composed of part firebrick. Brick is backed with several layers of insulation and a strong steel casing. Reinforcing metal lath is often used in wall construction. (c) Tangent tubes are favored in furnaces of many small and large boilers. In some designs tubes are staggered. This arrangement offers high heat-absorbing surface that is backed by solid block or plastic insulation and a strong steel external casing. (*From Bender, 1964.*)

is usually desirable that the final steam temperature stays essentially constant over a wide range of boiler loads. When the exit steam temperature becomes excessive, some units employ an attemperator or desuperheater in which compressed feedwater is sprayed into the superheated steam to lower its temperature. Many superheater sections are pendant-type units in which bundles of tubes are suspended in the exhaust-gas duct.

The reheat section of a large boiler is that portion of the boiler in which all of the exhaust steam from the high-pressure turbine is returned for additional superheat before it is sent to the intermediate-pressure turbine. The reheater is very similar to the superheater in appearance and location in the boiler.

4.4.4 Secondary Boiler Heating Surfaces

The secondary heat-transfer surfaces recover heat from the flue gas after it has transferred heat to the primary heat-transfer surfaces. In order to achieve a high boiler efficiency, it is desirable to reduce the temperature of the exhaust gas as much as possible. There are two kinds of secondary heat-transfer surface—the economizer and the air preheater. The economizer transfers heat from the flue gas to the incoming boiler feedwater while the air preheater (sometimes simply called the air heater) transfers the flue-gas energy to the incoming combustion air.

While a low exhaust-gas outlet temperature improves the boiler efficiency, the temperature should not be lowered below about 80 Celsius degrees (150 Fahrenheit degrees) above the dew-point temperature of the exhaust gas. If the temperature is lowered much more than this, condensation will probably occur on the surfaces of the cold exhaust-gas duct. Any condensation in the flue-gas system is extremely undesirable since the liquid will be strongly acidic and corrosive because of the sulfur dioxide and sulfur trioxide in the exhaust. Unfortunately, the presence of small quantities of sulfur dioxide in the flue gas drastically raises the dew-point temperature of the exhaust.

The economizer is normally a cross-flow heat exchanger in which the heat is transferred from the flue gas to the incoming feedwater. It has been estimated that an increase of 6 to 7 Celsius degrees (10 to 11 Fahrenheit degrees) in feedwater temperature obtained from heat recovery in the economizer increases the boiler efficiency by 1 percent. Not only must the exhaust gas leaving the economizer be at least 80 Celsius degrees (150 Fahrenheit degrees) above the dew-point temperature of the flue gas but the water leaving the economizer should be at least 30 Celsius degrees (about 50 Fahrenheit degrees) below the saturation temperature of the boiler feed. This prevents any boiling and two-phase flow with the accompanying high pressure drop in the economizer as the result of pressure fluctuations in the boiler.

It is common practice in the design of economizer systems to put the high-pressure water inside the economizer tubes. Since the gas-side, heat-transfer coefficient is low and is the controlling heat-transfer coefficient in the system, it

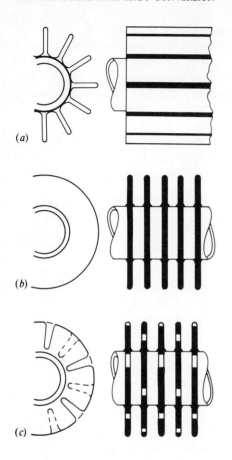

(a)

(b)

(c)

Figure 4.25 Economizer heat-transfer sections. (a) Welding continuous fins onto tubes lengthwise increases heat transfer. This design also improves anti-sag feature. Typical fins used today may be $\frac{1}{4}$ in thick, 2 in high. (b) Another extended heating surface consists of gilled ring tubing made of cast iron or steel. Fins may be square-shaped, totally independent, or a continuous spiral. (c) One design favors separate lentil-shaped steel lugs that are welded around the tube. They keep soot from accumulating and are easy to keep clean. (*From Bender, 1964.*)

is also common practice to use extended surfaces (fins) on the economizer tubes. Some typical economizer sections are shown in Fig. 4.25.

One problem associated with any coal-fired boiler system, particularly a pulverized-coal system, is the ash content of the flue gas and the resulting build-up of ash or slag deposits on the heat-transfer surfaces of the boiler, both the primary and the secondary surfaces. The buildup of these deposits is particularly severe for those heating surfaces employing extended surfaces. It is common practice in coal-fired boilers to incorporate devices, called soot blowers, to remove the ash deposits from the tubes. These systems can be composed of stationary nozzles or retractable lances with rotating nozzles. Most of the soot blowers are powered with high-pressure air or steam, and recently considerable research has been done using compressed water. The water is very effective in removing the ash deposits from the surface but it presents a thermal-shock problem. These soot-blower systems are shown schematically in Fig. 4.26.

The air preheater transfers heat from the exhaust gas to the cold combustion air. There are two broad classes of air preheaters—the regenerative heater and the recuperative heater. The recuperative heater is a plate-type or tubular heat

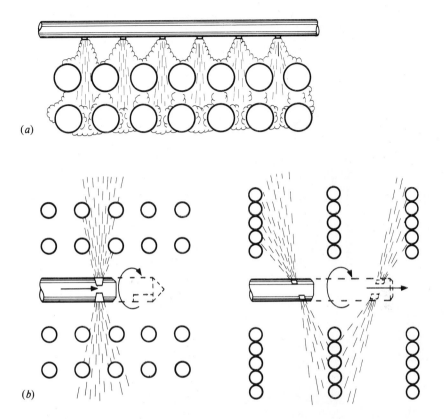

Figure 4.26 Soot-blower systems. (*a*) Stationary soot blowers employing straight nozzles clean banks of in-line tubes. With staggered tubes, the nozzles are offset for improved penetration. (*b*) Retractable blower lance, driven by two motors, traverses boiler at one speed, retracts at twice that speed. Result: about a 25 percent saving in cleaning time and in blowing medium. Speed of lance rotation is held throughout the cycle. (*From Bender, 1964.*)

exchanger operating as a counterflow or crossflow unit. A typical recuperative air preheater is shown in Fig. 4.27. A shot-cleaning system, rather than a soot-blower system, is commonly used to clean the flue-gas side of these exchangers. The shot-cleaning system operates by dropping porous, $\frac{1}{2}$-in steel shot from the top of the exchanger. The shot knocks off the ash deposits as it trickles through the tubing. The shot is then separated from the flyash and returned to the top of the exchanger by a pneumatic transfer system. A typical shot-cleaning system is shown in Fig. 4.28.

The regenerative air preheater, or Ljungstrum heater, employs a large rotor assembly with approximately half the element mounted in the exhaust-gas duct and the other half in the supply-air duct. The rotating element, which usually turns at 2 to 4 rev/min, contains many corrugated lamina that are alternately heated by the flue gas and cooled by the combustion air. The corrugated elements are commonly divided into upper and lower segments so that the coldest

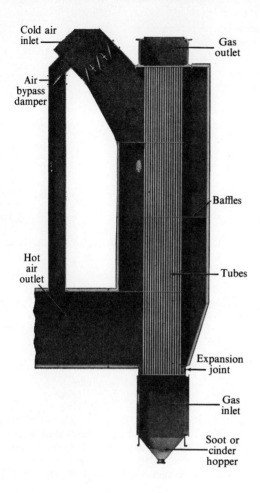

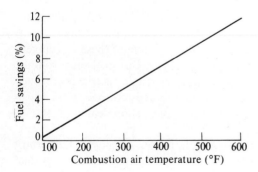

Figure 4.27 A cross-flow, recuperative air heater. (*From Bender, 1964.*)

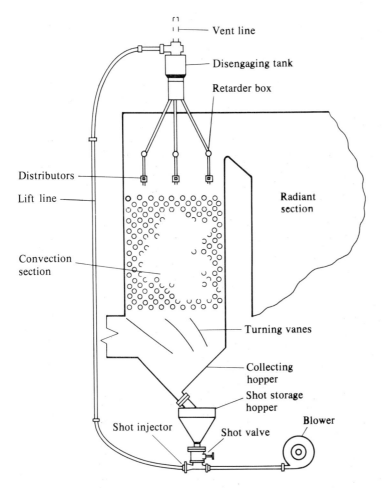

Figure 4.28 A shot-cleaning system. This technique calls for cascading metal balls on to heating surfaces. The cycle is automatic, utilizing pneumatic lift. *(From Bender, 1964.)*

portion can be replaced periodically. The coldest part of the air preheater is a likely place where condensation occurs, with its corresponding corrosion. Radial seals and diaphrams keep leakage to a minimum but some leakage still occurs (in both directions) because of gas entrainment in the rotating lamina. A typical regenerative heater is shown in Fig. 4.29.

The air preheaters are useful in other ways than just improving the overall efficiency of the unit. They reduce the time required for fuel ignition, thereby improving the fuel combustion. The maximum combustion-air temperature depends upon the type of furnace used. The inlet air for stoker furnaces is normally limited to a maximum temperature of 120 to 175°C (250 to 350°F) because this air is used to cool the stoker grates. The inlet air for a pulverized-coal furnace is limited to 290 to 315°C (550 to 600°F).

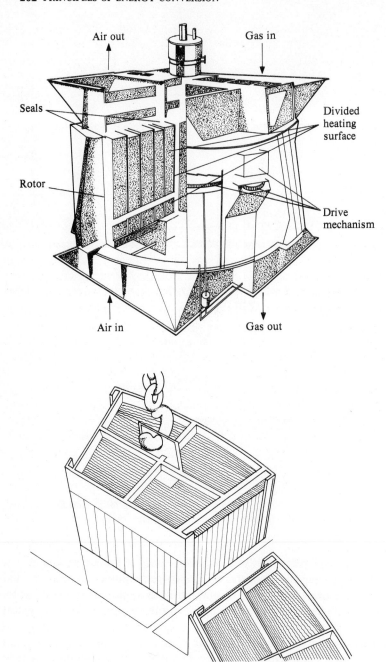

Figure 4.29 The regenerative air preheater. *(From Bender, 1964.)*

If one has a choice of secondary heat-transfer surfaces, either economizer surface or preheater surface, the optimum choice is dictated by the type of system used. For large pulverized-coal or cyclone systems, the preheater surface is usually cheaper than the economizer surface for a given heat recovery. The effectiveness of an economizer may also be limited by the use of many feedwater heaters. For small stoker units where air heating is limited and only one deaerating feedwater heater is used, the economizer surface is normally less expensive for a given heat recovery than the preheater surface.

4.4.5 Boiler Ratings and Performance

In the early part of the twentieth century, an attempt was made to relate the output of the boiler to the output of the steam prime mover, which at that time was the reciprocating steam engine. This produced a unit of boiler output, called a boiler horsepower, such that one boiler horsepower produced an equivalent of one horsepower of mechanical energy. At that time, one boiler horsepower was defined as the generation of 34.5 lbm of steam per hour from water at 100°C (212°F). This resulted in a value of 33,500 Btu/h per boiler horsepower. At the same time that the boiler horsepower was developed, another term, called the boiler rating, evolved. A boiler with a rating of 100 percent was supposed to develop one boiler horsepower for each 10 square feet of evaporative heating surface, excluding the superheating and economizer surfaces.

Both the boiler horsepower and the boiler rating terms are antiquated quantities as most modern steam generators have ratings in excess of 400 percent. Sometimes, one may encounter small or older boilers which are rated in terms of boiler horsepower. Most of the modern steam generators are rated in terms of steam capacity (usually lbm/h) along with the steam outlet pressure and temperature. There is no attempt, at present, to relate the turbine or generator output to the output of the boiler.

The figure of merit for operation of a boiler is the boiler or steam-generator efficiency η_{sg}. This quantity is defined as the ratio of the energy transferred to or absorbed by the working fluid in the boiler to the input chemical energy of the fuel. The boiler efficiency commonly ranges from 70 to 90 percent and is usually highest at low boiler loads. There are two ways to calculate the boiler efficiency—the direct method and the indirect method.

In the direct method of calculating the boiler efficiency, the total heat added to the working fluid in the economizer, evaporator, superheater, and reheat sections of the boiler is evaluated and this quantity is divided by the total fuel-input energy:

$$\text{Boiler efficiency} = \eta_{sg} = \frac{\text{total energy added to the working fluid}}{\text{total fuel-input energy}} \, 100$$

$$= \frac{\dot{m}_s(h_2 - h_1) + \dot{m}_r(h_4 - h_3)}{\dot{m}_f(\text{HHV})_{\text{fuel}}} \, 100 \qquad \% \qquad (4.9)$$

where $\dot{m}_f$ is the mass flow rate of the fuel, $\dot{m}_r$ is the mass flow rate of the steam from the reheater, $\dot{m}_s$ is the steam mass flow rate from the superheater, h_1 is the specific enthalpy of the inlet boiler feedwater, h_2 is the enthalpy of the steam leaving the superheater, h_3 is the enthalpy of the steam entering the reheater, and h_4 is the enthalpy of the steam leaving the reheater. Since it is very difficult to measure the fuel and fluid flow rates much more accurately than plus or minus 5 percent, the direct method of evaluating the steam-generator efficiency is not too accurate and is not used in common practice.

Most boiler-efficiency calculations are made using the indirect method of calculation. In this system, it is assumed that the total fuel-input energy is either transferred to the working fluid or it is lost in a number of ways, but that these losses can be readily determined. There are a total of six boiler heat losses and all of them are calculated in terms of energy lost per unit mass of fuel (kJ/kg or Btu/lbm). Using this system, the steam-generator efficiency becomes

$$\eta_{sg} = \frac{\text{higher heating value of fuel} - \text{total losses}}{\text{higher heating value of fuel}} 100$$

$$= \left[1 - \frac{\text{total losses}}{(\text{HHV})_{\text{fuel}}} \right] 100 \qquad (4.10)$$

where all the losses are based on a unit mass of fuel. The six major boiler losses are the dry-gas loss (DGL), the moisture loss (ML), the moisture in combustion air loss (MCAL), the incomplete-combustion loss (ICL), the unburned carbon loss (UCL), and the radiation and unaccounted loss (RL). These losses are determined by using the ultimate analysis of the fuel, the orsat analysis of the flue gas, the refuse analysis when the system is coal fired, along with the pressures and temperatures of the fluids in the steam generator. Since the pressures and temperatures can be experimentally measured more accurately than the steam flow rates, the indirect calculational method gives a more reliable estimate of the boiler efficiency than the direct method.

The dry-gas loss (DGL) is that portion of the boiler loss that can be attributed to the combustion air supplied to the steam generator. Since this is all sensible heat, it can be calculated from the actual dry air-fuel ratio and the gas temperatures. The mass of dry flue gas generated per unit mass of fuel is equal to the actual dry air-fuel ratio plus the unit fuel mass less the refuse mass fraction and the mass of water present and formed during the combustion process:

$$\text{DGL} = w_g c_p (T_{g,\,\text{out}} - T_{g,\,\text{in}})$$

$$= \left[\left(\frac{A}{F} \right)_{\text{act},\,m,\,d} + 1.0 - R - M - 9\text{H}_2 \right] c_p (T_{g,\,\text{out}} - T_{g,\,\text{in}}) \qquad \text{kJ/kg} \quad (4.11)$$

where $w_g = (A/F)_{\text{act},\,m,\,d} + 1.0 - R - M - 9\text{H}_2$, kg of dry flue gas/kg fuel
$\quad c_p$ = specific heat of flue gas (assumed to be the same as that of air)
$\quad\quad = 1.0048$ kJ/kg·°C
$\quad T_{g,\,\text{in}}$ = inlet air temperature, °C
$\quad T_{g,\,\text{out}}$ = outlet flue-gas temperature, °C

and R, M, and H_2 are the refuse, moisture, and hydrogen mass fractions as determined from the refuse and ultimate analyses.

The moisture loss (ML) includes the loss due to vaporizing the moisture in the fuel and the loss due to the latent heat of the moisture produced from the combustion of hydrogen in the fuel:

$$\text{ML} = (M + 9H_2)(h_s - h_w) \qquad \text{kJ/kg} \qquad (4.12)$$

where h_s = enthalpy of superheated steam at $T_{g,\text{out}}$ and a pressure of 1 lb/in² abs (the approximate partial pressure of the water vapor in the flue gas), kJ/kg

h_w = enthalpy of water at the inlet gas temperature $T_{g,\text{in}}$, kJ/kg

If $T_{g,\text{out}}$ exceeds 300°C,

$$h_s - h_w = 2442 + 2.093 T_{g,\text{out}} - 4.187 T_{g,\text{in}} \qquad (4.13)$$

If $T_{g,\text{out}}$ is less than 300°C,

$$h_s - h_w = 2492.6 + 1.926 T_{g,\text{out}} - 4.187 T_{g,\text{in}} \qquad (4.14)$$

The values of M and H_2 are the fuel mass fractions of moisture and hydrogen, respectively, from the as-burned ultimate fuel analysis, $(h_s - h_w)$ is the enthalpy difference in kilojoules per kilogram, and the temperatures are in degrees Celsius.

Another but much smaller moisture loss is the moisture in the combustion-air loss (MCAL). The latent heat is not involved in this calculation as the moisture enters and leaves the combustion process as a vapor:

$$\text{MCAL} = \left(\frac{A}{F}\right)_{\text{act, }m,\,d} \omega c_{p,\,g}(T_{g,\text{out}} - T_{g,\text{in}}) \qquad \text{kJ/kg} \qquad (4.15)$$

In Eq. (4.15), ω is the specific humidity of the entering air in kilograms of moisture per kilogram of dry air, and $c_{p,\,g}$ is the specific heat of water vapor, or 1.926 kJ/kg·°C.

The specific humidity can be evaluated from Fig. 4.30 as a function of the dry-bulb and wet-bulb temperatures in degrees Fahrenheit. It can also be calculated from Eq. (4.16) if the relative humidity ϕ and the total atmospheric pressure P_{atm} are known:

$$\omega = \frac{0.622 \phi P_{\text{sat}}}{P_{\text{atm}} - \phi P_{\text{sat}}} \qquad \text{kg water/kg dry air} \qquad (4.16)$$

In the above equation, P_{sat} is the saturation pressure of the water vapor at $T_{g,\text{in}}$ and ϕP_{sat} is the partial pressure of the water vapor in the air. The moisture in combustion-air loss is normally one of the smaller boiler losses, at least an order of magnitude lower than the moisture and dry-gas losses for most fuels.

The unburned-carbon loss (UCL) is the boiler loss associated with the appearance of carbon in the refuse. This loss is equal to the product of the mass

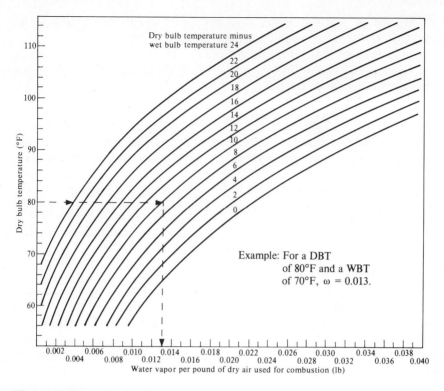

Figure 4.30 The specific humidity as a function of the dry-bulb and wet-bulb temperatures. *(From Lorenzi, 1952.)*

of unburned carbon per unit mass of fuel in the refuse (C_r) and the higher heating value of carbon $(HHV)_c$:

$$UCL = C_r(HHV)_c = 32,778C_r \quad kJ/kg \qquad (4.17)$$

The incomplete-combustion loss (ICL) is the energy lost as the result of the formation of carbon monoxide instead of carbon dioxide in the combustion process. This loss, like the unburned-carbon loss, is equal to the mass of carbon monoxide produced per unit mass of fuel times the higher heating value of carbon monoxide. The mass of carbon monoxide produced per unit mass of fuel can be obtained from the following equation:

$$kg \text{ CO formed/kg fuel} = \frac{\%CO}{\%CO + \%CO_2}\left(\frac{28.01}{12.01}\right)C_b \qquad (4.18)$$

where C_b is the mass of carbon burned per mass of fuel and $\%CO$ and $\%CO_2$ are the values directly from the orsat analysis. The incomplete-combustion loss (ICL) can be determined from the following equation:

$$\text{ICL} = \frac{\%\text{CO}}{\%\text{CO} + \%\text{CO}_2}\left(\frac{28.01}{12.01}\right)C_b(\text{HHV})_{\text{CO}}$$

$$= 23{,}630 C_b \frac{\%\text{CO}}{\%\text{CO} + \%\text{CO}_2} \qquad \text{kJ/kg} \qquad (4.19)$$

The radiation and unaccounted loss (RL) cannot be explicitly calculated, but is estimated from the data presented in Fig. 4.31. The data from this graph give the radiation loss as a fraction of the higher heating value of the fuel and this loss is a function of the actual steam output and the maximum possible output, in million British thermal units per hour, as well as the number of cooled walls in the steam generator. The steam output of the boiler can be estimated by multiplying the input power by an assumed value of boiler efficiency. Once this factor has been determined, the radiation loss (RL) is

$$\text{RL} = (\text{HHV})_{\text{fuel}} \text{ (factor from Fig. 4.31)} \qquad \text{kJ/kg} \qquad (4.20)$$

Once all the losses have been evaluated, the boiler efficiency can be determined from Eq. (4.10). This method is commonly used to evaluate the performance of steam-generator systems but it is not complete enough for some purposes. The Power Test Codes of the American Society of Mechanical Engineers (ASME) provide complete detailed instructions for determining the performance of steam generators. These instructions should be closely followed for contract tests and where precise results are desired.

Example 4.2 Using the data from Example 4.1, perform an energy balance for the system and calculate the boiler efficiency. Assume that the boiler has three sides that are water cooled and the system is operating at full load.

SOLUTION From Example 4.1, $T_{g,\text{in}} = 50°C$, $T_{g,\text{out}} = 288°C$, $M = 0.14$ kg/kg, $R = 0.1195$ kg/kg, $H_2 = 0.042$ kg/kg, $(A/F)_{\text{act},m,d} = 9.804$ kg/kg, $\omega = 0.04353$ kg/kg, $C_r = 0.0095$ kg/kg, $C_b = 0.58$ kg/kg, and $(\text{HHV})_{\text{fuel}} = 24{,}870$ kJ/kg fuel.

Dry-gas loss:

$$\text{DGL} = \left[\left(\frac{A}{F}\right)_{\text{act},m,d} + 1.0 - R - M - 9H_2\right]c_p(T_{g,\text{out}} - T_{g,\text{in}})$$

$$= [9.804 + 1.0 - 0.1195 - 0.14 - 9(0.042)](1.0048)(288 - 50)$$

$$= 2431.2 \text{ kJ/kg fuel}$$

Moisture loss:

$$\text{ML} = (M + 9H_2)(h_s - h_w)$$

$$= [0.14 + 9(0.042)](2837.9) = 1470.1 \text{ kJ/kg fuel}$$

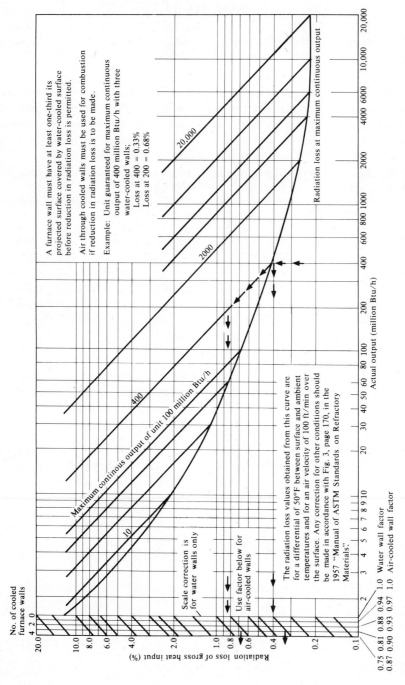

A furnace wall must have at least one-third its projected surface covered by water-cooled surface before reduction in radiation loss is permitted.

Air through cooled walls must be used for combustion if reduction in radiation loss is to be made.

Example: Unit guaranteed for maximum continuous output of 400 million Btu/h with three water-cooled walls;
Loss at 400 = 0.33%
Loss at 200 = 0.68%

20,000

2000

400

Maximum continous output of unit 100 million Btu/h

10

Radiation loss at maximum continuous output

Actual output (million Btu/h)

No. of cooled furnace walls
4 2 0

Radiation loss of gross heat input (%)

Scale correction is for water walls only

Use factor below for air-cooled walls

The radiation loss values obtained from this curve are for a differential of 50°F between surface and ambient temperatures and for an air velocity of 100 ft/min over the surface. Any correction for other conditions should be made in accordance with Fig. 3, page 170, in the 1957 "Manual of ASTM Standards on Refractory Materials."

0.75 0.81 0.88 0.94 1.0   Water wall factor
0.87 0.90 0.93 0.97 1.0   Air-cooled wall factor

Figure 4.31 Radiation loss from a boiler. (*American Boiler Manufacturers Association, as presented in "Steam/Its Generation and Use," 1972.*)

where $h_s - h_w = 2492.6 + 1.926T_{g,\,out} - 4.187T_{g,\,in}$

$$= 2492.6 + 1.926(288) - 4.187(50) = 2837.9 \text{ kJ/kg H}_2\text{O}$$

Moisture in combustion-air loss:

$$\text{MCAL} = \left(\frac{A}{F}\right)_{act,\,m,\,d} \omega c_{p,\,g}(T_{g,\,out} - T_{g,\,in})$$

$$= 9.804(0.04354)(1.926)(288 - 50) = 195.8 \text{ kJ/kg fuel}$$

Unburned-carbon loss:

$$\text{UCL} = 32{,}778C_r = 32{,}778(0.0095) = 311.4 \text{ kJ/kg fuel}$$

Incomplete-combustion loss:

$$\text{ICL} = 23{,}630C_b \frac{\%\text{CO}}{\%\text{CO} + \%\text{CO}_2} = \frac{23{,}630(0.58)(0.75)}{0.75 + 13.78}$$

$$= 707.4 \text{ kJ/kg fuel}$$

Radiation loss:

$$\text{Input power to boiler} = 600{,}000 \text{ kW}_e(8863 \text{ Btu}_{th}/\text{kW}_e\cdot\text{h})$$

$$= 5317.8 \text{ MBtu/h}$$

Assume that the boiler efficiency is 80%.
Output boiler power $= 0.8(5317.8) = 4254$ MBtu/h
Factor from Fig. 4.31 $= 0.21\%$
$\text{RL} = 0.0021(\text{HHV})_{fuel} = 0.0021(24{,}870) = 52.2$ kJ/kg fuel

Total losses $= \text{DGL} + \text{ML} + \text{MCAL} + \text{UCL} + \text{ICL} + \text{RL}$

$$= 2431.2 + 1470.1 + 195.8 + 311.4 + 707.4 + 52.2$$

$$= 5168.1 \text{ kJ/kg fuel}$$

$$\text{Steam-generator efficiency} = \eta_{sg} = \left[1.0 - \frac{\text{total losses}}{(\text{HHV})_{fuel}}\right](100)$$

$$= \left[1.0 - \frac{5168.1}{24{,}870}\right](100) = 79.22\%$$

The component and system layouts of some typical large power boilers are shown in Fig. 4.32.

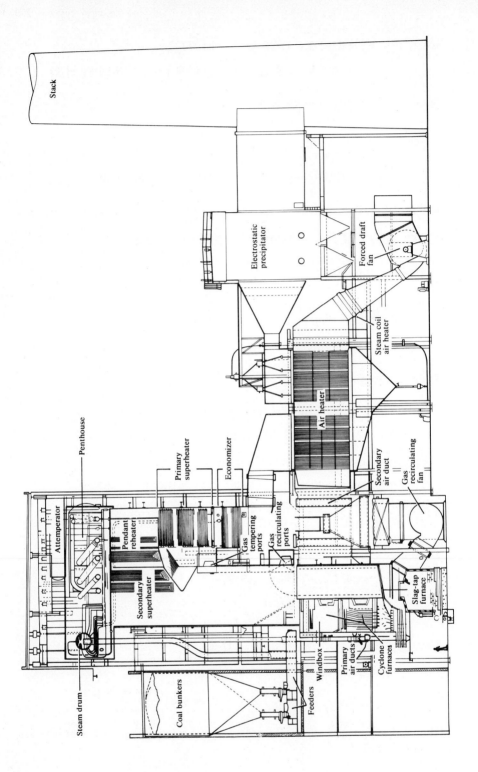

Stack

Electrostatic precipitator

Forced draft fan

Steam coil air heater

Air heater

Penthouse

Primary superheater

Economizer

Secondary air duct

Gas recirculating fan

Attemperator

Pendant reheater

Secondary superheater

Gas tempering ports

Gas recirculating ports

Slag-tap furnace

Steam drum

Coal bunkers

Feeders

Windbox

Primary air ducts

Cyclone furnaces

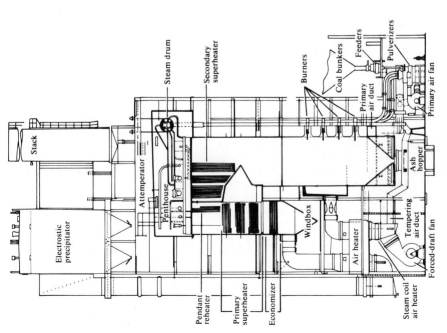

Figure 4.32 Typical radiant boilers and associated systems for cyclone and pulverized-coal combustion systems. (*Courtesy of Babcock and Wilcox, from "Steam/Its Generation and Use," 1972.*)

Stack

Electrostic precipitator

Attemperator

Penthouse

Steam drum

Secondary superheater

Pendant reheater

Primary superheater

Economizer

Windbox

Air heater

Steam coil air heater

Forced-draft fan

Tempering air duct

Ash hopper

Burners

Coal bunkers

Feeders

Pulverizers

Primary air fan

Primary air duct

PROBLEMS

4.1 A centrifugal pump gives the following performance: total pressure rise = 80 ft of water; water flow rate = 1000 gal/min; angular velocity = 1650 rev/min; mechanical efficiency = 82%. Calculate (a) the required motor power, in kilowatts, (b) the specific speed of the pump, and (c) the required power if the total pressure rise is increased to 100 ft of water.

4.2 A 600-MW$_e$ gas-fired power station burns Louisiana B natural gas (from Appendix F). The heat rate of the plant is 9400 Btu$_{th}$/kW$_e$·h. An orsat analysis of the flue gas yields: 9.09% CO_2, 4.11% O_2, and 0.58% CO. Calculate the volumetric and mass air-fuel ratios and the percentage of excess air. If this system uses two forced-draft fans, which are designed to pump ambient air at 50°C and 0.92 atm across a pressure differential of 112 mbars with an efficiency of 80%, find the horsepower of each of the fans and the outlet temperature of the air, assuming an adiabatic compression, and estimate the dew-point temperature of the exhaust gas if it leaves at 1 bar.

4.3 A 400-MW$_e$ power plant burns Clay County, Mo., coal having a moisture fraction of 0.12 and an ash fraction of 0.10, and operates with a thermal efficiency of 37%. An analysis of the refuse pit shows a heating value of 582 kJ/kg refuse. An orsat analysis of the flue gas gives: 14.91% CO_2, 3.67% O_2, and 0.15% CO. Find: (a) the coal rate, (b) the dilution coefficient and percentage of excess air, (c) the capacity of the f-d fan, in cubic feet per minute and in kilograms per minute, if the atmospheric conditions are 0°C and 0.92 bar, (d) the capacity of the i-d fan, in cubic feet per minute and in kilograms per minute, if the exhaust gas is at 260°C and 0.85 bar, and (e) the size of the motors, in kilowatts, required to drive the i-d and the f-d fans if there is a 50-mbar pressure differential across the fans.

4.4 During the test of a 510-hp Babcock and Wilcox boiler, the following data were obtained:

Coal rate = 995.8 kg/h   Refuse rate = 185.4 kg/h

Percentage of combustible in the refuse = 22.84%

Ultimate analysis: C = 60.31, H$_2$ = 4.06, O$_2$ = 6.71, N$_2$ = 1.13, S = 5.23, A = 22.56

Orsat analysis: 10.58% CO_2, 8.41% O_2, and 0.23% CO

Find: (a) the boiler efficiency if one boiler horsepower is 33,500 Btu/h, (b) the percentage of excess air, (c) the mass of flyash going up the stack, in kilograms per hour.

4.5 A 2284-MW$_e$ power plant burns Linn County, Mo., coal with M and A of 12% and 8%, respectively, and the heat rate for the system is 9000 Btu/kW·h. This plant has eight f-d fans and eight i-d fans. The f-d fans are designed to pump air at 40°C and 0.93 bar across a pressure rise of 35 mbar, and the i-d fans are designed to pump exhaust gas at 150°C and 0.88 bar across a pressure rise of 50 mbar. Assume that the mechanical efficiency of each fan is 80%. An analysis of the refuse pit shows that the higher heating value of the refuse is 3280 kJ/kg. An orsat analysis of the flue gas yields: 14.36% CO_2, 4.01% O_2, and 0.78% CO. Find the percentage of excess air for this system as well as the fan capacities, in cubic feet per minute, and the power requirements for the fans.

4.6 When St. Clair County, Ill., coal (with M = 10 and A = 12) is burned in a 600-MW$_e$ power plant with a heat rate of 9670 Btu$_{th}$/kW$_e$·h, the heating value of the refuse is 4685 kJ/kg and the orsat analysis gives: 12.45% CO_2, 5.45% O_2, and 1.43% CO.

(a) Find the coal rate for the unit in metric tons per hour and in short tons per hour.

(b) Find the mass (kg/h) and volumetric (ft^3/min) flow rates of combustion air required to burn the fuel if atmospheric air is at 0.95 bar and 40°C.

(c) Repeat part (b) for the exhaust gas if it leaves at 0.88 bar and 200°C.

(d) Find the horsepower requirements of the i-d and f-d fans if the pressure rise across each fan is 62 mbar and the fan efficiencies are 80%.

(e) Find the dew-point temperature of the gas.

4.7 Two hundred tons of coal per hour are burned in a boiler. The coal has the following ultimate analysis: C = 71.89%, H$_2$ = 6.69%, S = 1.91%, O$_2$ = 1.17%, N$_2$ = 2.68%, M = 8.82%, and A = 6.84%. The higher heating value of the refuse is 4333 kJ/kg and the orsat analysis is: 11.27% CO_2, 6.62% O_2, and 1.32% CO. Assume the air enters the f-d fan at 0.94 bar and 45°C and that the flue gas enters the i-d fan at 0.87 bar and 310°C. If the mechanical efficiency of each fan is 80%, find the required power of each fan, in kilowatts and in horsepower, and the capacity of each fan, in cubic feet per minute. Assume that the pressure rise across each fan is 43 mbar. Also, determine the percentage of excess air and the dew-point temperature of the exhaust gas and estimate the electrical power output of the plant, in megawatts, if the overall heat rate is 9750 Btu$_{th}$/kW$_e$·h.

4.8 If the relative humidity of the incoming air is 40 percent and all four walls of the furnace are water cooled, determine the boiler heat balance and the boiler efficiency for the system in Prob. 4.7.

4.9 The following data were taken during a boiler test:

Ultimate analysis of dry coal: 72.97% C, 4.83% H_2, 1.39% N_2, 5.70% O_2, 2.48% S, 12.63% A, and HHV = 30,724 kJ/kg

Moisture content as fired = 4.6%

Orsat analysis: 12.93% CO_2, 6.03% O_2, and 0.08% CO

Temperature of flue gas = 188°C

HHV of refuse = 2472 kJ/kg

Atmospheric conditions: 21°C, 0.92 atm, and relative humidity of 90%

The boiler is a 40-MBtu/h unit with water tubes on three sides of the unit which is operated at 10 MBtu/h during the test. Estimate the boiler efficiency.

4.10 The following data were collected during operation of a 600-MW_e power plant operating at a heat rate of 9650 $Btu_{th}/kW_e \cdot h$:

Coal (as fired): Williamson County, Ill., coal with $M = 8\%$ and $A = 12\%$

Orsat analysis: 13.8% CO_2, 4.8% O_2, and 0.7% CO

Combustion-air conditions: 90°F DBT and 81°F WBT

Outlet flue-gas temperature = 260°C

Percentage of combustible in the refuse = 17.8%

System power during the test = 400 MW_e

Determine:

(a) The coal rate in metric tons per hour and short tons per hour

(b) The percentage of excess air

(c) The energy losses for the boiler, including the radiation loss if only two of the furnace walls are cooled

(d) The boiler efficiency (using the indirect method)

(e) The dew-point temperature of the exhaust gas (assuming a maximum flue-gas pressure of 0.92 bar)

4.11 Calculate the boiler heat balance and the boiler efficiency from the following data:

Ultimate as-burned coal analysis: 10.48% A, 1.1% S, 4.1% H_2, 77.39% C, 1.62% N_2, 5.31% O_2, and HHV = 30,866 kJ/kg

Duration of test: 24 h

Fuel fired during the test = 382.18 metric t

Heating value of refuse = 1872 kJ/kg

Water supplied to boiler during test = 4,075,732 kg

Outlet steam conditions: 400°C and 58 bar

Inlet water temperature = 190°C

Fuel and air temperature = 33°C

Relative humidity of the inlet air = 50%

Outlet gas temperature = 226°C

Orsat analysis: 14.7% CO_2, 3.7% O_2, and 0.2% CO

4.12 The following data were obtained from a boiler evaporation test:

Boiler efficiency = 77%

Ultimate, as-burned coal analysis: 65% C, 4% H_2, 8% O_2, 2% S, 10% A, 1% N_2, 10% M, and HHV = 26,352 kJ/kg

Refuse analysis: 20% combustible

Orsat analysis: 14.18% CO_2, 3.55% O_2, and 1.42% CO

Combustion-air conditions: 27°C DBT and 22°C WBT

Flue-gas temperature = 250°C   Boiler steam temperature = 180°C

Find the individual boiler losses and the dilution coefficient, and estimate the dew-point temperature of the flue gas leaving the boiler if the flue gas has the same molecular weight as air.

4.13 An 800-MW_e power plant operates at an average heat rate of 9300 $Btu_{th}/kW_e \cdot h$. Two sides of the firebox are cooled and the system is fired with Coal County, Okla., coal with M and A of 6% and 10%, respectively. During a test with the boiler operating at 30% capacity, the following data were taken:

Refuse analysis: HHV = 7188 kJ/kg
Orsat analysis: 13.83% CO_2, 4.21% O_2, and 1.15% CO
Exhaust-gas temperature = 205°C
Air conditions: 32°C, 0.98 bar, and 50% relative humidity
Estimate the boiler efficiency and the percentage of excess air.

4.14 Data from a boiler test are as follows:
Ultimate analysis of the coal: 60% C, 4% H_2, 6% O_2, 1% N_2, 3% S, 12% M, and 14% A
Orsat analysis: 13.57% CO_2, 4.37% O_2, and 1.18% CO
Refuse analysis: HHV = 6770 kJ/kg
Air conditions: 21°C DBT and 16°C WBT
Flue-gas outlet temperature = 193°C
Calculate the percentage of excess air and the boiler efficiency if the radiation loss is 3 percent of the input energy. Use Dulong's formula to determine the HHV.

4.15 The Bruce Mansfield power plant of the Pennsylvania Power Company is a three-unit, pulverized-coal system with a total output of 2751 MW_e. Each unit has two f-d fans and six i-d fans. The f-d fans have a pressure rise of 80 mbar with the inlet air at 1 atm and 27°C. The i-d fans have a pressure rise of 85 mbar with 48°C and 0.9135 kg/m³ flue gas. Assume the system burns Elk County, Pa., coal with moisture and ash fractions of 0.03 and 0.09, respectively. The plant heat rate is 8800 $Btu_{th}/kW_e \cdot h$ and the HHV of the refuse is 1058 kJ/kg. The orsat analysis gives: 13.77% CO_2, 4.40% O_2, and 0.99% CO. Find:

    (*a*) The maximum coal rate for the plant at full power, in short tons per hour
    (*b*) The percentage of excess air
    (*c*) The horsepower of each of the six f-d fans
    (*d*) The horsepower of each of the eighteen i-d fans
    (*e*) The energy losses for the boiler, including the radiation loss (assuming that three of the furnace walls are cooled and that the units are operating at one-third full power), with the following data:

        Air conditions: 32°C DBT and 80% relative humidity
        Exit flue-gas temperature = 260°C
    (*f*) The boiler efficiency (using the indirect method)
    (*g*) The dew-point temperature of the flue gas

REFERENCES

Bender, R. J.: "Steam Generation," A *Power Magazine* Special Report, McGraw-Hill, Inc., New York, June, 1964.

Dillio, C. C., and E. P. Nye: "Thermal Engineering," International Textbook Company, Scranton, Pa., 1963.

Editors of *Power Magazine*: Axial Fans Reduce Operating Costs, in "Generation Planbook," McGraw-Hill, Inc., New York, 1976.

Fritsch, T. J.: World's Largest Boiler Feed Pump, *Proceedings of the American Power Conference*, 1974, vol. 36, Illinois Institute of Technology, Chicago, Ill.

Hays, T. C., B. C. Krippene, G. J. Clessuras, and G. E. McMackin: Development of Large Components for Large Steam Generators, *Proceedings of the American Power Conference*, 1973, vol. 35, Illinois Institute of Technology, Chicago, Ill.

Lorenzi, Otto de (ed.): "Combustion Engineering," Combustion Engineering-Superheater, Inc., New York, N.Y., 1952.

Mesko, J. E., and R. L. Gamble: Atmospheric Fluidized-Bed Steam Generators for Electric Power Generation, *Proceedings of the American Power Conference*, 1974, vol. 36, Illinois Institute of Technology, Chicago, Ill.

Skrotzki, B. G. A., and W. A. Vopat: "Power Station Engineering and Economy," McGraw-Hill Book Company, New York, 1963.

"Steam/Its Generation and Use," 38th ed., Babcock and Wilcox Company, New York, N.Y., 1972.

Wilhem, B. W., et al.: The Effect of Water Jet Lancing on Furnace Walls, *Proceedings of the American Power Conference*, 1975, vol. 37, Illinois Institute of Technology, Chicago, Ill.

NUCLEAR REACTOR DESIGN AND OPERATION

5.1 INTRODUCTION

This chapter covers the various power reactor systems currently in use in the United States as well as some of the problems associated with the thermal design and operation of these systems. It will be assumed that the student is familiar with the general reactor terminology and the operational principles presented in Chapter 3.

5.2 POWER REACTOR SYSTEMS

5.2.1 Light-Water Reactors

Introduction Almost all of the power reactors currently operating in the United States are light-water reactors (LWR). These reactors are thermal reactor systems in which the water serves as both the coolant and the moderator. The use of ordinary water in these systems means that they must employ uranium fuel which is enriched to 2 to 3 percent. All LWR systems employ cylindrical fuel rods mounted in a square fuel array similar to that shown in Fig. 3.10. There are two basic types of light-water reactors, including the pressurized-water reactor (PWR) and the boiling-water reactor (BWR). In 1973, the Atomic Energy Commission (forerunner of the Nuclear Regulatory Commission) issued its Standardization Policy which set the maximum power of any power reactor system at 3800 MW$_{th}$ plus the core power needed to drive the reactor auxiliaries.

Pressurized-water reactors The pressurized-water reactor is the most prevalent of all the reactor systems currently being built in this country. It is also the type of power system employed in the Naval Reactor Program. The PWR is offered by Westinghouse, Babcock and Wilcox, and Combustion Engineering.

In the PWR, there is no bulk boiling permitted in the reactor core as the water is pressurized to about 150 atm (2250 lb/in² abs). The hot water from the reactor flows to a steam generator where the heat is transferred to the water and steam in a secondary cooling loop operating at a lower pressure. The primary coolant then flows from the steam generator to the primary circulating pump where it is pumped back to the reactor. A typical PWR primary cooling system is shown in Fig. 5.1 along with the secondary steam system. The operational parameters for the "3800-MW$_{\text{th}}$" PWR systems are given in Table 5.1.

The PWR primary system shown in Fig. 5.1 contains one component not discussed previously—the pressurizer. This system is simply a pressure vessel with a heater at the bottom and a water spray at the top. The top of the pressure vessel is filled with steam at the primary-system pressure. If the primary-system pressure drops, the heater is activated to increase the steam content in the pressurizer which increases the pressure of the primary system. If the primary-

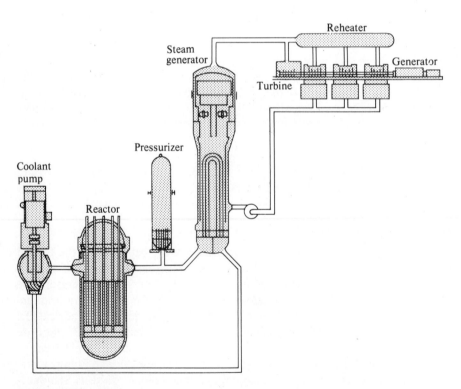

Figure 5.1 The primary and secondary cooling systems for a typical PWR. *(Courtesy of Westinghouse Electric Corporation.)*

Table 5.1 Operational and design parameters for " 3800-MW $_{th}$" pressurized-water reactor systems

NSSS vendor	Westinghouse	Combustion Engineering	Babcock and Wilcox
System designation	Westinghouse 3817	System 80	Babcock 241
Core power, MW$_{th}$	3817	3817	3815
Number of primary coolant pumps	4	4	4
Pump power, hp	8000	8991	6870
Primary system pressure, lb/in^2 abs	2250	2250	2250
Number of fuel-element assemblies	193	241	241
Active core length, in	164	150	143
Number of control rod assemblies	61	89	84
Reactor-inlet temperature, °F	559	564.5	573.5
Reactor-outlet temperature, °F	623	621.5	629.4
Coolant flow rate, 10^6 lbm/h	147	164	159
Number of primary loops	4	2	2
Secondary steam pressure, lb/in^2 abs	1100	1070	1135
Turbine steam temperature, °F	556 (sat)	553 (sat)	603 (sup)
Nominal electrical rating, MW$_e$	1300	1300	1322

system pressure is too high, cold pressurized water is sprayed into the steam volume to condense the steam and reduce the primary pressure.

Each PWR normally has two to four primary cooling loops but only one pressurizer. Each cooling loop contains a large steam generator and one or two primary pumps. The pressurizer for the Babcock and Wilcox reactor, which is typical of most reactors, is shown in Fig. 5.2.

Two different kinds of steam generators are employed in PWR systems. The Westinghouse and Combustion Engineering systems use a U-tube steam-generator design which produces saturated secondary steam for the turbine. Babcock and Wilcox employs a once-through counterflow steam generator in which the hot primary water from the reactor actually produces superheated steam in the secondary loop. This improves the turbine performance and the overall performance of the system. The two kinds of steam generators are shown in Fig. 5.3.

The reactor core for a PWR system is enclosed in a large pressure vessel. These vessels are up to 21 ft in diameter and more than 40 ft in length. The walls are constructed of 8 to 10 in of carbon steel with approximately $\frac{5}{16}$ in of stainless steel welded to the inside surface to provide corrosion protection from the hot water. All nozzles and major penetrations to the reactor pressure vessel are made above the reactor core to prevent the complete loss of coolant in the event of a failure of the primary-system piping. The cold water enters the reactor vessel above the fuel region, flows down the annular region between the core and the pressure vessel, cooling the vessel, and then flows up through the fuel elements

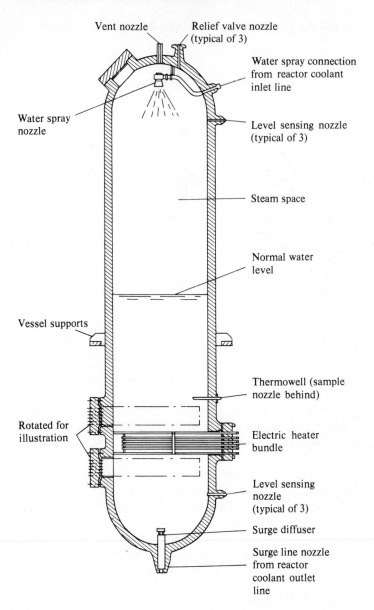

Vent nozzle

Relief valve nozzle
(typical of 3)

Water spray connection
from reactor coolant
inlet line

Water spray
nozzle

Level sensing nozzle
(typical of 3)

Steam space

Normal water
level

Vessel supports

Thermowell (sample
nozzle behind)

Rotated for
illustration

Electric heater
bundle

Level sensing
nozzle
(typical of 3)

Surge diffuser

Surge line nozzle
from reactor
coolant outlet
line

Figure 5.2 A typical pressurizer for the primary cooling loop of a PWR. *(Courtesy of Babcock and Wilcox, from "Steam/Its Generation and Use," 1972.)*

No.	Service
1	Primary inlet
2	Primary outlet
3	Auxiliary feedwater
4	Steam outlet
5	Blowdown
6	Liquid level
7	Primary manway
8	Secondary manway
9	Handhole
10	Upper feedwater
11	Lower feedwater

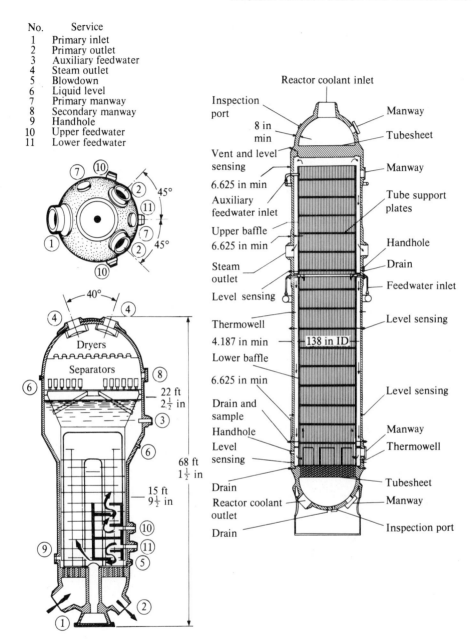

Figure 5.3 The large PWR steam generators. (*From Bevilacqua and Gibbons, 1974; and courtesy of Babcock and Wilcox, from "Steam/Its Generation and Use," 1972.*)

where the fission energy is transferred to the water. The hot water is then piped to the steam generator.

The control rods in a **PWR** enter from the top of the core and are actuated by a magnetic drive system. The absorber rods are pulled from the core by a magnetic jack or by a system containing an electromagnet. The reactor is "scrammed" by simply terminating the power to the magnets letting the rods drop into the core under the influence of gravity. The older **PWR** systems employed "cruciform rods" similar to that shown later in Fig. 5.9, while the newer systems use a "finger" array similar to that shown in Fig. 3.13. The newer system puts control elements ("fingers") into a number of different fuel assemblies. The basic **PWR** pressure vessel and internals are shown in Fig. 5.4.

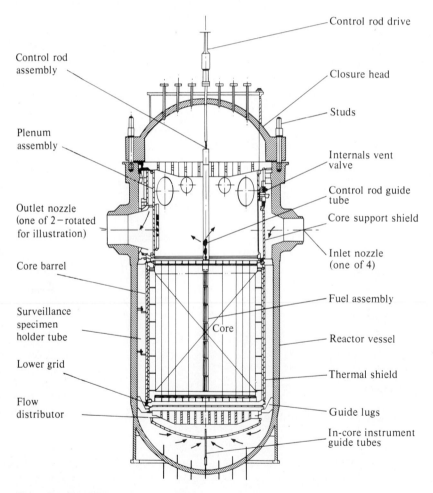

Figure 5.4 The PWR pressure vessel and internals. (*Courtesy of Babcock and Wilcox, from "Steam/Its Generation and Use," 1972.*)

Boiling-water reactors In the boiling-water power reactor (BWR), saturated steam is generated in the reactor core and this steam is dried and sent directly to the steam turbine. This system eliminates the large, expensive, secondary-steam generator although some of the early BWR systems used small ones for control purposes. These reactor systems have some problems with the carryover of radioactive material into the turbine and also with in-core vibration. The boiling-water power reactor is promoted by the General Electric Company.

The early BWR's were called dual-cycle flow systems because they generated most of the turbine steam directly in the reactor core, but they also generated a portion of the turbine steam in a secondary steam generator at lower pressure for control purposes. A schematic diagram of the dual-cycle BWR system is shown in Fig. 5.5.

In a BWR, the amount of steam bubbles in the reactor core essentially controls the reactor power. Since the water coolant also moderates (slows down) the fast neutrons, an increase in the steam fraction in the core displaces water in the core and reduces the reactor power unless the control rods are withdrawn. If the reactor steam is supplied to the turbine through a throttling valve and the valve is opened to increase the turbine power, the immediate effect is to lower the steam pressure in the reactor. The decrease in steam pressure increases the steam volume in the core and this results in a lower reactor power—just the opposite of the desired effect. This condition can be remedied by using a small secondary steam generator and placing the turbine throttle valve on the

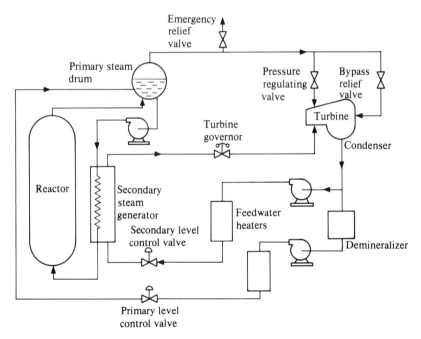

Figure 5.5 Schematic diagram of the dual-cycle BWR system. (*Courtesy of the General Electric Company.*)

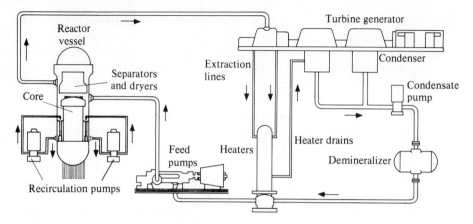

Figure 5.6 Schematic diagram of the single-cycle BWR system. *(Courtesy of the General Electric Company.)*

secondary steam supply line. In this case, opening the throttle valve reduces the secondary steam pressure and temperature, increasing the heat transfer from the primary water, and this reduces the reactor-inlet temperature. When the cold water enters the reactor, it condenses some of the in-core steam and increases the reactor power. This arrangement essentially gives the reactor an inherent load-following capability.

The latest BWR systems are single-cycle power reactors and require no secondary steam generators. The reactor power is controlled by varying the recirculation rate through the reactor core with a series of jet pumps. When the power demand increases, the water recirculation rate through the core is increased and this sweeps the steam bubbles from the core, increasing the reactor power level. The use of jet pumps around the periphery of the reactor core eliminates any moving parts inside the primary pressure vessel. A detailed illustration of the reactor is shown in Fig. 5.6. Some of the design parameters for the latest 3800-MW$_{th}$ single-cycle BWR system, as offered by the General Electric Company, are listed in Table 5.2.

The pressure vessel for the BWR system is considerably taller than the PWR vessel and the top of the BWR vessel is filled with steam-drying equipment (Fig. 5.7). The control rods enter from the bottom of the core since the top is filled with the steam equipment. The rods cannot rely on gravity for emergency insertion ("scram") as do the PWR systems. Instead, a hydraulic accumulator system is employed, and this system has enough stored energy to drive the rods into the core in the event of an emergency shutdown.

Care must be taken in the operation of any liquid-cooled reactor, but particularly in the design of a BWR system, to assure that the maximum heat flux in the reactor core does not exceed the value of the "departure from nucleate boiling" or "DNB." A typical boiling heat-transfer curve is shown in Fig. 5.8. At relatively low heat fluxes, the boiling heat-transfer process is called nucleate

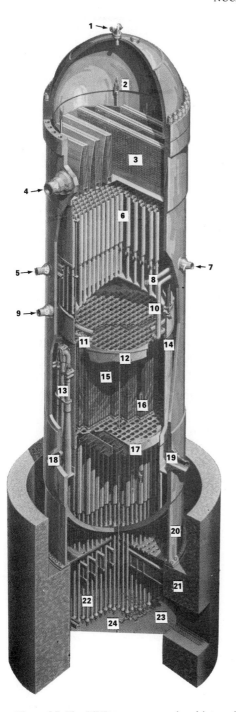

BWR/6
REACTOR ASSEMBLY

1. Vent and head spray
2. Steam dryer lifting lug
3. Steam dryer assembly
4. Steam outlet
5. Core spray inlet
6. Steam separator assembly
7. Feedwater inlet
8. Feedwater sparger
9. Low pressure coolant
   injection inlet
10. Core spray line
11. Core spray sparger
12. Top guide
13. Jet pump assembly
14. Core shroud
15. Fuel assemblies
16. Control blade
17. Core plate
18. Jet pump/recirculation
    water inlet
19. Recirculation water outlet
20. Vessel support skirt
21. Shield wall
22. Control rod drives
23. Control rod drive
    hydraulic lines
24. In-core flux monitor

Figure 5.7 The BWR pressure vessel and internals. *(Courtesy of the General Electric Company.)*

Table 5.2 Operational and design parameters for the "3800-MW$_{th}$" boiling-water reactor system

Reactor designation	GE/BWR/6
Core power, MW$_{th}$	3,833
Number of fuel elements	784
Number of control rods	193
Pressure vessel ID, in	251
Recirculation pipe size, in	24
Core flow rate, 10^6 lbm/h	113.5
Recirculation-pump flow, gal/min	44,900
Pump power, hp	7,290
Motor rating, hp	7,700

boiling. In this region, vapor bubbles are produced at the surface of the fuel element and break away, producing a lot of agitation. This produces very high heat-transfer rates (very high convective heat-transfer coefficients) and small temperature differences between the saturation temperature of the fluid and the fuel-element surface.

As the heat flux is raised, however, the steam-bubble formation rate increases, increasing the heat-transfer rate until the entire surface is mostly covered by vapor. When the surface is completely covered with vapor, the vapor acts as an insulating layer and the surface temperature increases dramatically. This mode

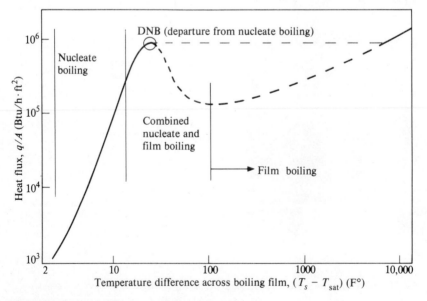

Figure 5.8 Typical boiling heat-transfer relationships.

of boiling heat transfer is called film boiling and unless the heat flux is reduced, the surface will actually melt. The transition from nucleate to film boiling essentially takes place at the departure from nucleate boiling.

Since a nuclear reactor has no theoretical maximum temperature, power, or heat-flux limit as long as a supercritical assembly can be maintained, care must be taken in the design and operation of a reactor to assure that the maximum

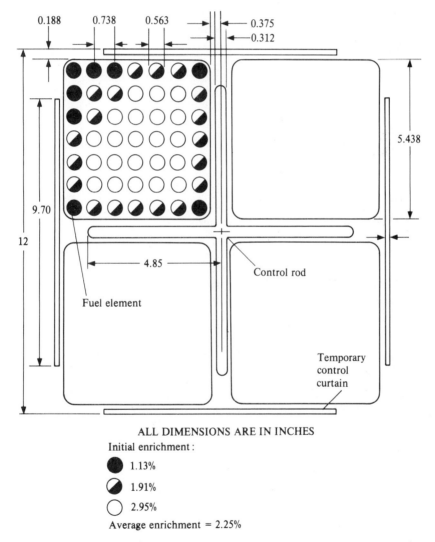

ALL DIMENSIONS ARE IN INCHES
Initial enrichment:

● 1.13%

◑ 1.91%

○ 2.95%

Average enrichment = 2.25%

Figure 5.9 A typical in-core section of an older BWR fuel module consisting of four fuel elements surrounding a cruciform control rod. Newer fuel modules have an 8 × 8 array of fuel rods. *(Courtesy of the General Electric Company.)*

heat flux is below the DNB value. This consideration is important in any liquid-cooled system but it is particularly true for the boiling-water reactor since the phase change is already taking place in the core.

Most of the boiling-water reactors are designed to operate at a maximum heat flux which is well below the value of DNB. To accomplish this, it may be necessary to reduce the power generation in those fuel tubes at the outer edge of the fuel element next to the water gap where the control rod is inserted. A typical fuel-element and control rod arrangement is shown in Fig. 5.9. When the cruci-form control rods are withdrawn, the thermal-neutron flux peaks in the water gap produced by the rod withdrawal. This high neutron flux can cause excessive heating in the fuel rods at the edge of the element. To reduce the power genera-tion in these rods, lower fuel enrichments are employed in these rods than in the center of the fuel element. The different rod cross sections in the fuel rods of the upper right element, in Fig. 5.9, represent different fuel enrichments.

The operating boiling-water reactor systems have developed some unantic-ipated generic problems recently that have somewhat dimmed the utilization of these systems. In-core vibration brought about by the boiling process compli-cates the retrieval of reliable data from the in-core instrumentation. A few of the BWR systems have also experienced cracks in the water recirculation lines, resulting in expensive shutdowns and repairs.

5.2.2 High-Temperature, Gas-Cooled Reactors

Two gas-cooled commercial power reactors have been built in the United States. One of these systems was a small 40-MW_e reactor (Peach Bottom-I) built for the Philadelphia Electric Company. The other reactor is a 300-MW_e power reactor (the Ft. St. Vrain reactor) built by the Public Service Company of Colorado. These systems are cooled with helium and moderated with graphite and are called high-temperature, gas-cooled reactors (HTGR). The entire reactor core is constructed of nonmetallic materials (primarily graphite), so that very high tem-peratures can be tolerated in this system. In addition, the large amount of graphite in the core has such a high heat capacity that it can absorb all of the fission-product decay heat without the problem of core meltdown. This reactor type is being promoted by Gulf General Atomics, a division of the Gulf Oil Corporation.

The HTGR is completely contained within a prestressed-concrete pressure vessel (PCPV) along with the helium coolant and the steam-generating and superheating equipment. A schematic diagram of the HTGR power system is shown in Fig. 5.10. The proposed reactor for a "3800-MW_{th}" system is shown in Fig. 5.11, and the pertinent operational parameters are presented in Table 5.3.

It will be noted that this system produces high-temperature (955°F), high-pressure (2400 lb/in^2 gage) steam for the turbine inlet. These steam conditions are comparable to those from a fossil-fueled system. The resulting thermal efficiency of the HTGR plant is around 38 percent, which compares favorably with the

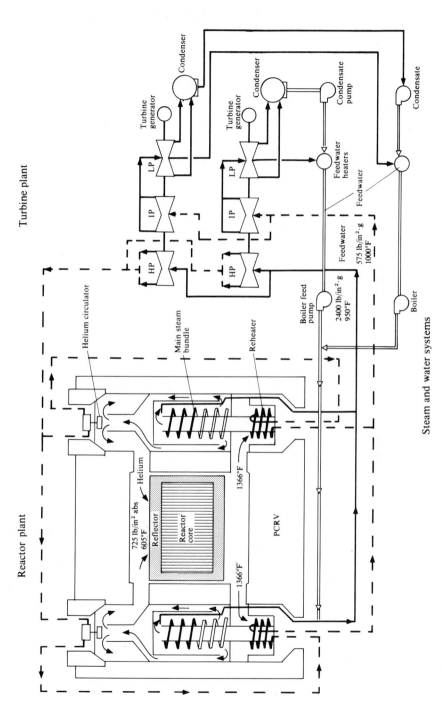

Figure 5.10 Schematic diagram for the HTGR power system. (*From Waage et al., 1974.*)

227

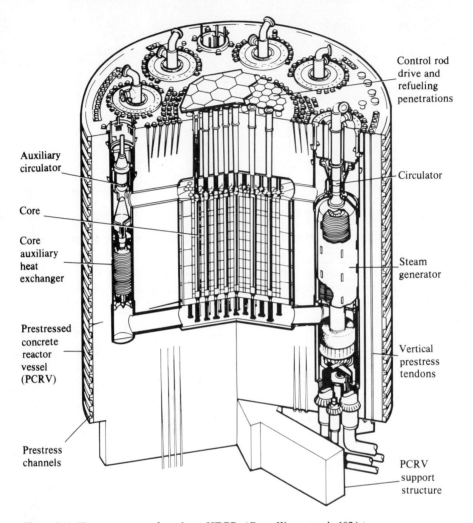

Control rod
drive and
refueling
penetrations

Auxiliary
circulator

Circulator

Core

Core
auxiliary
heat
exchanger

Steam
generator

Prestressed
concrete
reactor
vessel
(PCRV)

Vertical
prestress
tendons

Prestress
channels

PCRV
support
structure

Figure 5.11 The arrangement for a large HTGR. *(From Waage et al., 1974.)*

efficiency of the best fossil units. This efficiency is much higher than the typical thermal efficiency values of 32 to 33 percent for the LWR plants.

The HTGR has many attractive features and, by 1974, six of the 1100-MW$_e$ systems had been ordered by various utilities. However, problems with the helium cooling system have caused major revisions and a corresponding escalation of the cost so that all six of the HTGRs on order were cancelled by early 1976. As a result, the future prospects for the HTGR are somewhat clouded. Considerable research and design have been directed toward the possibility of adapting the HTGR to a Brayton-cycle (gas-turbine) system in order to achieve higher thermal efficiencies.

Table 5.3 Operational parameters for the "3800-MW$_{th}$" high-temperature gas-cooled reactor

Core power, MW$_{th}$	3800
Net thermal output, MW$_{th}$	3764
Primary coolant (helium) pressure, lb/in^2 abs	725
Helium-inlet temperature, °F	605
Helium-outlet temperature, °F	1366
Throttle-steam conditions, lb/in^2 gage/°F	2400/950
Reheat-steam conditions, lb/in^2 gage/°F	575/1000
Net electrical output, MW$_e$	1450
Net thermal efficiency, %	38.2

5.2.3 Liquid-Metal, Fast-Breeder Reactors

Barring a government-imposed moratorium or a technical breakthrough by another reactor type, the next generation of power reactors appear to be the liquid-metal, fast-breeder reactors (LMFBR). This system is a sodium-cooled fast reactor and the program has been under way for some time. The fast flux test facility (FFTF) is almost completed at Hanford, Wash., and design and limited construction is currently under way on the Clinch River breeder reactor (CRBR) power system at Oak Ridge, Tenn. This reactor is a 975-MW$_{th}$ (375-MW$_e$) power prototype and is supposed to supply electricity for the TVA grid. The schematic diagram for the CRBR is shown in Fig. 5.12.

In this system, as in any sodium-cooled power reactor, the fission energy is transferred to the primary sodium; from the primary sodium to sodium in a secondary loop in the intermediate heat exchanger (IHX); and finally to the water-steam system. This precludes the possibility of a sodium-water reaction with the radioactive sodium. A plan view of the reactor vessel and internal components are shown in Fig. 5.13 and the pertinent reactor operating parameters are listed in Table 5.4.

Table 5.4 Operational parameters for the Clinch River breeder reactor

Core power, MW$_{th}$	975
Number of fuel rods/assembly	198
Number of core assemblies	150
Core height/diameter, ft	3.0/6.2
Breeding ratio	1.2
Doubling time, year	23
Primary sodium-inlet/outlet temperatures, °F	730/995
Secondary sodium-inlet/outlet temperatures, °F	651/936
Throttle-steam conditions	1450/900
Net thermal efficiency, %	38.4
Net electrical output, MW$_e$	375

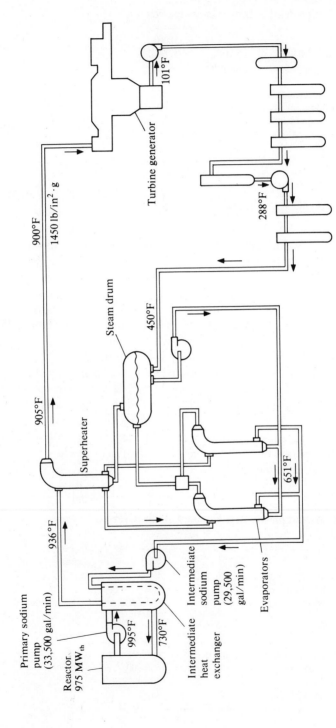

Figure 5.12 The schematic flow diagram for the CRBR. (*From U.S. Atomic Energy Commission, 1974.*)

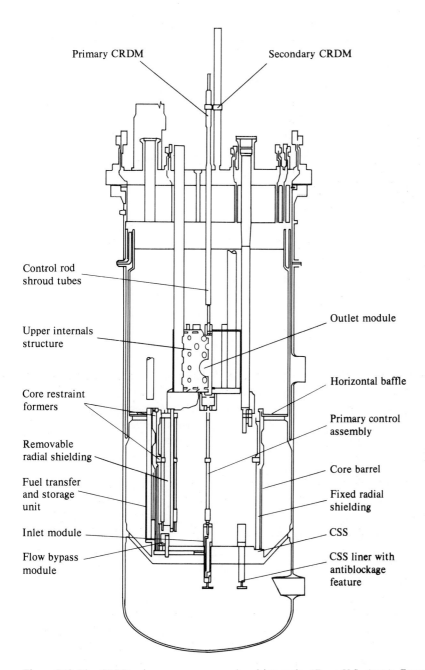

Figure 5.13 The CRBR primary pressure vessel and internals. (*From U.S. Atomic Energy Commission, 1974.*)

There have been a number of sodium-cooled fast reactors constructed and operated previously with mixed results. The experimental breeder reactor II (EBR-II) and the southeast fast oxide reactor (SEFOR) have been operated successfully while the EBR-I and the Enrico-Fermi reactors have experienced partial core meltdowns. Despite possible technical problems, the main threat is from a politically motivated moratorium. In 1977, President Carter has expressed a desire to eliminate or indefinitely delay the LMFBR program, including the Clinch River breeder reactor, and has vetoed an appropriations bill for this program. Despite the political problems in this country, the LMFBR program is being actively pursued in England, France, Germany, and Japan, among others.

5.3 REACTOR THERMAL DESIGN

5.3.1 Reactor In-Core Power Distributions

In the design or operation of any nuclear reactor, there is normally some maximum operating temperature (e.g., the melting point of the fuel) or heat flux (e.g., the value of DNB) that limits the reactor power. The maximum temperature or heat flux is a function of the fission rate per unit volume or the power density in the fuel material. The maximum-possible safe reactor power that can be attained is realized when the entire reactor core is operating at the limiting temperature or heat flux. Likewise, the safest operating condition occurs at lower powers when the entire core is operating at the same temperature or the same heat flux in order to achieve as large a margin of safety as possible.

The local fission rate per unit volume is equal to the product of the local atomic density of the fuel N_{fuel}/V, the average microscopic fission cross section $\bar{\sigma}_f$, and the local value of the neutron flux causing fission ϕ. Since the fission cross section is essentially constant, the local power density varies as the product of the atomic concentration of the fuel and the neutron flux of the neutrons causing fission if the reactor has a variable fuel loading. If the reactor is uniformly loaded, the local power density is the same as the distribution of the neutrons causing fission, i.e., the fast neutron flux in a fast reactor and the thermal neutron flux in a thermal reactor.

A uniform temperature or heat flux can be achieved in a reactor core by employing either a nonuniform fuel distribution or by using absorber rods to shape the power. Most LWR systems move the partially burned fuel from the outer edge of the reactor to the center of the core and insert the new fuel at the edge of the core in an effort to raise the flux in the outer edge in order to level-out and improve the radial power distribution. The use of absorber rods to shape the reactor power is not as efficient as varying the reactor fuel loading.

The use of a variable fuel loading is very difficult to analyze because of the fact that the local neutron flux and the local fuel loading are not independent variables. As the fuel concentration is increased in a given region the power increases, but not proportionally to the fuel loading, because the addition of more fuel into a given region of the core increases the absorption cross section which, in turn, decreases the neutron flux in that region. The local fuel concentration at a given point can be increased by putting more fuel material in that volume and/or by varying the enrichment of the fuel. An analysis of the nonuniform fuel-loading condition is beyond the scope of this text and is even beyond the scope of most textbooks on reactor physics.

While the uniformly loaded reactor systems are not common in power reactors, it is common in low-powered research reactors. These systems can be studied analytically and allows the student to obtain some idea of the method of analysis. In fact, this text treats the simplest type of reactor, which is the uniformly loaded, bare reactor system, and starts with the neutron-flux distributions as developed from reactor physics.

In a bare, uniformly loaded reactor, the neutron flux is a maximum at the center of the assembly and drops off at the boundaries because of neutron leakage. It is common practice to extrapolate the flux distribution to zero at some point outside the core, called the extrapolated core boundary. Using the simple approximation of diffusion theory from reactor physics, in which the neutrons are treated as gas atoms, the neutron-flux distributions can be obtained for some simple bare reactor geometries.

For an infinite slab reactor, as shown in Fig. 5.14, the neutron flux as a function of x is given by the following equation:

$$\phi(x) = \phi_{\max} \cos\left(\frac{\pi x}{a_0}\right) \tag{5.1}$$

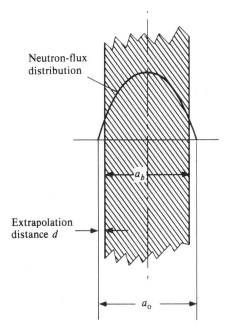

Neutron-flux distribution

a_b

Extrapolation distance d

a_0

Figure 5.14 The neutron-flux distribution in an infinite, bare, uniformly loaded slab reactor.

where a_0 is the extrapolated core boundary as shown in Fig. 5.14; a_0 is equal to the actual core dimension a_b plus twice the extrapolation distance d. Since the power and neutron-flux distributions for a uniformly loaded reactor are the same, the power distribution for an infinite bare slab reactor is

$$P(x) = P_{max} \cos\left(\frac{\pi x}{a_0}\right) \tag{5.2}$$

For a bare, infinite, uniformly loaded cylindrical reactor, whose cross section is shown in Fig. 5.15, the neutron flux, as a function of r, is given by the following equation:

$$\phi(r) = \phi_{max} J_0\left(2.405 \frac{r}{r_0}\right) \tag{5.3}$$

where J_0 is the zeroth-order Bessel function of the first kind and r_0 is the extrapolated core radius. In this case, the extrapolated core radius r_0 is equal to the sum of the reactor core radius r_b and the extrapolation distance d. The properties of the zeroth-order Bessel function of the first kind are such that $J_0(0) = 1.0$ and $J_0(2.405) = 0.0$. The power distribution has the same general equation and is equal to

$$P(r) = P_{max} J_0\left(2.405 \frac{r}{r_0}\right) \tag{5.4}$$

For a uniformly loaded, bare spherical reactor, the neutron-flux distribution is similar to that shown in Fig. 5.15 and is given by the following equation:

$$\phi(r) = \frac{A_0}{r} \sin\left(\frac{\pi r}{r_0}\right) \tag{5.5}$$

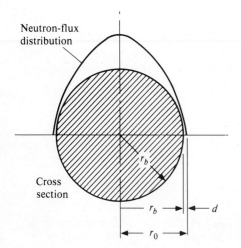

Neutron-flux distribution

Cross section

Figure 5.15 The neutron-flux distribution in an infinite, bare, uniformly loaded cylindrical reactor.

where A_0 is a constant but it is not the maximum flux and r_0 is the extrapolated core radius or the actual radius r_b plus the extrapolation distance d. The value of A_0 is obtained by setting the value of the flux at $r = 0$ to the maximum neutron flux in the core ϕ_{max}:

$$\phi(\text{at } r = 0) = \phi_{max} = \lim_{r \to 0} \frac{A_0}{r} \sin\left(\frac{\pi r}{r_0}\right) = \frac{0}{0} \quad \text{(indeterminate)}$$

Applying L'Hospital's rule,

$$\phi(\text{at } r = 0) = \phi_{max} = \lim_{r \to 0} \frac{A_0 \pi}{r_0} \cos\left(\frac{\pi r}{r_0}\right) = \frac{A_0 \pi}{r_0} \tag{5.6}$$

or

$$\phi_{max} = \frac{A_0 \pi}{r_0} \quad \text{and} \quad A_0 = \frac{\phi_{max} r_0}{\pi} \tag{5.7}$$

Substituting Eq. (5.7) into Eq. (5.5) gives

$$\phi(r) = \frac{\phi_{max} r_0}{\pi r} \sin\left(\frac{\pi r}{r_0}\right) \tag{5.8}$$

Since the power and flux distributions are the same in a uniformly loaded reactor,

$$P(r) = \frac{P_{max} r_0}{\pi r} \sin\left(\frac{\pi r}{r_0}\right) \tag{5.9}$$

The reactor design engineer needs the ratio of P_{max} to P_{ave} in order to evaluate the maximum temperature and/or heat flux in the reactor core. In order to determine this power ratio, the average power P_{ave} must be obtained over the volume of the reactor core. The average power in the core can be found from the following equation:

$$P_{ave} = \frac{\displaystyle\int_{vol} P(V)\, dV}{\displaystyle\int_{vol} dV} \tag{5.10}$$

For the infinite slab reactor, $V = Ax$ and $dV = A\, dx$, where A is the cross-sectional area of the slab (in this case it is infinity). Substituting the expression for dV into Eq. (5.10), the area cancels and Eq. (5.10) becomes

$$P_{ave} = \frac{\displaystyle\int_{-a_b/2}^{a_b/2} P_{max} \cos(\pi x/a_0)\, dx}{\displaystyle\int_{-a_b/2}^{a_b/2} dx} \tag{5.11}$$

Integrating this equation and substituting the limits gives

$$\frac{P_{max}}{P_{ave}} = \frac{\pi a_b/2a_0}{\sin(\pi a_b/2a_0)} \tag{5.12}$$

The maximum possible value (the upper limit) of the power ratio, $(P_{max}/P_{ave})_{max}$, occurs when the extrapolation distance d is small compared to the core dimension a_b. When $a_b \to a_0$, Eq. (5.12) reduces to

$$\left(\frac{P_{max}}{P_{ave}}\right)_{max} \text{(infinite bare slab)} = \frac{\pi}{2} = 1.57 \qquad (5.13)$$

While the power engineer will never encounter an infinite slab reactor, the same results can be used to estimate the maximum-to-average power ratio for a bare, rectangular parallelepiped reactor. In this system, it is common practice to assume that the neutron-flux and the power distributions are variable separable, or

$$\phi(x, y, z) = \phi_{max} \cos\left(\frac{\pi x}{a_0}\right) \cos\left(\frac{\pi y}{b_0}\right) \cos\left(\frac{\pi z}{c_0}\right) \qquad (5.14)$$

and

$$P(x, y, z) = P_{max} \cos\left(\frac{\pi x}{a_0}\right) \cos\left(\frac{\pi y}{b_0}\right) \cos\left(\frac{\pi z}{c_0}\right) \qquad (5.15)$$

where a_0, b_0, and c_0 are the respective extrapolated core dimensions in the x, y, and z directions. Integrating either of these last two expressions over the extrapolated core boundaries in each direction gives a factor of 1.57 in each direction:

$$\left(\frac{\phi_{max}}{\phi_{ave}}\right)_{max \text{(parallelepiped)}} = \left(\frac{P_{max}}{P_{ave}}\right)_{max} = (1.57)^3 = 3.87 \qquad (5.16)$$

For an infinite bare cylindrical reactor, $V = \pi r^2 L$ or $dV = 2\pi r L \, dr$. Substituting this expression into Eq. (5.10) along with the power distribution given by (5.4) gives

$$P_{ave} = \frac{\int_0^{r_b} P_{max} J_0(2.405r/2r_0) 2\pi r L \, dr}{\int_0^{r_b} 2\pi r L \, dr} \qquad (5.17)$$

Integrating this equation and substituting the limits yields

$$\frac{P_{max}}{P_{ave}} = \frac{2.405 r_b / 2 r_0}{J_1(2.405 r / r_0)} \qquad (5.18)$$

The maximum-to-average power ratio occurs when the extrapolation distance is negligible and $r_b \to r_0$. In this case, Eq. (5.18) reduces to

$$\left(\frac{P_{max}}{P_{ave}}\right)_{max \text{(cylinder)}} = \frac{2.405/2}{J_1(2.405)} = \frac{1.2025}{0.5191} = 2.32 \qquad (5.19)$$

For a bare, uniformly loaded right circular cylinder, the neutron flux and power are assumed to be variable and separable, so that

$$\phi(r, z) = \phi_{max} J_0 \left(2.405 \frac{r}{r_0} \right) \cos \left(\frac{\pi z}{c_0} \right) \tag{5.20}$$

and

$$P(r, z) = P_{max} J_0 \left(2.405 \frac{r}{r_0} \right) \cos \left(\frac{\pi z}{c_0} \right) \tag{5.21}$$

Integrating either of these two expressions over all the extrapolated core boundaries ($r_b \rightarrow r_0$ and $c_b \rightarrow c_0$) gives

$$\left(\frac{\phi_{max}}{\phi_{ave}} \right)_{max} = \left(\frac{P_{max}}{P_{ave}} \right)_{max} = \left(\frac{P_{max}}{P_{ave}} \right)_{max, \, cyl} \left(\frac{P_{max}}{P_{ave}} \right)_{max, \, slab}$$

$$= (2.32)(1.57) = 3.64 \tag{5.22}$$

This value is lower (and better) than the value of 3.87 obtained for the rectangular parallelepiped reactor.

For a bare, uniformly loaded spherical reactor, $V = 4\pi r^3/3$ and $dV = 4\pi r^2 \, dr$. Substituting this expression for dV into Eq. (5.10) along with Eq. (5.9) for the power distribution of a sphere as a function of r gives

$$P_{ave} = \frac{\displaystyle\int_0^{r_b} (P_{max} r_0/\pi r) \sin (\pi r/r_0) 4\pi r^2 \, dr}{\displaystyle\int_0^{r_b} 4\pi r^2 \, dr} \tag{5.23}$$

Integrating the equation by parts and substituting the limits yields

$$\frac{P_{max}}{P_{ave}} = \frac{(\pi r_b/r_0)^2}{3[-\cos (\pi r_b/r_0) + (r_0/\pi r_b) \sin (\pi r_b/r_0)]} \tag{5.24}$$

The maximum possible power ratio occurs when the extrapolation distance is negligible and $r_b \rightarrow r_0$:

$$\left(\frac{P_{max}}{P_{ave}} \right)_{max(sphere)} = \frac{\pi^2}{3} = 3.29 \tag{5.25}$$

For a bare, uniformly loaded sphere, the value of 3.29 given by Eq. (5.25) is considerably lower and better than the values for either the right, circular cylinder or the rectangular parallelepiped.

Again, it should be emphasized that the maximum power ratios, as given by Eqs. (5.16), (5.22), and (5.25) are indeed maximum values and the actual ratio of maximum-to-average power will be significantly less. For instance, if $r_b = 0.9r_0$ for the bare, spherical system, the value of P_{max}/P_{ave} becomes 2.513 instead of 3.29.

Example 5.1 If the maximum thermal-neutron flux in a 4-m-diameter bare, uniformly loaded sphere is 10^{17} n/m²·s and the flux at the core boundary is 2×10^{16} n/m²·s, find the average thermal-neutron flux and the extrapolation distance d.

SOLUTION

Given: $\phi_{max} = 10^{17}\,n/m^2 \cdot s$, $\phi(at\ r = r_b) = 2 \times 10^{16}\,n/m^2 \cdot s$, $r_b = 2\,m$

$$\phi(at\ r = r_b) = \frac{\phi_{max}r_0}{\pi r_b} \sin\left(\frac{\pi r_b}{r_0}\right) = 0.2 \times 10^{17}$$

$$= 10^{17}\left(\frac{r_0}{\pi r_b}\right)\sin\left(\frac{\pi r_b}{r_0}\right)$$

$$\frac{\sin\left(\pi r_b/r_0\right)}{\left(\pi r_b/r_0\right)} = 0.2 \qquad \text{(a transcendental equation)}$$

Since $r_0 > r_b$, $\pi r_b/r_0 < \pi$ rad

By trial and error, $\pi r_b/r_0 = 2.595$ rad

$r_0 = r_b\,\pi/2.595 = 2\pi/2.595 = 2.421\,m$

Extrapolation distance $= d = r_0 - r_b = 2.421 - 2.0 = 0.421\,m$

$$\frac{P_{max}}{P_{ave}} = \frac{\phi_{max}}{\phi_{ave}} = \frac{(\pi r_b/r_0)^2}{3[-\cos{(\pi r_b/r_0)} + (r_0/\pi r_b)\sin{(\pi r_b/r_0)}]}$$

$$= \frac{(2.595)^2}{3[-\cos{(2.595)} + (1/2.595)\sin{(2.595)}]}$$

$$= 2.128$$

Average thermal-neutron flux $= \phi_{ave} = \dfrac{\phi_{max}}{2.128} = 4.699 \times 10^{16}\,n/m^2 \cdot s$

Barring coolant temperature changes, the design engineer desires a maximum-to-average power ratio of unity in the reactor core to achieve a uniform maximum temperature or heat flux. If the coolant temperature increases significantly as it passes through the core, it is desirable to increase the power density in the cold part of the reactor, if possible. If the radial power distribution cannot be shaped to achieve a uniform maximum temperature, it may be desirable to orifice the coolant flow so that those elements operating at higher than average power receive a higher coolant flow than those operating at lower power densities.

The maximum-to-average power ratio in a nuclear reactor can be improved (decreased) by putting a reflector around a bare core. The reflector essentially reduces the neutron leakage by returning neutrons to the reactor core as the result of neutron scattering. Not only does the reflector return leakage neutrons to the core but it also reduces the kinetic energy of the fast and intermediate neutrons. This makes these neutrons more effective in inducing fission and producing fission energy at the outer regions of the core.

The effect of adding a reflector to a bare thermal reactor is shown in Fig. 5.16. It will be noted that the reflector is very effective in raising the neutron

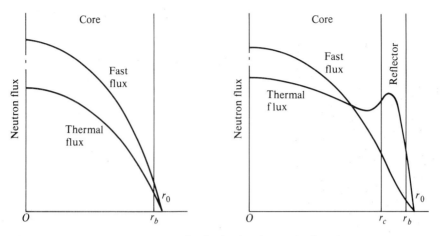

Figure 5.16 The thermal neutron-flux distribution in a bare and reflected reactor core.

flux and hence the power density at the edges of the core. In some systems, the addition of a reflector may actually cause the peak power to occur at the edge of the reactor core rather than at the center of the reactor.

The addition of a reflector to a bare reactor core accomplishes two things. It reduces the critical mass of the reactor by decreasing the neutron leakage rate. If the same fuel concentration is used the reactor can be smaller, or if the same dimensions are maintained the fuel concentration or enrichment can be lowered. The addition of a reflector also causes a marked improvement in the ratio of maximum-to-average power in the core by raising the neutron flux and hence the power generation at the core boundary. A typical value of P_{max}/P_{ave} for a reflected assembly is about 2.4.

5.3.2 Fuel-Element Temperatures and Heat Fluxes

General Once the power distribution in the reactor core has been determined, it is a relatively simple matter to estimate the maximum temperatures and heat flux in the core. Two basic types of fuel elements are commonly employed in nuclear reactors. The flat-plate element is used in many research reactors and the pellet-in-tube element is used in most power reactors. Both elements are treated in this section.

In the design of any fuel element, it is normally assumed that all of the fission energy is generated in the fuel material and that it is distributed uniformly throughout the volume of the fuel material in a given plate or tube. This is actually a conservative assumption as the heat-generation rate in the fuel is highest at the outer edges of the fuel plate or tube because of the neutron-flux depression in the fuel material due to its high neutron-absorption cross section. Assuming that the volumetric heat-generation rate q' is uniform a given fuel element gives higher temperatures than actually occur.

Normally, the maximum temperature and maximum heat flux are found at the location of the maximum heat-generation rate or power density. The maximum temperatures may occur downstream of the location of the maximum power density if the coolant undergoes a significant temperature rise as it passes through the core. The maximum volumetric heat-generation rate in the core q' is given by the following equations:

$$q' = \left(\frac{P_{th}}{V_{fuel_m}}\right)\left(\frac{P_{max}}{P_{ave}}\right) \quad kW/m^3 \tag{5.26}$$

or

$$q' = \left(3413\frac{P_{th}}{V_{fuel_e}}\right)\left(\frac{P_{max}}{P_{ave}}\right) \quad Btu/h \cdot ft^3 \tag{5.27}$$

where P_{th} is the thermal power of the reactor in kilowatts, V_{fuel_m} is the volume of the fuel material in cubic meters, and V_{fuel_e} is the fuel volume in cubic feet.

Once the internal heat-generation rate has been determined, the temperature distribution in the fuel element can be determined with the use of the heat conduction equation and the convection heat-transfer relationship. For constant thermal conductivity k, the heat conduction equation is

$$\nabla^2 T + \frac{q'}{k} = \frac{1}{\alpha}\frac{\partial T}{\partial t} \tag{5.28}$$

where T is the temperature, α is the thermal diffusivity of the material, t is the time, and ∇^2 is the Laplacian operator. Since the design engineer is normally concerned with steady-state values, $\partial T/\partial t = 0$ and Eq. (5.28) reduces to

$$\nabla^2 T + \frac{q'}{k} = 0 \tag{5.29}$$

For cartesian coordinates,

$$\nabla^2 T = \frac{\partial^2 T}{\partial x^2} + \frac{\partial^2 T}{\partial y^2} + \frac{\partial^2 T}{\partial z^2} \tag{5.30}$$

For cylindrical coordinates,

$$\nabla^2 T = \frac{\partial^2 T}{\partial r^2} + \frac{1}{r}\frac{\partial T}{\partial r} + \frac{1}{r^2}\frac{\partial^2 T}{\partial \theta^2} + \frac{\partial^2 T}{\partial z^2} \tag{5.31}$$

For spherical coordinates,

$$\nabla^2 T = \frac{d^2 T}{dr^2} + \frac{2}{r}\frac{\partial T}{\partial r} + \frac{1}{r^2}\frac{\partial^2 T}{\partial \theta^2} + \frac{\cos\theta}{r^2\sin\theta}\frac{\partial T}{\partial \theta} + \left(\frac{1}{r^2}\sin\theta\right)\frac{\partial^2 T}{\partial \psi^2} \tag{5.32}$$

Plate-type fuel elements The plate-type or slab fuel element is composed of a thin slab of fuel material clad on each side with structural material or cladding. The structural material is metallurgically bonded to the fuel material so that there is a very small temperature drop across the fuel-clad interface. The fuel

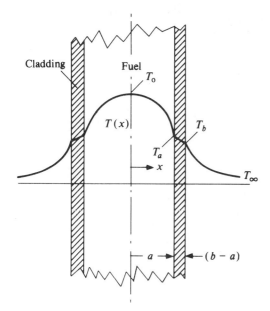

Figure 5.17 Temperature distribution in a plate-type fuel element.

plates are thin enough so that a single plate can be approximated by a one-dimensional infinite slab as pictured in Fig. 5.17. The origin of the axis is established at the centerline of the fuel and the values of $x = \pm a$ mark the fuel-clad boundaries, and the values of $x = \pm b$ mark the outer boundaries of the cladding material.

In the fuel material of the infinite slab, Eq. (5.29) reduces to

$$\frac{d^2 T}{dx^2} + \frac{q'}{k_f} = 0 \qquad (5.33)$$

Assuming a symmetrical temperature distribution in the fuel material (maximum temperature at the center of the fuel) gives the following relationship in the fuel:

$$T_x - T_a = \frac{q'(a^2 - x^2)}{2k_f} \qquad (5.34)$$

where T_a is the temperature of the fuel at the fuel-clad interface and k_f is the thermal conductivity of the fuel material. Thus the temperature at the center of the fuel T_0 is the maximum temperature and is equal to

$$T_0 - T_a = \frac{q' a^2}{2k_f} \qquad (5.35)$$

The steady-state heat-transfer rate q through the fuel-clad interface, the cladding, and the coolant film is a constant and is equal to

$$q = -k_f A \left(\frac{dT}{dx}\right)_{\text{at } x = a} = aAq' \qquad \text{W or Btu/h} \qquad (5.36)$$

where A is the surface area normal to the flow of heat. Using this equation, along with the assumption that there is no heat generation in the cladding, gives

$$T_a - T_b = \frac{q(b-a)}{Ak_c} = \frac{q'a}{k_c}(b-a) \tag{5.37}$$

where T_b is the surface temperature of the element and k_c is the thermal conductivity of the cladding material. Again, employing Eq. (5.36), along with the Newton equation for convective heat transfer, gives the following temperature drop across the fluid film:

$$T_b - T_\infty = \frac{q}{hA} = \frac{q'a}{h} \tag{5.38}$$

where T_∞ is the bulk temperature of the coolant and h is the surface conductance of the coolant.

The total temperature drop from the center of the fuel to the coolant is obtained by adding Eqs. (5.35), (5.37), and (5.38). This gives

$$T_0 - T_\infty = q'a\left(\frac{a}{2k_f} + \frac{b-a}{k_c} + \frac{1}{h}\right) \tag{5.39}$$

Cylindrical fuel elements The cylindrical fuel element shown in Fig. 5.18 is commonly employed in most of the current power reactors. The fuel rods are less than $\frac{1}{2}$ in in diameter and over 12 ft long. Consequently, the heat flow from

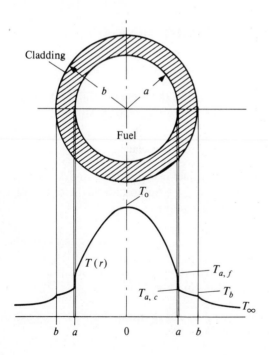

Figure 5.18 Temperature distribution in a cylindrical fuel element.

the fuel rods can be approximated by a one-dimensional, infinite cylinder. For a one-dimensional cylinder, the steady-state heat conduction equation (5.29) reduces to

$$\frac{d^2 T}{dr^2} + \frac{1}{r}\frac{dT}{dr} = \frac{-q'}{k_f}$$ (5.40)

The heat flow through the cylindrical fuel element is complicated by an additional thermal resistance term that is not encountered in the slab element. In the slab element, the fuel and cladding are metallurgically bonded to each other in a hot rolling process that produces a very low thermal resistance. In the fabrication of the cylindrical elements, on the other hand, the ceramic fuel pellets are placed inside the cladding tube. This means that there is a small gap between the pellet and the wall of the cladding. This gap produces a significant thermal resistance because of the poor contact between the two surfaces. For the purpose of this study, it will be assumed that the heat flow through this contact resistance can be found by using the Newton heat-flow equation and employing an effective surface conductance h for the flow of heat. Despite the gap that exists between the pellet and the tube, and the actual gap changes as the reactor operates over a long period of time, it will be assumed that the outer radius of the fuel pellet and the inner radius of the tube are both equal to a.

Equation (5.40) can be integrated twice to determine the temperature drop in the fuel pellet. For the condition of symmetrical cooling, this equation gives

$$T_r - T_{a,f} = \frac{q'(a^2 - r^2)}{4k_f}$$ (5.41)

where T_r is the temperature at any fuel radius and $T_{a,f}$ is the temperature at the surface of the fuel pellet. The maximum temperature drop in the fuel pellet is then obtained by setting $r = 0$. This gives

$$T_0 - T_{a,f} = \frac{q'a^2}{4k_f}$$ (5.42)

At steady state, all of the heat that is generated in the fuel pellets must be transferred through the contact resistance, through the cladding and then to the coolant. Since it is assumed that all of the fission energy is generated in the fuel, the standard heat-transfer equations are applicable outside the fuel material. The steady-state heat transfer is equal to the product of the fuel volume and the volumetric heat-generation rate q':

$$q = q'\pi a^2 l$$ (5.43)

Applying Eq. (5.43) to the Newton convection equation gives the temperature drop between the outer surface of the fuel pellet $T_{a,f}$ and the inside surface of the cladding $T_{a,c}$:

$$T_{a,f} - T_{a,c} = \frac{q}{2\pi a l h_a} = \frac{q'a}{2h_a}$$ (5.44)

Substituting Eq. (5.33) into the equation for conduction heat flow through a cylindrical wall gives the following equation for the temperature drop through the tube wall with an inner radius of a and an outer radius of b:

$$T_{a,c} - T_b = \frac{q \ln (b/a)}{2\pi k_c l} = \frac{q'a^2}{2k_c} \ln \frac{b}{a} \tag{5.45}$$

The temperature drop across the coolant film can be obtained in the same manner as Eq. (5.34):

$$T_b - T_\infty = \frac{q}{2\pi b l h_b} = \frac{q'a^2}{2bh_b} \tag{5.46}$$

The maximum temperature drop across the cylindrical fuel rod is obtained by adding Eqs. (5.42), (5.44), (5.45), and (5.46). This gives

$$T_0 - T_\infty = \frac{q'a^2}{2} \left(\frac{1}{2k_f} + \frac{1}{ah_a} + \frac{1}{k_c} \ln \frac{b}{a} + \frac{1}{bh_b} \right) \tag{5.47}$$

If the maximum fuel temperature $T_{0,\max}$ is desired, it can be estimated from either (5.39) or (5.47) along with the exit coolant temperature $T_{\infty,\max}$. In some systems, particularly the boiling-water reactors, the maximum heat flux at the surface of the cylindrical fuel element is the condition that limits the reactor power. This quantity can be obtained by dividing the maximum heat-transfer rate in the core by the surface area of the cladding. This gives

$$\left(\frac{q}{A} \right)_{\max} = \frac{q'\pi a^2 l}{2\pi b l} = \frac{q'a^2}{2b} \tag{5.48}$$

Example 5.2 A cylindrical fuel element in a BWR is composed of zircoloy-2 tubes filled with UO_2 pellets. The tubes have an OD of 12.5 mm and an ID of 10.8 mm. If the thermal conductivity of the fuel and cladding are 1.8 and 14.0 W/m·°C, respectively, the thermal conductance of the pellet-clad interface is 5500 W/m²·°C, the surface temperature is 520°C, and the heat flux at the cladding surface is 950,000 W/m², find the maximum temperature in the fuel pellet and the average thermal-neutron flux in the fuel. Assume the UO_2 has a density of 10 g/cm³, a U-235 enrichment of 2.2%, and assume that the average coolant temperature is 305°C.

SOLUTION

$a = 10.8/2 = 5.4$ mm $= 0.0054$ m $\qquad b = 0.0125/2 = 0.00625$ m

Heat-transfer rate through the clad $= q = \left(\dfrac{q}{A} \right) A = \left(\dfrac{q}{A} \right)_b (2\pi b l)$

$q/l = 950,000(2\pi)(0.00625) = 37,306$ W/m

Volume of fuel $= \pi a^2 l = (0.0054)^2 \pi l = 0.0000916 l$ m^3 $= V_{fuel_m}$

$$q' = \frac{q/l}{V_{fuel}/l} = \frac{37{,}306}{0.0000916} = 4.072 \times 10^8 \text{ W/m}^3$$

$$T_0 - T_b = \frac{q'a}{2}\left[\frac{a}{2k_f} + \frac{1}{h_a} + \frac{a(\ln b/a)}{k_c}\right]$$

$$k_f = 1.8 \text{ W/m·°C} \qquad h_a = 5500 \text{ W/m}^2\text{°C} \qquad k_c = 14.0 \text{ W/m·°C}$$

$$T_0 - T_b = (4.072 \times 10^8)\left(\frac{0.0054}{2}\right)$$

$$\times \left[\frac{0.0054}{2(1.8)} + \frac{1}{5500} + 0.0054\frac{\ln 0.00625/0.00540}{14}\right]$$

$$= (1.099 \times 10^6)(0.0015 + 0.0001818 + 0.0000564)$$

$$= (1.099 \times 10^6)(0.0017382) = 1910 \text{ Celsius degrees}$$

$$T_0 = T_b + 1910 = 520 + 1910 = 2430\text{°C}$$

For a unit meter of tubing:

$$\text{Fission rate} = \bar{\sigma}_f N_{fuel}\bar{\phi}_{th}$$

$$= (3.1 \times 10^{10} \text{ fissions/W·s})(37{,}306 \text{ W})$$

$$= 1.1565 \times 10^{15} \text{ fissions/s}$$

Average U-235 fission cross section $= \bar{\sigma}_f$

$$= \frac{582[293/(305 + 273)]^{1/2}}{1.128} = 367.4 \text{ barns} = 3.674 \times 10^{-26} \text{ m}^2$$

$$N_{fuel} = \frac{(V_{fuel})(\rho_{fuel})[238/(238 + 32)](\text{Avagadro's number})(\text{enrichment})}{235}$$

$$= \frac{(91.6 \text{ cm}^3)(10 \text{ g/cm}^3)(238/270)(0.6023 \times 10^{24})(0.022)}{235}$$

$$= 4.553 \times 10^{22} \text{ atoms}$$

$$\bar{\phi}_{th} = \frac{1.1565 \times 10^{15}}{(4.553 \times 10^{22})(367.4 \times 10^{-28})}$$

$$= 6.9140 \times 10^{17} \text{ n/m}^2\text{·s}$$

5.4 REACTOR OPERATIONS

5.4.1 Reactor Kinetics

A term commonly encountered in the operation of any nuclear reactor is the reactor period t_r. The reactor period is defined as the time required for the reactor power or the neutron flux to *increase* by a factor of e, the base of natural logarithms. In a supercritical reactor, the multiplication factor k is greater than unity, the reactivity ρ is positive, and the reactor period t_r is positive. In a critical reactor, the multiplication factor is exactly unity, the reactivity is zero, and the reactor period is equal to infinity. In a subcritical reactor, the multiplication factor is less than unity, the reactivity is negative, and the reactor period is negative.

An important parameter in the determination of the reactor period is the lifetime of the average neutron in the reactor because this is also equal to the average generation time for the neutrons in the chain reaction. The average-neutron lifetime is defined as the time between the instant that the fission occurs that produces the neutron until this average neutron is lost from the reactor core as the result of absorption or leakage. In a thermal reactor system, the lifetime of the thermal neutrons is equal to the sum of the production time l_{pr}, the slowing-down time l_s, and the thermal-diffusion time l_{th}, or

$$\text{Average-neutron lifetime} = \bar{l} = l_{pr} + l_s + l_{th} \tag{5.49}$$

The production time l_{pr} is the time from the instant that the fission occurs until the fission neutron appears. For most of the reactor neutrons, the production time is very short, within 10^{-14} s. These neutrons are called prompt neutrons and, because of the extremely short production time, this time is usually neglected when calculating the average lifetime. A small number of fission neutrons, called delayed neutrons, have a long production time because they are produced as the result of radioactive decay of certain fission products. These fission products are halogen atoms (isotopes of iodine and bromine) and are called delayed-neutron precursors. Although these neutrons comprise less than 1 percent of all the fission neutrons, the production time of these neutrons is so long that they are very important in evaluating the average neutron lifetime in the reactor.

The slowing-down time l_s is relatively short, ranging from around 10^{-5} s in a large graphite or heavy-water moderated reactor to less than 10^{-8} s in a small fast reactor. This time is normally neglected in most thermal reactor systems because it is several orders of magnitude lower than the thermal-diffusion time. In a fast reactor, however, the average lifetime of prompt neutrons is essentially the same as the slowing-down time since the production time is very short and the thermal-diffusion time is zero.

The thermal-diffusion time l_{th} is the average time that the thermal neutrons spend in a thermal reactor until they are either absorbed or they leak from the core. This time ranges from about 10^{-4} s for LWR systems to around 10^{-3} s for the large graphite or heavy-water moderated systems. If the thermal-neutron leakage is small, the thermal-diffusion time in a homogeneous assembly is equal to

$$l_{th} = \frac{\lambda_a}{v_{th}} = \frac{1}{\Sigma_a v_{th}} \tag{5.50}$$

where Σ_a is the total macroscopic neutron-absorption cross section and v_{th} is the average velocity of the thermal neutrons.

For prompt neutrons, the average lifetime is called the prompt-neutron lifetime l_p. Since l_{pr} is very small, the prompt-neutron lifetime for fast reactors is equal to l_s, while that for thermal reactors is approximately equal to l_{th}. For fast-reactor systems,

$$10^{-8} < l_p < 10^{-7} \text{ s}$$

For thermal-reactor systems,

$$10^{-4} < l_p < 10^{-3} \text{ s}$$

The lower values in each case represent values for small, highly loaded reactors where the leakage and/or the absorption cross section is high.

For the delayed neutrons, the delayed-neutron lifetime l_d is approximately equal to the production time because it is much longer than either the slowing-down time or the thermal-diffusion time. The production time of these neutrons is equal to the average lifetime of the precursor isotopes. There are six groups of delayed-neutron precursors with half-lives ranging from a minimum of 0.23 s to a maximum of 55.72 s. According to Eq. (2.23), the average lifetime of any radioisotope is equal to the reciprocal of the decay constant of the isotope. Thus, the production time for the ith group of delayed neutrons is

$$l_{d_i} = l_{pr_i} = \frac{(T_{1/2})_i}{\ln 2} \tag{5.51}$$

where $(T_{1/2})_i$ is the half-life of the ith group of delayed-neutron precursors. The production times and hence the delayed-neutron lifetimes range from 0.332 s for the first or shortest-lived group to 80.4 s for the sixth or longest-lived group.

Table 5.5 lists the average kinetic energy, the half-lives of the delayed-neutron precursors and the fraction β_i of the total fission neutrons produced by each of the delayed-neutron precursors. These values are tabulated for each of the three common fuel isotopes. It will be noted that uranium-235 has the largest delayed-neutron fraction while plutonium-239 has the smallest fraction.

The average lifetime of all the fission neutrons can be estimated by summing the products of the fraction of each group of neutrons and their respective average lifetimes. Since the total fraction of delayed neutrons is equal to β, where β is the summation of β_i over all six delayed-neutron groups, the fraction of fission neutrons that are prompt neutrons is $(1 - \beta)$. The average lifetime of all the fission neutrons is given by the following equation:

$$\bar{l} = (1 - \beta)l_p + \sum_{i=1}^{6} \beta_i l_{d_i} = (1 - \beta)l_p + \sum_{i=1}^{6} \frac{\beta_i (T_{1/2})_i}{\ln 2} \tag{5.52}$$

Table 5.5 Delayed-neutron parameters for the common fuel isotopes

Group, i	Energy for uranium-235 fission, MeV	Half-life, s	Total neutrons from fission, %		
			Uranium-235	Uranium-233	Plutonium-239
Prompt	≈ 2.00	< 0.001	99.359	99.736	99.790
1	0.25	55.72	0.021	0.023	0.007
2	0.56	22.72	0.140	0.079	0.063
3	0.43	6.22	0.126	0.066	0.044
4	0.62	2.30	0.253	0.073	0.069
5	0.42	0.61	0.074	0.014	0.018
6		0.23	0.027	0.009	0.009
	Total delayed-neutron fraction =		0.641	0.264	0.210

The maximum value of $(1 - \beta)l_p$ is less than 0.001 s while the value of the second term or the summation of $\beta_i(T_{1/2})_i/(\ln 2)$ ranges from 0.0832 s for uranium-235 to 0.0327 s for plutonium-239. Thus, the delayed neutrons are very effective in increasing the average-neutron lifetime even though there are very few of them. In fact, the average-neutron lifetimes are essentially independent of the prompt-neutron lifetimes and depend only on the type of fuel used in the reactor. As will be discussed later, however, there are some modes of operation where the prompt-neutron lifetime is very important.

If it is assumed that all the neutrons have one lifetime, it is relatively simple to develop an equation relating the transient response of the reactor to the reactor multiplication factor k or the system reactivity ρ. The average-neutron lifetime is also the average-neutron generation time since a fission cannot occur until the neutron is absorbed in the fission process.

The rate of change of neutrons with time, dn/dt, is

$$\frac{dn}{dt} = \frac{n_1 - n}{l} = n\left(\frac{n_1}{n} - 1\right)\Big/l \tag{5.53}$$

where n_1 is the number of neutrons produced in one generation and n is the number produced in the preceding generation. The ratio of n_1/n is the definition of the multiplication factor k. Since $(k - 1)$ is approximately equal to the reactivity ρ for values of k close to unity, Eq. (5.53) reduces to

$$\frac{dn}{dt} = \frac{n(k - 1)}{l} = \frac{n\rho}{l} \tag{5.54}$$

For small step changes in k or ρ, this equation can be integrated to yield

$$n = n_0 e^{\rho t/l} \tag{5.55}$$

where n is the number of neutrons at some arbitrary time t and n_0 is the number that was present at some time, $t = 0$. The neutron population n in the reactor core is directly proportional to the neutron flux and also to the reactor power. Consequently, Eq. (5.55) may also be written as either

$$\phi = \phi_0 e^{(k-1)t/\bar{l}} = \phi_0 e^{\rho t/\bar{l}} \tag{5.56}$$

or

$$P = P_0 e^{(k-1)t/\bar{l}} = P_0 e^{\rho t/\bar{l}} \tag{5.57}$$

In practice, Eqs. (5.56) and (5.57) have a fairly limited range of application because all the fission neutrons do not have the same lifetime. In a uranium-235-fueled reactor, these equations are applicable for reactivity values in the range of ± 0.001. The reactor period t_r is defined as the time required for the power or neutron flux to increase by a factor of e. Substituting this definition into the previous equations yields

$$n = n_0 e^{t/t_r} \quad \text{or} \quad \phi = \phi_0 e^{t/t_r} \quad \text{or} \quad P = P_0 e^{t/t_r} \tag{5.58}$$

where

$$t_r = \frac{\bar{l}}{\rho} \tag{5.59}$$

As mentioned above, these equations are applicable for uranium-235-fueled reactors ($\bar{l} = 0.0832$ s) for step changes in reactivity ranging from -0.001 to $+0.001$. Using the upper limit of $\rho = +0.001$ gives a stable reactor period of 83.2 s. A rigorous calculation of the stable period for the same reactivity insertion gives a value of about 60 s.

Figure 5.19 is a plot of the stable reactor period for uranium-235-fueled reactors as a function of a step change of reactivity and also as a function of the prompt-neutron lifetime. A rigorous solution of the reactor transient equations for a step change in reactivity involves the solution of a seventh-order equation

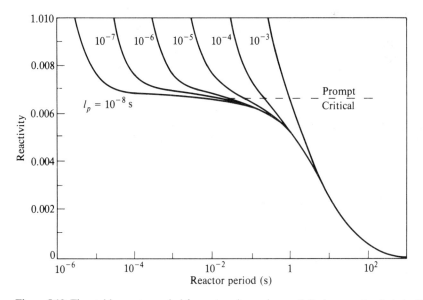

Figure 5.19 The stable reactor period for a step change in reactivity in a reactor fueled with U-235.

to find the seven roots. There is one root for each of the six groups of delayed neutrons plus the prompt-neutron group. These roots, a_i, yield a general flux equation of the following form:

$$\phi(t) = A_0 e^{a_0 t} + A_1 e^{a_1 t} + A_2 e^{a_2 t} + A_3 e^{a_3 t} + A_4 e^{a_4 t} + A_5 e^{a_5 t} + A_6 e^{a_6 t} \quad (5.60)$$

For positive insertions of reactivity, one root, a_0, and one coefficient, A_0, are positive while all the other roots and coefficients are negative. All the terms involving the negative roots decay exponentially with time and the stable period is $1/a_0$. For negative reactivity insertions, all the roots are negative and all the coefficients are positive. For small negative changes, the stable period is $1/a_0$ while the stable period for large negative reactivity changes, such as that which occurs in a reactor scram, is -80.4 s. This corresponds to the lifetime of the longest-lived group of delayed-neutron precursors.

Before considering the response of the actual reactor to step changes in reactivity, consider the importance of the delayed neutrons to reactor control. As was discussed earlier, for a uranium-235-fueled reactor and reactivity values ranging between ± 0.001, the stable reactor period is $t_r = \bar{l}/\rho$. If there were no delayed neutrons, the average-neutron lifetime would be equal to the prompt-neutron lifetime l_p, and the stable and only period becomes $t_r = l_p/\rho$. A small fast reactor with a prompt-neutron lifetime of 10^{-8} s would achieve a reactor period of 10^{-5} s for a step increase in reactivity of $+0.001$. This means that the reactor would go from a thermal power of 1 W to a power of more than 5×10^{15} MW in a 1-s time interval. This response approaches or exceeds that of an atomic bomb. The presence of delayed neutrons increases the actual period to about 60 s, which means that the power actually increases from 1 to 2.718 W in a 1-min interval. If it were not for the presence of delayed neutrons, reactor control would be essentially impossible.

As discussed above, the actual response to a step change in reactivity is difficult to calculate explicitly. However, some general trends can be examined. For a step insertion of positive reactivity into a critical reactor at some arbitrary time, $t = 0$, the power will start to increase with an initial period of $t_r = l_p/\rho$, but then the rate of increase slows until the stable period is attained. This response is shown in the semilog plot of power versus time in Fig. 5.20. The exponential relationship such as that given by Eq. (5.58) plots as a straight line on semilog graph paper.

For large negative step changes in reactivity, such as occur during a reactor scram when the control rods are dropped into the core, there is an initial rapid transient drop in power and then the power eventually decays with a stable period of -80 s. The reactor response to a small step decrease in reactivity is shown in Fig. 5.21.

For large positive values of reactivity insertions, the prompt-neutron lifetime l_p becomes very important. If the amount of positive reactivity is equal to the delayed-neutron fraction or $\beta = \rho$, the multiplication factor k is approximately $(1 + \beta)$. This means that the delayed neutrons could be discarded and the reactor would still be critical. When this happens, the reactor is said to be prompt

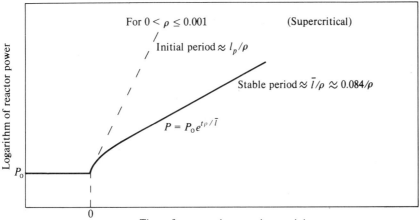

Figure 5.20 Reactor power following a step increase in reactivity in a critical reactor fueled with U-235.

critical because the reactor could be critical on only prompt neutrons. If the system reactivity exceeds β, the stable reactor period approaches

$$t_r < \frac{l_p}{\rho - \beta} \qquad (5.61)$$

The prompt-critical condition is normally avoided in all power reactors but it is especially severe in small fast reactors because of the very short prompt-neutron lifetimes. The super-prompt-critical condition is normally attained in

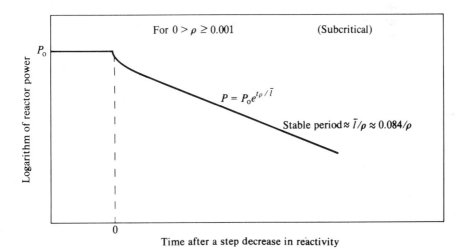

Figure 5.21 Reactor power following a small step decrease in reactivity in a critical reactor fueled with U-235.

the detonation of atomic bombs to achieve a high energy release or "yield" before the assembly is blown apart. There are a few pulsed research reactors that operate above the prompt-critical condition in order to achieve very high power pulses. These systems depend on the temperature increase to reduce the maximum power to tolerable levels before meltdown.

Some opponents of fast reactors point to the response of fast reactor systems to the prompt-critical condition as a major disadvantage. However, it should be pointed out that for small changes in reactivity, the response of fast and thermal reactors are about the same because the delayed neutrons produce about the same value of stable reactor period. Moreover, it is much more difficult to achieve the prompt-critical condition in a fast reactor system as compared to thermal reactors. Actually, the possibility that any reactor system might detonate like an atomic bomb is zero. However, there is a possibility that a fast transient could cause a reactor meltdown or the sudden vaporization of the reactor coolant which could result in considerable damage to the reactor complex.

Throughout this discussion, the delayed-neutron fractions were assumed to be constant for a given fuel, as listed in Table 5.5. Actually, this is true for only heterogeneous reactor systems. In homogeneous reactors where the fuel solution is circulated outside the reactor core, some of the delayed neutrons will be produced in the external piping where they have no effect on the system reactivity. Thus, the effective delayed-neutron fractions and the prompt-critical value of reactivity are somewhat less than the actual delayed-neutron fraction β.

There are several units of reactivity in common use today. The "in-hour" of reactivity, Ih, is a small unit of reactivity and is equal to the amount of reactivity that will put a critical reactor on a stable period of 1 h or 3600 s. The term in-hours is a contraction of "inverse hours," and for small changes in reactivity the stable period is

$$t_r = \frac{3600}{\text{Ih}} \quad \text{s} \quad \text{(for Ih} \leq 50) \tag{5.62}$$

The other common unit of reactivity is the "dollar of reactivity." One dollar of reactivity inserted into a critical reactor will make the reactor prompt critical. For heterogeneous reactors, a dollar's worth of reactivity is equal to the delayed-neutron fraction but it is less than this quantity for a homogeneous reactor. One "cent of reactivity" is equal to one-hundredth of a dollar.

Example 5.3 Find the stable reactor period and the excess reactivity, in percentage of reactivity and in in-hours of reactivity, if five cents of reactivity are inserted into a critical thermal reactor fueled with Pu-239. How long will it take the critical reactor to go from a power of 100 W to 3800 MW after the reactivity insertion? Also evaluate the time required to go from 3800 to 500 MW if the reactor is scrammed and the initial transient is neglected.

SOLUTION Using the values from Table 5.5,

The delayed-neutron fraction for Pu-239 $= \beta = 0.210\% = 0.0021$

$$\text{Five cents of reactivity} = 0.05\beta = 0.05(0.0021) = 0.000105$$

$$\text{Excess reactivity} = \rho = 0.000105 = 0.0105\%$$

Average-neutron lifetime for Pu-239 $= \Gamma$

$$\approx \sum_{i=1}^{6} \frac{\beta_i(T_{1/2})_i}{\ln 2}$$

$$= [55.72(0.00007) + 22.72(0.00063) + 6.22(0.00044) + 2.3(0.00069)$$
$$+ 0.61(0.00018) + 0.23(0.00009)]/(\ln 2)$$

$$= 0.0327 \text{ s}$$

Stable reactor period $= t_r = \dfrac{\Gamma}{\rho} = \dfrac{0.0327}{0.000105} = 311.4 \text{ s}$

In-hours of reactivity $= \dfrac{3600}{t_r} = \dfrac{3600}{311.4} = 11.56 \text{ Ih}$

Time required to go from 100 W to 3800 MW:

$$P = P_0 e^{t/t_r} \qquad \text{or} \qquad t = t_r[\ln (P/P_0)]$$

$$t = 311.4\left(\ln \frac{3800 \times 10^6}{100}\right) = 311.4(\ln 38,000,000)$$

$$= 5435 \text{ s} = 90.58 \text{ min} = 1.51 \text{ h}$$

The time required to go from 3800 to 500 MW if the reactor is scrammed and the initial transient is neglected can be determined as follows. The stable reactor period is equal to the lifetime of the longest-lived group of delayed neutrons:

$$t_r = -\frac{(T_{1/2})_l}{\ln 2} = -\frac{55.72}{\ln 2} = -80.38 \text{ s}$$

$$t = t_r \ln \frac{P}{P_0} = -80.38 \ln \frac{500}{3800} = -80.38(\ln 0.2669)$$

$$= 163 \text{ s} = 2.72 \text{ min}$$

It should be stressed here that this is not a realistic value because the initial transients will significantly reduce the calculated value.

5.4.2 Reactivity Coefficients

Reactivity coefficients are very important in the operation of reactors because they indicate how the system reactivity ρ changes with respect to a change in certain operating parameters such as power, temperature, pressure, etc. Actually, the reactivity coefficient of greatest interest is the power coefficient of reactivity α_P, which gives the rate of change of reactivity with power:

$$\alpha_P = \frac{\partial \rho}{\partial(\text{power})} \qquad (5.63)$$

For ease of control, the power coefficient of reactivity should be large in absolute magnitude and negative in arithmetic sign. With this condition, the reactor will be very stable because an increase in power causes a decrease in reactivity which, in turn, limits the reactor power. A positive power coefficient, on the other hand, gives positive feedback and a diverging power condition. It is usually difficult to predict the value of the power coefficient directly because it is a function of many variables. Instead, some of the other reactivity coefficients are evaluated which are closely related to the power coefficient.

One of the more important reactivity coefficients is the temperature coefficient of reactivity α_T. This coefficient gives the rate of change of reactivity with reactor temperature:

$$\alpha_T = \frac{\partial \rho}{\partial T} \qquad (5.64)$$

As is the case for all reactivity coefficients for ease of reactor control, α_T should be large in magnitude and negative in sign. If α_T is positive, an unstable condition exists because an increase in temperature would cause an increase in reactivity which would, in turn, cause a corresponding increase in power and temperature, producing another increase in reactivity, etc.

An important temperature phenomenon encountered in the operation of some reactors is doppler broadening. Doppler broadening effectively causes a neutron cross-section peak or resonance, commonly found in the intermediate energy range, to decrease in magnitude but increase in energy width. This effect is caused by the relative motion between the incident neutrons and the target nuclei. Above absolute zero, all nuclei in a solid are vibrating about a fixed lattice position and, as the temperature increases, the molecular (nuclear) kinetic energy increases directly with the absolute temperature. If the incident neutron reacts with the target nucleus while they are moving in the same direction, the apparent kinetic energy of the incident neutron will be less than its absolute energy, and if the neutron and target nucleus collide while they are going in opposite directions, the apparent kinetic energy of the reacting neutron will be more than its absolute kinetic energy. Thus, some of the neutrons having energies slightly above or below the resonance energy band will appear to fall within the resonance range due to the relative motion of the target nucleus. The doppler broadening effectively increases the neutron reaction rate with a given nuclear resonance because of an increase in temperature.

The doppler broadening effect can be very important in the control of some reactors, particularly fast reactors. If most of the material resonances are fission resonances, an increase in temperature increases the fission rate because of the doppler broadening of those resonances. The increased fission rate produces a positive temperature coefficient of reactivity and this makes the reactor very difficult to control unless there are other effects, such as fuel expansion, etc., that make the overall temperature coefficient negative.

Typical values of the temperature coefficient of reactivity range from $-10^{-4} \Delta\rho$ per Celsius degree to $-10^{-8} \Delta\rho$ per Celsius degree. Homogeneous reactors normally have values of α_T approaching $-10^{-4} \Delta\rho$ per Celsius degree, because any increase in reactor temperature causes the fuel solution to expand which removes fuel from the reactor core. Fast heterogeneous reactors, on the other hand, commonly have an α_T of about $-10^{-8} \Delta\rho$ per Celsius degree and the doppler coefficient may be a prime component of the overall temperature coefficient of reactivity in these systems.

For operational purposes, the overall temperature coefficient of reactivity is further divided into a prompt component of the temperature coefficient and a delayed component. The prompt temperature coefficient is that portion of the overall temperature coefficient which is associated with fuel effects, such as fuel expansion, doppler broadening, etc. The delayed temperature coefficient is that part of the overall coefficient that is associated with the temperature effects on the coolant, moderator, and other nonfuel materials in the core.

If the prompt temperature coefficient of reactivity is positive, it can cause operational problems during fast positive transients when the reactor period is very short. In fact, the experimental breeder reactor I (EBR-I), which had a negative overall temperature coefficient, actually had an unknown positive prompt coefficient due to bowing of the fuel rods as the result of fuel expansion. During some fast transient tests that were carried out just before the reactor was scheduled for dismantling, the positive prompt temperature coefficient produced an unexpectedly high power excursion that culminated in a core meltdown.

Another important reactivity coefficient in the operation of liquid-cooled reactors is the void coefficient of reactivity α_V. The void coefficient gives the rate of change of reactivity with the percentage of voids in the core:

$$\alpha_V = \frac{\partial\rho}{\partial(\% \text{ voids})} \tag{5.65}$$

Coolant vaporization in any liquid-cooled reactor essentially introduces voids into the reactor core. Thus, the void coefficient of reactivity is particularly important in the normal operation of BWR systems. For stable operation, the void coefficient should be large in absolute magnitude and negative in sign. While this coefficient is particularly important in boiling-water systems, it is also important in the operation of any liquid-cooled reactor. A power excursion in any of these systems can produce localized or general boiling of the coolant. This sudden vaporization of the coolant can aggravate or reduce the power excursion, depending on whether the void coefficient is positive or negative.

5.4.3 Fission-Product Poisoning

General poisoning effects During the operation of any nuclear fission reactor, the system reactivity decreases as the system is operated. This is due to the fact that the fuel is being consumed and also because fission products are accumulating in the core and they absorb some of the neutrons. While all fission products absorb some neutrons and consequently poison the reactor, some of them have extremely high microscopic absorption cross sections and present some interesting operating problems. This section investigates the effect of the buildup of fission products on the reactivity and operation of the reactor. The effect of fuel burnup will be studied in the next general section.

An examination of the modified, one-group critical equation (3.28) shows that the only terms strongly influenced by the buildup of fission-product poisons are the thermal diffusion length L and the thermal utilization factor f. The thermal diffusion length appears in the thermal nonleakage probability function. This term, $(1 + L^2 B^2)$, is the denominator of the critical equation. In most reactors, the thermal-neutron leakage is quite small and $L^2 B^2$ is much smaller than unity. For most systems, this means that the variation of the multiplication factor with respect to the buildup of fission products essentially occurs only in the thermal utilization factor.

The thermal utilization factor is defined as the fraction of thermal neutrons absorbed by the fuel. As fission products accumulate in the core, they absorb some of the thermal neutrons, reducing f and k. The equation for the multiplication factor can be written as follows:

$$k = K_1 f \tag{5.66}$$

where
$$K_1 = \frac{\eta p \varepsilon}{[1 + L^2 B^2] e^{B^2 \tau}}$$

For a "clean" reactor, which is one in which there are no fission products in the core, the total thermal-neutron absorption rate in the reactor core is equal to

$$(\Sigma_a \bar{\phi}_{\text{th}} V)_{\text{total}} = (\Sigma_a \bar{\phi}_{\text{th}} V)_{\text{fuel}} + (\Sigma_a \bar{\phi}_{\text{th}} V)_{\text{mod}}$$
$$+ (\Sigma_a \bar{\phi}_{\text{th}} V)_{\text{cool}} + (\Sigma_a \bar{\phi}_{\text{th}} V)_{\text{rods}}$$
$$+ (\Sigma_a \bar{\phi}_{\text{th}} V)_{\text{clad}} + (\Sigma_a \bar{\phi}_{\text{th}} V)_{\text{struct}} + \cdots \tag{5.67}$$

Lumping the absorptions by the nonfuel materials (moderator, coolant, etc.) into one term and assuming that the average-neutron flux in all the materials is about the same (not a very good assumption in some heterogeneous reactors), the absorption-rate equation (5.67) reduces to

$$\Sigma_{a,\text{ total}} = \Sigma_{a,\text{ fuel}} + \Sigma_{a,\text{ other}} \tag{5.68}$$

The thermal utilization factor for the clean reactor f_c becomes

$$f_c = \frac{\Sigma_{a,\text{ fuel}}}{\Sigma_{a,\text{ total}}} = \frac{\Sigma_{a,\text{ fuel}}}{\Sigma_{a,\text{ fuel}} + \Sigma_{a,\text{ other}}} = \frac{1}{1 + \delta} \tag{5.69}$$

where
$$\delta = \frac{\Sigma_{a,\text{ other}}}{\Sigma_{a,\text{ fuel}}}$$

Substituting Eq. (5.69) into Eq. (5.66) gives

$$k_c = K_1 f_c = \frac{K_1}{1 + \delta} \tag{5.70}$$

and the resulting reactivity for the clean reactor ρ_c becomes

$$\rho_c = \frac{k_c - 1}{k_c} = 1 - \frac{1}{k_c} = 1 - \frac{1 + \delta}{K_1} = \frac{K_1 - 1 - \delta}{K_1} \tag{5.71}$$

In a "dirty" reactor, which is one containing fission products, the thermal utilization factor is reduced because of the absorption by the fission products. In this case, Eq. (5.68) becomes

$$\Sigma_{a, \text{total}} = \Sigma_{a, \text{fuel}} + \Sigma_{a, \text{other}} + \Sigma_{a, \text{fission prod}} \tag{5.72}$$

The thermal utilization factor for the dirty reactor f_d is

$$f_d = \frac{1}{1 + \delta + \theta} \tag{5.73}$$

where θ is called the "fission-product poisoning" and is given by the following equation:

$$\theta = \frac{\Sigma_{a, \text{fission prod}}}{\Sigma_{a, \text{fuel}}} \tag{5.74}$$

The multiplication factor for the dirty reactor k_d becomes

$$k_d = K_1 f_d = \frac{K_1}{1 + \delta + \theta} \tag{5.75}$$

and the reactivity for the dirty reactor ρ_d is

$$\rho_d = \frac{k_d - 1}{k_d} = 1 - \frac{1}{k_d} = \frac{K_1 - 1 - \delta - \theta}{K_1} \tag{5.76}$$

The change in reactivity that occurs in going from the "clean" to the "dirty" condition $\Delta \rho_{\text{fp}}$ is equal to

$$\Delta \rho_{\text{fp}} = \rho_d - \rho_c = \frac{K_1 - 1 - \delta - \theta}{K_1} - \frac{K_1 - 1 - \delta}{K_1} = \frac{-\theta}{K_1} \tag{5.77}$$

If the fuel is the principal absorber in the core, $\delta \ll 1.0$, which means that K_1 approaches unity. Thus the change in reactivity due to the buildup of fission products in the core approaches $-\theta$, the fission-product poisoning term. While this derivation is carried out for thermal reactors, the result is also applicable to fast reactor systems.

Thermal-reactor poisons There are two fission products that are important in the operation of thermal reactors because of their high microscopic neutron-absorption cross sections. These two thermal-reactor poisons are xenon-135 and samarium-149. To determine the atomic density and the fission-product poisoning due to these two isotopes, the production and depletion rates of the various isotopes appearing in the fission-product decay chains must be examined. On the average, the first or primary fission products produced from fission undergo three stages of beta decay before they reach a stable configuration. The first isotope appearing in the fission-product chain is called the primary fission product for that particular chain. All the other products appearing in the radioactive-decay chain are called secondary fission products.

The production rate of a primary fission product is directly proportional to the fission rate. The constant of proportionality is called the fission yield constant γ_i, and this quantity is experimentally determined. Since there are two fission products produced in the fission process, the summation of all values of γ_i total two. The typical double-humped yield curve for uranium-235 fission is shown in Fig. 5.22. The production rate of a primary fission product is equal to the yield constant times the fission rate, or $\gamma_i N_{fuel} \bar{\sigma}_f \bar{\phi}_{th}$.

The primary fission product is lost as the result of radioactive decay and as the result of neutron burnup. The radioactive-decay rate of any radioisotope is equal to the product of the number of atoms N_i and the decay constant λ_i, or simply $\lambda_i N_i$. The neutron-burnup rate of the isotope is equal to the product, $N_i \bar{\sigma}_{a,i} \bar{\phi}_{th}$. The production rate of any secondary fission product is equal to the decay rate of the precursor product in the decay chain.

The xenon-135 fission-product isotope has the highest microscopic neutron cross section of any known material with $\bar{\sigma}_{a,th}$ equal to 3.2×10^6 b. This isotope is both a primary and a secondary fission product and is also radioactive with a half-life of 9.2 h. The fission-product decay chain containing xenon-135 is shown below:

Fission   Fission
↓   ↓

$$^{135}_{52}\text{Te} \xrightarrow[\text{2 min}]{\beta^-} {}^{135}_{53}\text{I} \xrightarrow[\text{6.7 h}]{\beta^-} {}^{135}_{54}\text{Xe} \xrightarrow[\text{9.2 h}]{\beta^-} {}^{135}_{55}\text{Cs} \xrightarrow[\text{2000 years}]{\beta^-} {}^{135}_{56}\text{Ba (stable)}$$

The design or operating engineer is normally interested in the maximum values of fission-product poisoning. During reactor operation, this occurs when equilibrium conditions are attained. To find the equilibrium xenon-135 concentration, the equilibrium concentrations of the preceding products in the chain must be determined first. The rate of change of the number of tellurium-135 atoms with time is

$$\frac{d(\text{Te})}{dt} = \gamma_{\text{Te}} N_{fuel} \bar{\sigma}_f \bar{\phi}_{th} - \lambda_{\text{Te}}(\text{Te}) - (\text{Te})\bar{\sigma}_{a,\text{Te}} \bar{\phi}_{th} \tag{5.78}$$

At equilibrium, $d(\text{Te})/dt$ is zero and the burnup term in the above equation happens to be negligible when compared to the radioactive decay term. Substituting these relationships into (5.78) gives the equilibrium number of tellurium atoms $(\text{Te})_\infty$:

$$(\text{Te})_\infty = \frac{\gamma_{\text{Te}} N_{fuel} \bar{\sigma}_f \bar{\phi}_{th}}{\lambda_{\text{Te}}} \tag{5.79}$$

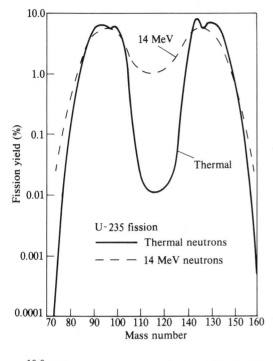

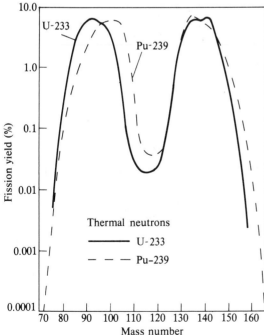

Figure 5.22 The mass distribution of fission products from the fission of U-235, U-233, and Pu-239. (*From "Steam/Its Generation and Use,"* 1972.)

A similar approach can be used to develop an equation for the secondary fission product, iodine-135. The rate of change of iodine-135 is given by the following equation:

$$\frac{d(I)}{dt} = \lambda_{Te}(Te) - \sigma_{a,I}(I)\bar{\phi}_{th} - \lambda_I(I) \tag{5.80}$$

At equilibrium in the reactor, $d(I)/dt$ is zero and $(Te) = (Te)_\infty$. As in the case for tellurium, $\sigma_{a,I}$ is so small (about 7 b) that the neutron-burnup term is several orders of magnitude smaller than the radioactive-decay rate. The equilibrium number of iodine atoms $(I)_\infty$ in the reactor becomes

$$(I)_\infty = \frac{\gamma_{Te} N_{fuel} \bar{\sigma}_f \bar{\phi}_{th}}{\lambda_I} \tag{5.81}$$

The differential equation for the variation of xenon-135 atoms with time is somewhat more complicated than those for tellurium and iodine because xenon-135 is both a primary and a secondary fission product:

$$\frac{d(Xe)}{dt} = \lambda_I(I) + \gamma_{Xe} N_{fuel} \bar{\sigma}_f \bar{\phi}_{th} - \lambda_{Xe}(Xe) - (Xe)\sigma_{a,Xe}\bar{\phi}_{th} \tag{5.82}$$

The neutron-burnup term in this equation cannot be neglected because of the extremely large value of the microscopic neutron-absorption cross section of xenon-135 unless the neutron flux is very low (less than 10^{15} n/m^2·s). In fact, if the average-neutron flux exceeds 10^{19} n/m^2·s, the neutron-burnup term in Eq. (5.82) overshadows the xenon decay-rate term. At equilibrium, $d(Xe)/dt$ is zero and the iodine will have already attained equilibrium concentration, $(I)_\infty$, because of the shorter half-life. Thus, the equilibrium number of xenon-135 atoms, $(Xe)_\infty$, in the core is

$$(Xe)_\infty = \frac{(\gamma_{Te} + \gamma_{Xe}) N_{fuel} \bar{\sigma}_f \bar{\phi}_{th}}{\sigma_{a,Xe}\bar{\phi}_{th} + \lambda_{Xe}} \tag{5.83}$$

The equilibrium poisoning value of xenon-135, $\theta_{Xe,\infty}$, is obtained from the following equation:

$$\theta_{Xe,\infty} = \frac{\Sigma_{a,Xe,\infty}}{\Sigma_{a,fuel}} = \frac{Xe_\infty \sigma_{a,Xe}/V}{N_{fuel}\sigma_{a,fuel}/V} \tag{5.84}$$

Substituting Eq. (5.83) into Eq. (5.84) gives

$$\theta_{Xe,\infty} = \frac{\sigma_{a,Xe}(\sigma_f/\sigma_a)_{fuel}\bar{\phi}_{th}(\gamma_{Te} + \gamma_{Xe})}{\sigma_{a,Xe}\bar{\phi}_{th} + \lambda_{Xe}} \tag{5.85}$$

For a uranium-235-fueled thermal reactor, $\gamma_{Te} = 0.056$, $\gamma_{Xe} = 0.003$, $\sigma_f = 582$ b, and $\sigma_a = 683$ b, so that Eq. (5.85) reduces to

$$\theta_{Xe,\infty,U\text{-}235} = \frac{1.7 \times 10^{-23}\bar{\phi}_{th}}{3.2 \times 10^{-22}\bar{\phi}_{th} + 2.1 \times 10^{-5}} \tag{5.86}$$

Table 5.6 Equilibrium xenon-poisoning values in a uranium-235-fueled thermal reactor

Average operating thermal flux $\bar{\phi}_{th}$, n/m²·s	10^{15}	10^{16}	10^{17}	10^{18}	10^{19}
Equilibrium xenon poisoning $\theta_{Xe, \infty, U\text{-}235}$	0.0008	0.007	0.032	0.050	0.053

Again, it will be noted in Eq. (5.86) that if the average thermal-neutron flux in the core is 10^{15} n/m²·s, or less, the first term in the denominator is negligible when compared to the second term and consequently the equilibrium xenon poisoning is directly proportional to the average thermal flux. However, when the thermal flux exceeds 10^{19} n/m²·s, the first term in the denominator becomes the dominate term and the poisoning value is then independent of the neutron flux and is constant at 0.053 or 5.3 percent or about 10 dollars of reactivity. The equilibrium xenon-poisoning values in a uranium-235-fueled reactor is shown as a function of the average thermal-neutron flux in the core in Table 5.6. Most power reactors operate with average thermal-neutron fluxes ranging between 10^{17} and 5×10^{18} n/m²·s.

It is interesting to examine the variation of xenon poisoning during the normal operation of the reactor. If the reactor is a high flux (high-powered) system, the xenon-poisoning value may approximate the plot in Fig. 5.23. Because the yield of tellurium is larger than the yield of xenon, most of the xenon comes from the decay of iodine. An increase in reactor power causes an immediate increase in the thermal flux which, in turn, causes an immediate increase

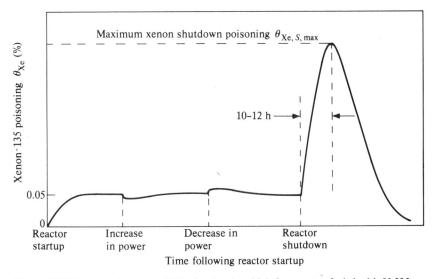

Figure 5.23 The operating xenon-135 poisoning in a high-flux reactor fueled with U-235.

in the neutron-burnup term. There is a lag, however, in the increased production rate of xenon (from the decay of tellurium and iodine) and consequently the xenon-production rate increases rather slowly. This means that there will be an immediate but temporary decrease in the reactor xenon-poisoning value. Similarly, a slight decrease in the reactor power will produce an immediate but temporary increase in the reactor xenon poisoning.

It will be noted in Fig. 5.23 that the xenon poisoning in a high-flux reactor goes through a maximum following reactor shutdown. A maximum in xenon poisoning occurs only if the iodine-decay term exceeds the radioactive-decay term for xenon. For a uranium-235-fueled thermal reactor a maximum poisoning value appears only if the average thermal flux exceeds $3.20 \times 10^{15} \, \text{n/m}^2 \cdot \text{s}$. This maximum poisoning value occurs because the iodine decays at a faster rate than the xenon, following reactor shutdown. When the two decay rates are the same, the poisoning is a maximum. The maximum shutdown value of xenon poisoning, $\theta_{\text{Xe, S, max}}$, in a high-flux thermal reactor is normally reached 10 to 12 h following reactor shutdown. The maximum shutdown xenon-135-poisoning values are tabulated in Table 5.7 as a function of the average operating thermal flux.

The maximum shutdown xenon poisoning can sometimes be a major problem in the operation of a high-flux reactor. Consider the xenon poisoning shown in Fig. 5.24 for a reactor with an average thermal-neutron flux of $10^{18} \, \text{n/m}^2 \cdot \text{s}$. If the reactor control system has a maximum reactivity worth of 15 percent, the reactor may be restarted until the maximum xenon poisoning reaches 15 percent. After this time, the xenon poisoning prevents startup of the reactor even though all the control rods are withdrawn from the core. This condition continues until the xenon poisoning has decayed to 15 percent, at which time the reactor can be restarted. This means there is some time interval during which the reactor cannot be operated. While this condition may not be too bad for many systems, it may be intolerable for a nuclear-powered submarine sitting on the ocean floor.

Table 5.7 Maximum xenon-poisoning values in a uranium-235-fueled thermal reactor, following shutdown

Average thermal-neutron flux $\overline{\phi}_{\text{th}}$, n/m²·s	Time following shutdown for maximum xenon poisoning, h	Maximum shutdown xenon poisoning $\theta_{\text{Xe, S, max}}$
5×10^{15}	0.30	0.00378
10^{16}	0.98	0.00751
5×10^{16}	4.39	0.0255
10^{17}	6.40	0.0420
5×10^{17}	9.82	0.1512
10^{18}	10.88	0.2830
5×10^{18}	11.11	1.334
10^{19}	11.20	2.884
∞	11.28	∞

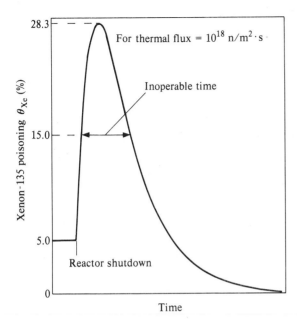

Figure 5.24 The shutdown xenon-135 poisoning in a high-flux thermal reactor fueled with U-235.

Another xenon operational problem encountered in some large, high-powered reactors is called the "xenon instability." This problem is caused by the increased production of xenon in that region of the reactor core where the power density is the highest. As the xenon builds up in that region, it poisons that part of the core reducing the thermal-neutron flux and the power generation in that volume. As a result, the location of the maximum power in the core of these systems tends to move around because of the xenon instability.

The other fission product that is important in the operation of thermal reactors is samarium-149. This isotope is a secondary fission product that appears at the end of the fission-product decay chain and hence is stable. This isotope has a microscopic neutron-absorption cross section of 66,000 b for 0.0253 eV neutrons. The fission-product chain that produces samarium-149 is shown below:

$$
\begin{array}{c}
\text{Fission} \\
\downarrow \\
{}^{149}_{60}\text{Nd} \xrightarrow[\text{1.7 h}]{\beta^-} {}^{149}\text{Pm} \xrightarrow[\text{47 h}]{\beta^-} {}^{149}\text{Sm (stable)}
\end{array}
$$

As in the case of xenon-135, only the equilibrium operating values will be developed here. The rate equation for the primary fission-product isotope neodymium-149 is

$$
\frac{d(\text{Nd})}{dt} = \gamma_{\text{Nd}} N_{\text{fuel}} \bar{\sigma}_f \bar{\phi}_{\text{th}} - \lambda_{\text{Nd}}(\text{Nd}) - (\text{Nd})\bar{\sigma}_{a,\,\text{Nd}} \bar{\phi}_{\text{th}} \tag{5.87}
$$

As was the case for the precursors in the xenon chain, the neutron-burnup term in this equation is negligible compared to the radioactive decay-rate term and, at equilibrium, the fission production rate is equal to the decay rate. Thus, the equilibrium concentration of neodymium-149 is

$$(Nd)_\infty = \frac{\gamma_{Nd} N_{fuel} \bar{\sigma}_f \bar{\phi}_{th}}{\lambda_{Nd}} \tag{5.88}$$

The rate equation for promethium-149, the first secondary product in the chain, is

$$\frac{d(Pm)}{dt} = \lambda_{Nd}(Nd) - \lambda_{Pm}(Pm) - (Pm)\bar{\sigma}_{a,\,Pm}\bar{\phi}_{th} \tag{5.89}$$

Again, neglecting the neutron-burnup term, the above equation gives the equilibrium concentration of promethium-149 as follows:

$$(Pm)_\infty = \frac{\gamma_{Nd} N_{fuel} \bar{\sigma}_f \bar{\phi}_{th}}{\lambda_{Pm}} \tag{5.90}$$

The rate equation for samarium-149 is

$$\frac{d(Sm)}{dt} = \lambda_{Pm}(Pm) - (Sm)\bar{\sigma}_{a,\,Sm}\bar{\phi}_{th} \tag{5.91}$$

In this equation, the burnup term cannot be neglected because it is the only way that samarium is lost. The equilibrium samarium-149 concentration in the reactor core is

$$(Sm)_\infty = \frac{\gamma_{Nd} N_{fuel} \sigma_f}{\sigma_{a,\,Sm}} \tag{5.92}$$

It will be noted in Eq. (5.92) that the number of samarium atoms in the core is independent of the thermal-neutron flux. The equilibrium samarium-poisoning value is found to be equal to

$$\theta_{Sm,\,\infty} = \frac{\Sigma_{a,\,Sm}}{\Sigma_{a,\,fuel}} = \gamma_{Nd}\left(\frac{\sigma_f}{\sigma_a}\right)_{fuel} \tag{5.93}$$

For uranium-235, γ_{Nd} is 0.014, σ_a is 683 b, and σ_f is 582 b, which gives an equilibrium samarium poisoning of 0.012 or about two dollars of reactivity. While the equilibrium samarium poisoning is independent of the neutron flux and hence is the same in all reactors, it takes a long time to reach the equilibrium poisoning value in a low-powered reactor.

The poisoning due to samarium-149 actually becomes more important following shutdown of the reactor because it is stable. Following shutdown, all the neodymium-149 and the promethium-149 atoms in the reactor decay to samarium-149. Consequently, the samarium poisoning does not go through a maximum and then to zero like the xenon poisoning but, instead, it builds up to some asymptotic value. The maximum samarium poisoning is a function of the neutron flux because the equilibrium concentrations of neodymium-149 and promethium-149 are directly proportional to the average thermal flux in the core. For a uranium-235 fueled reactor, the maximum shutdown value of samarium poisoning is

$$\theta_{Sm, S, max, U-235} = 0.012 + 2 \times 10^{-20} \, \bar{\phi}_{th} \tag{5.94}$$

One high-flux research or production reactor (the high-intensity flux reactor, HIFR, at Oak Ridge) normally operated for a two-week period before refueling. If this system experienced an unscheduled shutdown partway through the run, it had to be restarted within 20 min or the xenon poisoning prevented startup. By the time the xenon had decayed to tolerable levels, the samarium poisoning had increased to the point where it prevented startup of the reactor. In order to restart this system, a portion of the core had to be refueled.

5.4.4 Fuel Requirements for a Nuclear Reactor

Almost all nuclear reactors must be initially loaded with more fuel than the actual critical mass to compensate for fuel burnup, for the effect of reactivity coefficients, and for the fission-product poisoning. In the case of fuel burnup, the mass of fuel needed to compensate for neutron burnup can usually be estimated from the reactor power and the operating time. In the case of the reactivity coefficients and the fission-product poisoning, the change in reactivity can normally be estimated from the various relationships.

The amount of fuel that must be added to a reactor to compensate for neutron burnup depends on the kinetic energy of the neutrons causing fission, the type of fuel used, the operating period of the reactor, the average power of the reactor, and the conversion or breeding ratio of the reactor. Approximately 1 g of any of the three fissionable isotopes, completely fissioned, produces 1 MW·day of thermal energy. Actually, it takes more than 1 g of these fuels to produce this much energy because some of the fuel is destroyed in the radiative-capture and other nonfission absorption reactions.

The mass of fuel, in grams, consumed during the operation of a reactor depends on the average thermal power of the reactor P_{ave}, in megawatts, over the operating period of t_{op} days. The burnup mass m_{bu}, in grams, is

$$m_{bu} = K_2 P_{ave} t_{op} \tag{5.95}$$

The constant K_2 is a function of the type of fuel used and the average kinetic energy of the neutrons causing fission. The values of K_2 are listed in Table 5.8.

In a reactor loaded with fully enriched fuel, Eq. (5.95) can be used to evaluate the fuel-burnup requirements for the reactor. However, if the core contains any fertile material at all, some new fuel atoms will be generated and burned as the reactor operates, and this reduces the required burnup mass. It is possible to use a simplified approach to estimate the fuel requirements for a uranium-235-fueled reactor. This method assumes that the average reactor power P_{ave}, the average thermal-neutron flux $\bar{\phi}_{th}$, and the conversion ratio of the reactor r are

Table 5.8 The mass of fuel consumed per thermal megawatt-day of energy release, K_2

Fissionable isotope	Thermal reactors	Fast reactors
U-233	1.14	1.05
U-235	1.23	1.11
Pu-239	1.49	1.10

all constant during reactor operation. Solving the fuel-burnup differential equation yields the data presented in Fig. 5.25. Using this method for a uranium-235-fueled reactor operating for a total of t_{op} days, the amount of uranium-235 that must be added to the core to compensate for neutron burnup m_{bu} can be obtained from the following equation:

$$m_{bu} = P_{ave}(1.23t_{op} - rS) \qquad (5.96)$$

where S is the value obtained from the graph in Fig. 5.25.

The increase in the reactivity of a reactor caused by the addition of the fuel needed for burnup is evaluated by adding this burnup mass to the critical mass of the actual reactor and then calculating the value of the multiplication factor k from the critical equations. For the same burnup mass, the increase in reactivity caused by the addition of the extra burnup mass is larger for reactors with small critical masses than for those that have high critical masses (like fast reactors).

Additional fuel must be initially loaded into the reactor to overcome the effect of the reactivity coefficients in going from the ambient condition to the operating condition. If all the reactivity coefficients are negative, the system reactivity decreases as the reactor is brought to power. This means that the cold reactor will have a higher reactivity than the hot reactor.

Unlike the burnup mass calculations, the amount of extra fuel that must be added to a cold reactor to compensate for the reactivity coefficients cannot be explicitly evaluated. Instead, the decrease in system reactivity can be evaluated and the mass of fuel needed to compensate for these coefficients must be determined from the criticality equations.

If the average power coefficient of reactivity α_P is known for a particular reactor, the excess reactivity $\Delta\rho_{pc}$ needed to overcome the power coefficient of reactivity is

$$\Delta\rho_{pc} = \bar{\alpha}_P P_{op} \qquad (5.97)$$

where P_{op} is the operating power of the reactor. In a fluid-cooled reactor in which there is no phase change, the power coefficient of reactivity is closely related to the temperature coefficient of reactivity. Thus, the change in reactivity $\Delta\rho_{tc}$ caused by the temperature increase is equal to the product of the average

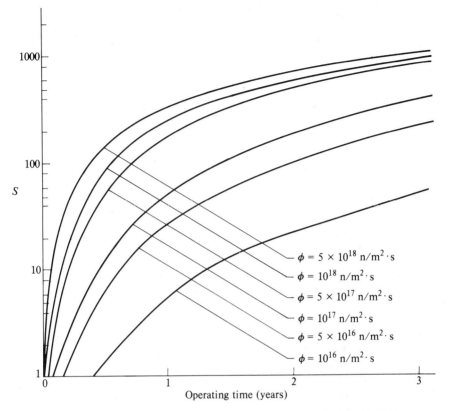

Figure 5.25 Fuel-burnup parameters for a thermal converter reactor fueled with U-235.

temperature coefficient $\bar{\alpha}_T$ and the temperature rise (usually the rise in coolant temperature), or

$$\Delta\rho_{tc} = \rho_{hot} - \rho_{cold} = \bar{\alpha}_T(T_{op} - T_{amb}) \tag{5.98}$$

where T_{op} is the operating temperature of the coolant and T_{amb} is the ambient temperature.

If there is any boiling in the reactor core, there will be an additional decrease in system reactivity if the void coefficient of reactivity $\bar{\alpha}_V$ is negative. The change in reactivity due to the vapor bubbles is equal to the product of the average void coefficient $\bar{\alpha}_V$ and the percentage of "voids" in the core:

$$\Delta\rho_{vc} = \bar{\alpha}_V(\%\ \text{voids}) \tag{5.99}$$

The total change in the reactivity of a BWR system due to the reactivity coefficients is

$$\Delta\rho_{rc} = \rho_{op} - \rho_{amb} = \bar{\alpha}_T(T_{sat} - T_{amb}) + \bar{\alpha}_V(\%\ \text{voids}) \tag{5.100}$$

If the operating value of reactivity ρ_{op} is zero, the cold or ambient reactivity ρ_{amb} must be

$$\rho_{amb} = -\bar{\alpha}_T(T_{sat} - T_{amb}) - \bar{\alpha}_V(\% \text{ voids}) \tag{5.101}$$

In addition to fuel-burnup and reactivity coefficients, extra fuel must be added to a critical reactor to compensate for the decrease in reactivity caused by the buildup of fission products. Fission products "poison" all reactors but they are particularly important in the operation of high-flux thermal systems because of the presence of xenon-135 and samarium-149. As discussed earlier, the decrease in reactivity in going from the "clean" reactor with no fission products to the "dirty" system is approximately equal to the value of the fission-product poisoning. In high-powered thermal reactors in which the flux exceeds 10^{19} n/m$^2 \cdot$s, the decrease in reactivity is approximately equal to

$$\Delta\rho_{fp} = (\rho_{\text{dirty}} - \rho_{\text{clean}})_{\max} = -\theta_{Xe, \infty} - \theta_{Sm, \infty} = -0.053 - 0.012 = -0.065 \tag{5.102}$$

This means that the clean multiplication factor must be greater than 1.069 to compensate for the buildup of these products. If the average thermal flux in the core is less than 10^{19} n/m$^2 \cdot$s, the xenon-135 poisoning will be less than 5.3 percent and should be calculated from Eq. (5.86). Again, it should be stressed that the numerical values in the above equations are applicable only for uranium-235-fueled thermal reactors.

Some nuclear reactors, particularly military systems and other systems where continuity of operation is important, must have the capability of being started at any time. This means that the reactor must have enough excess reactivity that it can override the buildup of fission-product poisoning following reactor shutdown. If the reactor is a high-flux system, this means that the reactor must have a lot of excess reactivity to override the buildup of xenon and samarium.

The maximum excess reactivity for most reactor systems occurs for the cold, clean reactor if all the reactivity coefficients are negative. At the end of the reactor operating period, the system reactivity should be equal to zero or equal to the reactivity for fission-product override if this quantity is included in the design of the reactor system. This means that the reactivity of the cold, clean reactor is

$$\rho_{\text{cold, clean}} = \Delta\rho_{\text{burnup}} + \Delta\rho_{\text{void}} + \Delta\rho_{\text{steady-state fp}} + \Delta\rho_{\text{fp override}} + \Delta\rho_{\text{temp}} \tag{5.103}$$

For a high-flux thermal reactor, Eq. (5.103) becomes

$$\rho_{\text{cold, clean}} = \Delta\rho_{\text{burnup}} - \bar{\alpha}_T \Delta T - \bar{\alpha}_V(\% \text{ voids}) + 0.065 + \Delta\rho_{\text{fp override}} \tag{5.104}$$

Since $k = 1/(1 - \rho)$, the cold, clean value of reactivity can be used to evaluate the cold, clean multiplication factor. Once this value is known, the criticality relationships can be used to estimate the required fuel loading for the reactor.

The reactor control system must be able to handle all this excess reactivity with a significant margin of safety. Some reactors control the excess reactivity with a large number of control rods. At the beginning of reactor operation, almost all of the rods are partially inserted into the core and are then withdrawn as the reactor is operated and the reactivity decreases. These rods can be programmed to shape the power distribution in the reactor core to improve the maximum-to-average power ratio. When all the rods have been withdrawn and the reactor power starts to drop, it is time to refuel the core.

Some nuclear reactors employ a " burnable poison " to reduce the reactivity requirement of the control system. In a high-powered reactor system, most of the excess fuel is used to overcome the problem of fuel burnup. Theoretically, the burnable poison is supposed to compensate for the decrease in reactivity caused by the depletion of the fuel. Boron is commonly used as the burnable poison and may be incorporated in either the fuel element or in the reactor coolant. Some PWR systems employ a boric acid solution in the water coolant.

Unfortunately, boron has a slightly higher absorption cross section than the uranium-235 fuel so that it burns at a faster rate than the fuel. This means that the overall reactivity of the system initially rises until the two burnout rates are the same and then the reactivity drops. The maximum system reactivity that must be handled by the control system with a solid burnable poison commonly occurs near the mid-life of the reactor system.

Example 5.4 A 3800-MW_{th} BWR operates with a critical mass of 3 ton of U-235 and a conversion ratio of 0.8. The average core temperature of the coolant (the saturation temperature of the water) is 340°C and the average in-core steam fraction is 20%. This system is designed to be completely refueled at 18-month intervals while operating at an average load factor of 80%. The system has an average temperature coefficient of $+3 \times 10^{-6} \Delta\rho$ per Celsius degree and an average void coefficient of $-3 \times 10^{-4} \Delta\rho$ per void %. Assuming that the multiplication factor of the reactor is equal to the square root of the ratio of the actual fuel mass to the critical mass, find the initial U-235 loading, the value of the cold-clean and maximum reactivities, and the multiplication factors. Also find the maximum reactivity worth of the control system if it must have the capability of reducing the system multiplication factor to the point where the reactivity is -5% at any time. Assume that no provision is made for overriding the fission-product poisons.

SOLUTION

Average fission cross section $= \bar{\sigma}_f = \dfrac{582[293/(273 + 340)]^{1/2}}{1.128}$

$$= 356.7 \text{ b}$$

Number of fuel atoms in the critical mass $= N_{fuel, c}$

$$= \frac{3(2000)(0.4536)(0.6023 \times 10^{27})}{235} = 6.975 \times 10^{27}$$

Neutron flux (maximum) based on the critical mass $= \bar{\phi}_{th,\,max}$

$$= \frac{3.1 \times 10^{16} P_{th}}{\sigma_f N_{fuel,\,c}} = \frac{3.1 \times 10^{16}(3800)}{(356.7 \times 10^{-28})(6.975 \times 10^{27})}$$

$$= 4.735 \times 10^{17} \, n/m^2 \cdot s$$

Average-neutron flux over the operating period $= 0.8(4.905 \times 10^{17})$

$$\bar{\phi}_{th,\,ave} = 3.788 \times 10^{17} \, n/m^2 \cdot s$$

(Use this value for the burnup calculations, but use the maximum value for the xenon poisoning.)

Fuel burnup calculations:

$$P_{ave} = 0.8 P_{max} = 0.8(3800) = 3040 \, MW_{th}$$

$$t_{op} = 365.25(1.5) = 547.9 \text{ days}$$

$$r = 0.8$$

S (from Fig. 5.25) $= 290$

$$m_{bu} = P_{ave}(1.23 t_{op} - rS) = 3040[1.23(547.9) - 0.8(290)]$$

$$= 1.343 \times 10^6 \, g = 1.481 \text{ ton}$$

$$k_{bu} = \left(\frac{m_{bu} + m_c}{m_c}\right)^{1/2} = \left(\frac{1.481 + 3}{3}\right)^{1/2} = 1.222$$

$$\rho_{burnup} = \frac{k_{bu} - 1}{k_{bu}} = \frac{1.222 - 1}{1.222} = 0.1818$$

Reactivity compensation for temperature coefficient:

$$\Delta\rho_{tc} = \rho_{hot} - \rho_{cold} = \bar{\alpha}_T(T_{op} - T_{amb}) = 3 \times 10^6(340 - 20) = 0.00096$$

This is positive because the reactor is overmoderated and the water acts as a poison. Water expansion improves the system reactivity.

Reactivity compensation for fission-product poisoning:

$$\Delta\rho_{fp} = \rho_{clean} - \rho_{dirty} = \theta_{Sm,\,\infty} + \theta_{Xe,\,\infty}$$

$$= 0.012 + \frac{1.7 \times 10^{-23}\bar{\phi}_{max}}{3.2 \times 10^{-22}\bar{\phi}_{max} + 2.1 \times 10^{-5}}$$

$$= 0.012 + \frac{(1.7 \times 10^{-23})(4.735 \times 10^{17})}{(3.2 \times 10^{-22})(4.735 \times 10^{17}) + 2.1 \times 10^{-5}}$$

$$= 0.012 + 0.04666 = 0.05866$$

Reactivity for the operating, clean reactor $= \rho_{op} = \rho_{burnup} - \Delta\rho_{fp}$

$$\rho_{op,\,clean} = 0.1818 + 0.05866 = 0.2405$$

Reactivity compensation for void coefficient:

$$\Delta\rho_{vc} = \rho_{op} - \rho_{hot} = \bar{\alpha}_V(\% \text{ voids}) = -3 \times 10^{-4}(20) = -0.006$$

Reactivity for the hot (no voids), clean reactor $= \rho_{op} - \Delta\rho_{vc}$

$$\rho_{hot, \text{ clean}} = 0.2405 - (-0.006) = 0.2465$$

$$k_{hot, \text{ clean}} = \frac{1}{1 - \rho} = \frac{1}{1 - 0.2465} = 1.327$$

Reactivity for the cold, clean reactor $= \rho_{hot, \text{ clean}} - \Delta\rho_{tc}$

$$\rho_{hot, \text{ clean}} - \Delta\rho_{tc} = 0.2467 - (+0.00096) = \rho_{cold, \text{ clean}}$$

$$\rho_{cold, \text{ clean}} = 0.2457$$

$$k_{cold, \text{ clean}} = \frac{1}{1 - 0.2457} = 1.326$$

The maximum reactivity and multiplication factors are the values for the hot, clean reactor with no voids:

$$\rho_{max} = \rho_{hot, \text{ clean}} = 0.2467$$

$$k_{max} = k_{hot, \text{ clean}} = 1.327$$

Total fuel loading $=$ (critical mass)$(k_{max})^2 = 3(1.327)^2 = 5.283$ ton U-235
Required worth of control system $= \rho_{max} + 0.05 = 0.2467 + 0.05 = 0.2967$

PROBLEMS

5.1 A bare, cubical reactor, 100 cm on a side, has a neutron flux at the center of any outside face which is 0.2 times the flux at the center of the core. Find the maximum-to-average power ratio in the reactor core and determine the extrapolation distance for the reactor.

5.2 A bare, spherical homogeneous reactor has a core with a diameter of 50 cm and operates at a thermal power of 2 MW_{th}. If the power at the edge of the reactor is 12% of the maximum power, find the maximum power density, in watts per cubic centimeter, in the reactor.

5.3 A bare, spherical reactor has a core diameter of 2 m and operates at a thermal power of 1000 MW_{th}. If the power density at the edge of the core is 8% of the maximum power density, find the value of the maximum power density, in watts per cubic centimeter, and the extrapolation distance.

5.4 A bare, spherical, homogeneous thermal reactor has a core with a diameter of 2 m. The reactor operates at a thermal power of 300 MW_{th} with an average moderator temperature of 260°C. If the maximum thermal-neutron flux is 5×10^{18} n/m^2·s and the flux at the core boundary is 1.75×10^{18} n/m^2·s, find the mass of U-233 needed to fuel the reactor, the maximum and average power densities, in watts per cubic centimeter, and the average specific power, in kilowatts per kilogram of fuel.

5.5 A bare, spherical, homogeneous thermal reactor has a core with a diameter of 1 m. The reactor operates at a thermal power of 50 MW_{th} with an average moderator temperature of 100°C. If the maximum flux is 10^{19} n/m^2·s and the minimum flux is 2×10^{18} n/m^2·s, find the mass of Pu-239 required to fuel the reactor and find the average and maximum power densities in kilowatts per cubic meter.

5.6 A reactor fuel plate is fabricated with 1.5 mm of 10 weight percent UO_2 and 90 weight percent aluminum alloy as the fuel with 0.4 mm of aluminum cladding on each side. Assume the conductivity of the fuel and the aluminum to be 173 and 208 W/m·°C respectively, and the densities are 3.3 and 2.7 g/cm^3. The uranium is 20% enriched and the peak thermal-neutron flux is 10^{18} n/m^2·s. The surface conductance at the outside surface of the plate is 4260 W/m^2·°C and the coolant temperature is 180°C. Determine the peak heat flux at the surface of the element and also find the temperature at the center of the fuel, plus the temperature at the plate surface at that same location.

5.7 A boiling-water reactor fuel rod is filled with UO_2 ($k = 1.8$ W/m·°C) pellets that have a diameter of 9.5 mm and a cladding thickness of 0.5 mm of zirconium alloy ($k = 24.2$ W/m·°C). The surface conductance between the surface of the fuel pellet and the internal surface of the cladding tube is 5680 W/m^2·°C. The outside film conductance is 12,000 W/m^2·°C and the boiling water has a temperature of 340°C. If the maximum allowable fuel temperature is 2500°C, find the maximum power output of the rod, in kilowatts per foot, and the maximum heat flux at the surface of the rod, in watts per square meter.

5.8 A plate-type fuel element is composed of a 1-mm layer of a UO_2 dispersion in an aluminum matrix, which is clad on each side with a 0.5-mm layer of aluminum. The fuel dispersion contains 25% UO_2, by weight, and the uranium is enriched to a level of 90%. The thermal conductivities of the fuel and cladding may be assumed to be 204 W/m·°C. The density of pure UO_2 is about 10 g/cm^3 while that of aluminum is 2.7 g/cm^3. The thermal flux in the fuel region is 10^{18} n/m^2·s. The plate is cooled on both sides with water at 38°C and the convective film coefficient is 40,000 W/m^2·°C. Calculate the maximum fuel and cladding temperatures and determine the surface temperature of the cladding.

5.9 A BWR core contains 560 fuel elements and operates at a thermal power of 2436 MW$_{th}$. Each fuel element consists of a 7 by 7 array of fuel rods and the maximum-to-average power ratio in the core is 2.4. The UO_2 fuel pellets ($k = 2$ W/m·°C) have an OD of 12.4 mm and an active length of 3.66 m and the cladding is 1 mm of zircaloy-2 ($k = 14.7$ W/m·°C). The surface conductance between the fuel pellet and the cladding is 4000 W/m^2·°C, the boiling heat transfer coefficient is 6000 W/m^2·°C, and the saturation temperature of the water is 360°C. Estimate the maximum heat flux at the cladding surface and the maximum temperature in the core.

5.10 A 3%-enriched uranium fuel rod is 19.1 mm in diameter and is clad with 0.76 mm of aluminum. Assume that the thermal conductivities of the fuel and cladding materials are 29.4 and 200 W/m·°C, respectively. If the fuel element is cooled with air at 35°C and the average convective heat-transfer coefficient is 115 W/m^2·s, find the maximum neutron flux that can be tolerated as well as the maximum aluminum temperature when the maximum fuel temperature is limited to 425°C. Assume that the density of the uranium is 18.7 g/cm^3, that the temperature of the graphite moderator is 100°C, and assume that the neutron flux is constant throughout the fuel material.

5.11 A fuel rod of UO_2 ($k = 2$ W/m·°C) has a diameter of 9.5 mm and is clad with 0.4 mm of 304 stainless steel ($k = 15.6$ W/m·°C). The surface conductance between the surface of the fuel pellet and the internal surface of the cladding is 6000 W/m^2·°C. The rod is to develop 20 kW$_{th}$/m at the point under consideration. The outside film coefficient is 7000 W/m^2·s and the boiling water has a temperature of 340°C. Calculate the temperature at the center of the rod and determine the heat flux at the surface of the cladding.

5.12 Determine the relative energy of a 2-eV neutron as it collides with a U-233 nucleus which is moving away from the neutron at the instant of collision at the most probable velocity corresponding to a temperature of 2250°C. Also calculate the relative energy of the neutron if the nucleus is moving toward the neutron at the instant of contact.

5.13 Compute the relative kinetic energy for a 0.1-eV neutron if it collides with a Pu-239 nucleus which is moving toward the neutron at the most probable velocity corresponding to a temperature of 2500°C, and again if the nucleus is moving away from the neutron at the instant of contact.

5.14 Estimate the percentage increase in reactivity in cents or reactivity in "in-hours" that will result in a stable reactor period of $2\frac{1}{2}$ min in reactors fueled by each of the three fuel isotopes, U-233, U-235, and Pu-239.

5.15 Considering only the stable negative period due to the longest delayed-neutron group, estimate the minimum time required to reduce the reactor power so that the average-neutron flux drops from 10^{18} to 10^{10} n/m²·s.

5.16 Using the data in Table 5.5 find the stable reactor period and the increase in reactivity, in percentage and in cents, that correspond to an insertion of 40 Ih into each of two reactors, one fueled with U-233 and the other fueled with Pu-239.

5.17 Estimate the percentage increase in reactivity and the cents of reactivity that correspond to a reactivity insertion of 60 Ih in reactors fueled by each of the fuel isotopes.

5.18 What is the reactivity (in cents and in-hours) of a carbon-moderated, Pu-239-fueled reactor when the power is suddenly increased, resulting in a reactor period of 12 min?

5.19 If ten "cents" of reactivity are inserted into a critical reactor which is fueled with U-235 or U-233 or Pu-239 (solve for each fuel), find the stable reactor period and the reactivity insertion in percentages and in in-hours.

5.20 Due to meteorite damage, it is found that the automatic control system of a nuclear reactor on board a spacecraft is inoperable and during a time of 15 min a constant value of reactivity of -10^{-4} has existed. The astronauts are instructed to manually restore the power level to that prior to the malfunction. If the power is to be restored within 5 min, find the amount of excess reactivity required. Assume that the reactor is fueled with U-235, and find the percentage drop in the power while the malfunction was in effect.

5.21 A high-flux research reactor is fueled with fully enriched U-235 and has an operational cycle such that it operates at a power of 200 MW$_{th}$ for a 1-week period before refueling. At the time of shutdown, the reactor has an average thermal-neutron flux of 10^{19} n/m²·s and the average temperature coefficient of reactivity is -3.6×10^{-5} $\Delta\rho$ per Celsius degree. If the reactor operates at a maximum temperature of 190°C, estimate the mass of fuel left in the reactor when it is shut down for refueling. Find the cold, clean value of multiplication factor, the total excess reactivity that must be contained by the control system, and the initial fuel loading in the reactor if the multiplication factor is approximately equal to the square root of the ratio of the actual fuel mass divided by the critical mass.

5.22 A 1200-MW$_e$ BWR has an overall thermal efficiency of 33% and a critical mass of 100 ton of 2.2%-enriched UO$_2$. The system has an average temperature coefficient of reactivity of -10^{-6} $\Delta\rho$ per Celsius degree and an average void coefficient of reactivity of -7×10^{-5} $\Delta\rho$ per void %. The plant is designed to operate at a pressure of 100 bar with a nominal steam fraction in the core of 15%. The average conversion ratio is 0.8 and the system is designed to operate for a period of 18 months before refueling. If the multiplication factor k is approximately equal to the square root of the ratio of the actual fuel loading to the critical mass, find (a) the maximum system reactivity that must be controlled, (b) the total UO$_2$ loading for the system, (c) the initial average thermal-neutron flux in the reactor core, and (d) the average thermal-neutron flux at the end of 18 months. Base the fission-product poisoning on the flux obtained by using the critical mass.

5.23 An 800-MW$_e$ boiling-water reactor has a thermal efficiency of 32%, a critical mass of 90 ton of 2.4%-enriched UO$_2$, and is designed to operate at an 80% load factor for 300 days before refueling. The system has an average temperature coefficient of reactivity of -5.4×10^{-6} $\Delta\rho$ per Celsius degree and a void coefficient of reactivity of -5×10^{-4} $\Delta\rho$ per void %. The plant is designed to operate at a pressure of 68 bar with a nominal steam fraction in the core of 18% and a conversion ratio of 0.85. Assuming that the multiplication factor is approximately equal to the square root of the ratio of the actual fuel loading to the critical mass, find (a) the total loading of UO$_2$ required for this system, (b) the maximum system reactivity that must be controlled during startup, (c) the initial average thermal-neutron flux in the reactor, and (d) the average thermal-neutron flux at the end of 300 days.

REFERENCES

Bevilacqua, F., and J. F. Gibbons: System 80: Combustion Engineering's Standard 3,800 MWT PWR, *Proceedings of the American Power Conference*, 1974, vol. 36, pp. 70–97, Illinois Institute of Technology, Chicago, Ill.

El-Wakil, M. M.: "Nuclear Heat Transport," International Textbook Company, Scranton, Pa., 1971.

Foster, A. R., and R. L. Wright, Jr.: "Basic Nuclear Engineering," 3d ed., Allyn and Bacon, Inc., Boston, Mass., 1977.

Geurts, J. R., D. R. Patterson, and E. P. Schlinger: BWR Plant Standardization Utilizing GESSAR and STRIDE, *Proceedings of the American Power Conference*, 1974, vol. 36, pp. 107–116, Illinois Institute of Technology, Chicago, Ill.

Glasstone, S., and A. Sesonske: "Nuclear Reactor Engineering," D. Van Nostrand Company, Inc., New York, 1963.

MacMillan, J. H., and T. O. Johnson: Once Through Performance at 3,800 MWT—The Babcock 241 Standardized Nuclear Steam System, *Proceedings of the American Power Conference*, 1974, vol. 36, pp. 117–124, Illinois Institute of Technology, Chicago, Ill.

"Steam/Its Generation and Use," 38th ed., Babcock and Wilcox Company, New York, 1972.

Toth, G. P., and R. J. Nath: Westinghouse 3,817-MWT NSSS, *Proceedings of the American Power Conference*, 1974, vol. 36, pp. 98–106, Illinois Institute of Technology, Chicago, Ill.

U.S. Atomic Energy Commission: "Reactor Physics Constants," ANL 5800, 2d ed., Argonne National Laboratory, July, 1963.

U.S. Atomic Energy Commission: Clinch River Breeder Reactor Project, *Proceedings of the Breeder Reactor Corporation*, October 1974 Information Session, Monroeville, Pa.

Waage, J. M., et al.: General Atomics 1,500 MWe High-Temperature Gas-Cooled Reactor, *Proceedings of the American Power Conference*, 1974, vol. 36, pp. 125–134, Illinois Institute of Technology, Chicago, Ill.

ENVIRONMENTAL IMPACT OF POWER PLANT OPERATION

6.1 INTRODUCTION

The design, location, construction, and operation of electrical power generation facilities have been markedly affected by the concern for the environment. The power engineer must have a sincere concern for the environment but he also must be concerned with producing enough power to meet the public demand at as low a cost as possible. Meeting all three of these considerations, i.e., the three "E's" (energy, ecology, and economy), is the major technological challenge of this century. While ecology is very important, many of the self-proclaimed "environmentalists" are not true environmentalists but are simply antinuclear or even antipower advocates. These people fight the siting and construction of new power plants with long and costly litigation producing serious delays. The same people are the first to complain about the brownouts, blackouts, and the rate increases produced by the delays.

Federal and state governments have passed environmental legislation that limits the amount of solid and gaseous pollutants that can be discharged to the atmosphere along with the amount of thermal energy that can be discharged to natural waters. According to the "Clean-Air Act of 1970," any new steam generating units or any plant construction or modification started after August 17, 1971, that has a thermal input of more than 250 million Btu/h must meet the following standards as a minimum.

Particulate matter:

The maximum two-hour average permitted is 0.1 lbm of particulate matter per million Btu of input energy.

The visible emissions from the stack shall not exceed 20 percent opacity, except for two minutes in any one hour, when emissions may be as great as 40 percent opacity, as determined by the standard Ringlemann Charts. These charts are simply lined plastic cards that the inspector compares to the effluent.

Sulfur oxides (primarily sulfur dioxide):

The maximum two-hour average for an oil-fired system is 0.80 lbm per million Btu of input thermal energy.

The maximum two-hour average for a coal-fired system is 1.20 lbm per million Btu of input thermal energy.

Nitrogen oxides (nitric oxide and nitrogen dioxide):

The maximum two-hour average for a gas-fired system is 0.20 lbm per million Btu of input thermal energy.

The maximum two-hour average for an oil-fired system is 0.30 lbm per million Btu of input thermal energy.

The maximum two-hour average for a coal-fired system is 0.70 lbm per million Btu of input thermal energy.

These rules are currently enforced by the Environmental Protection Agency (EPA). In the generation of electricity, these rules are applicable to any system producing more than around 25 MW_e of electrical energy.

6.2 PARTICULATE EMISSIONS

Of the three general atmospheric pollutants emitted from fossil-fueled power plants, the particulates are probably the easiest of the pollutants to control. Particulate matter is normally classified either according to the size of the particles or their source. Particles having a diameter of less than 1 μm (10^{-6} m) are generally classified as dust and these particles are small enough that they do not settle to the ground but behave as aerosols. Particles with a diameter of more than 10 μm normally settle to the ground. Smoke is usually composed of stable suspensions of particles that have a diameter of less than 10 μm and are visible only in the aggregate. Fumes are very small particles resulting from chemical reactions and are normally composed of metals or metallic oxides. The ash particles that are found in the flue gas of a fossil-fuel-fired boiler with diameters of 100 μm or less are called flyash while larger particles are called cinders. The size range of various particles, fumes, aerosols, etc., are shown in Fig. 6.1.

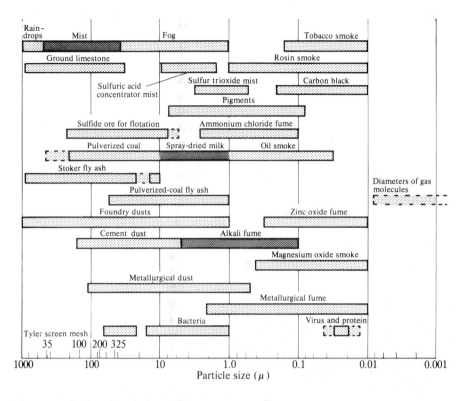

Figure 6.1 The size of particulate pollutants. *(From Faith, 1959.)*

There are many ways of removing particulates from the flue gas. The performance parameter of a particular removal system is called the collection efficiency of the system and is defined by the following equation:

$$\text{Collection efficiency} = \frac{\text{mass of dust removed}}{\text{mass of dust present}} (100) \qquad (6.1)$$

The collection efficiency varies considerably from system to system but normally ranges from 50 percent for some of the simpler mechanical systems to as high as 99 percent or more for an electrostatic precipitator.

Some typical mechanical systems for the removal of particulates are shown in Fig. 6.2. These include (1) a sudden decrease in gas velocity, (2) an abrupt change in the direction of gas flow, (3) the impingement of the gas stream on a series of baffles, and (4) the use of centrifugal force as in the cinder-vane fan. The first three systems in Fig. 6.2 are called "cinder catchers" and are commonly used on stoker and small cyclone furnaces where crushed coal is burned rather than the very fine pulverized coal. Cyclone and mechanical separators, similar to that shown in Fig. 6.3, are also employed when the greater collection efficiencies are desired. The collection efficiency of the cinder catchers

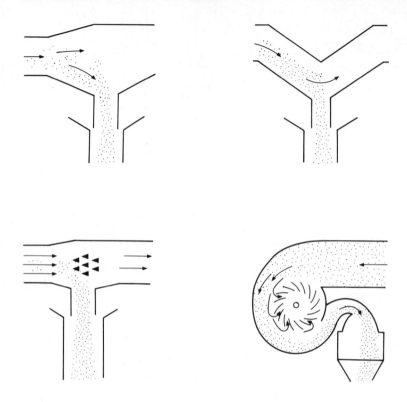

Figure 6.2 Mechanical dust-collection systems. *(From Skrotski and Vopat, 1960.)*

and the cinder-vane fan (of Fig. 6.2) range from 50 to 75 percent. The cyclone separators give collection efficiencies as high as 85 percent.

Wet scrubbers, similar to that shown in Fig. 6.4, are commonly used to remove particulates from process gases in the chemical and the grain-milling industries, but they have not been commonly used to remove ash from the flue gas. Some of the problems associated with the use of wet scrubbers in the flue gas system are that the gas is cooled so much that it must be reheated before it is put up the stack; the resulting pressure drop across the scrubber is quite high; the water, which is used as the scrubber fluid, becomes contaminated with sulfurous and sulfuric acid if the exhaust contains any sulfur oxides, and this can cause corrosion problems. The collection efficiency of wet scrubbers is around 90 percent.

While wet scrubbers are not commonly used to remove particulate matter from the flue gas, more and more power plants are employing wet scrubbers in the flue gas system to remove the sulfur oxides from the gas. While it might be expected that sulfur dioxide and particulates could be removed in the same wet scrubber system, a separate particulate-removal system is commonly employed

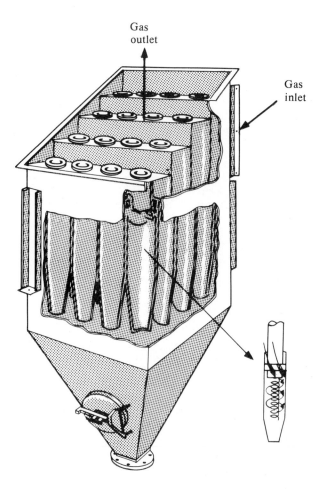

Gas
outlet

Gas
inlet

Figure 6.3 A cyclone particulate-removal system. *(Courtesy of Western Precipitation Company.)*

ahead of the sulfur dioxide scrubber because the simultaneous removal of both pollutants adversely affects the performance of the sulfur dioxide removal system.

In the past, most large pulverized-coal and large cyclone furnaces used a large electrostatic precipitator to remove the particulate matter from the gas. A schematic diagram of one of these systems is shown in Fig. 6.5. In the electrostatic precipitator, highly charged (30,000 to 60,000 V) wires are suspended in a gas-flow passage between two grounded plates. The particles in the gas stream acquire a charge from the negatively charged wires and are then attracted to the grounded plates. The grounded plates are periodically rapped by a steel plug which is raised and dropped by an electromagnet and the collected dust particles fall into hoppers below the plates.

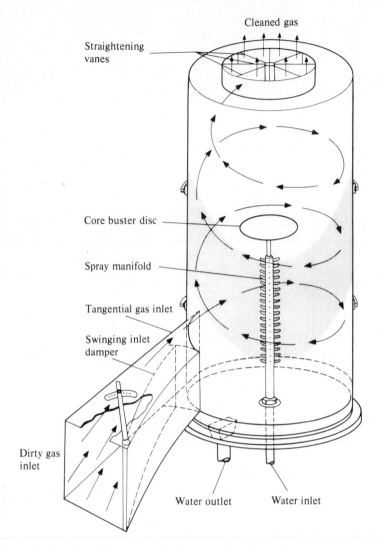

Cleaned gas

Straightening vanes

Core buster disc

Spray manifold

Tangential gas inlet

Swinging inlet damper

Dirty gas inlet

Water outlet

Water inlet

Figure 6.4 A cyclonic wet scrubber system. *(From Faith, 1959.)*

Care must be taken in the operation of any electrostatic precipitator to assure that there is no possibility that a large charge of unburned gas enters the precipitator. If such a charge does enter the unit, power should be immediately terminated because there is constant arcing between the wires and plates and this can trigger an explosion.

While electrostatic precipitators have been commonly used throughout the power industry and have collection efficiencies as high as 99 percent, they do not work well when the flyash has a high electrical resistivity. High-resistivity flyash

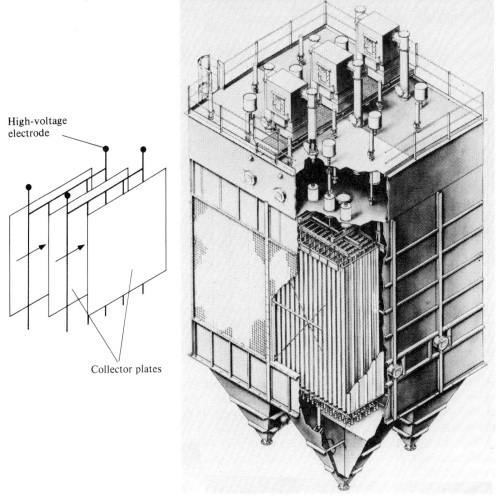

High-voltage
electrode

Collector plates

Figure 6.5 The conventional electrostatic precipitator. *(Courtesy of Babcock and Wilcox and Research-Cottrell, Inc.)*

commonly results from the combustion of low-sulfur coal. One solution to this problem that has actually been used is to inject sulfur trioxide (SO_3) gas into the exhaust gas to improve the conductivity of the flyash. This solution is hardly an environmentally acceptable solution because of the increased emission of sulfur oxides.

Because of the problem encountered with electrostatic precipitators and systems burning low-sulfur coal and because of the increased consumption of low-sulfur coal, more and more utilities are resorting to the use of baghouse filters for the removal of particulate matter. A schematic diagram of this system is shown in Fig. 6.6.

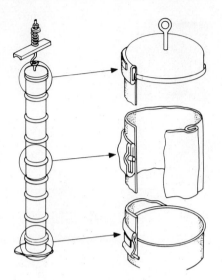

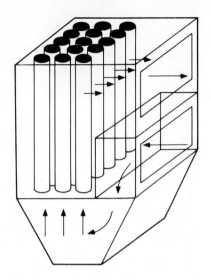

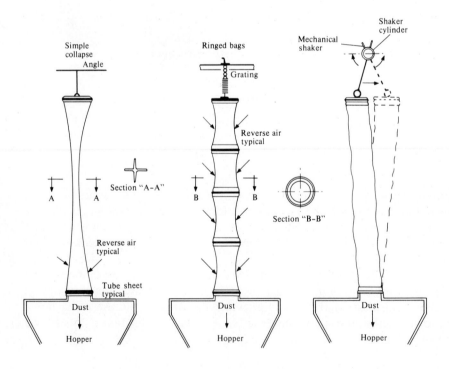

Figure 6.6 The baghouse filter system. *(From Helfritch and Beach, 1976.)*

A major problem with baghouse filters is the high maintenance cost associated with the cloth bags. The bags, which represent up to 20 percent of the erected cost of the baghouse, have an average life of 18 to 36 months. There are two basic types of bag materials but all must be suited for use in a high-temperature dust-laden gas stream. The two types of bag materials are the felted bags like those fabricated from Teflon^r and the woven bags fabricated from glass fibers with various coatings. The baghouse unit is usually cleaned by flushing the system with a reversed flow of atmospheric air. The felted bag is difficult to clean and creates a large pressure drop, while the woven bag loses its dust precoating when cleaned and this produces a temporary drop in the collection efficiency.

Baghouses are inherently large structures and require a minimum of 70 ft^2/MW$_e$. Despite the maintenance problems and the large collection volumes required for these systems, baghouse filters are finding wider and wider acceptance in coal-fired power systems. One estimate is that 30 percent of the coal-fired units will convert to baghouse filters by 1980 and that 75 percent of the planned new coal-fired units will be equipped with baghouses after 1980.

6.3 GASEOUS POLLUTANTS

The emission of sulfur oxides, primarily sulfur dioxide with some sulfur trioxide, is a primary source of atmospheric pollution from fossil-fired electrical generation plants. It is estimated that more than twenty million tons of sulfur dioxide are discharged to the atmosphere each year. One way to meet the sulfur dioxide emission requirements of the Clean-Air Act of 1970 is to burn low-sulfur coal. Many utilities have opted for this solution, even though it normally entails a significant transportation cost as most of it is mined in Wyoming and Montana.

In order to comply with the sulfur dioxide limitations of the Clean-Air Act, the maximum allowable sulfur mass fraction that a coal can have is directly proportional to the higher heating value (HHV) of the fuel. Assuming that all the sulfur in the coal is converted to sulfur dioxide and sulfur trioxide, the relationship for the allowable sulfur mass fraction is

$$\text{Maximum allowable sulfur mass percentage} = 6 \times 10^5 (\text{HHV}) \quad \% \quad (6.2)$$

where the higher heating value is in British thermal units per pound-mass. Thus, a coal with a higher heating value of 12,000 Btu/lbm (27,910 kJ/kg) can have a maximum sulfur mass fraction of 0.72 percent and still not violate the federal law. Actually, some of the sulfur is extracted in the ash so the allowable sulfur fraction can be somewhat greater than 0.72 percent.

In some cases, it may be possible to remove the sulfur from the fuel before it is burned and a considerable amount of research has been directed in this direction to achieve it economically. The sulfur can be removed from the fuel oil before it is burned, by several different processes, but it adds significantly to the fuel cost, as shown in Fig. 6.7. A lot of research is being done in the area of coal

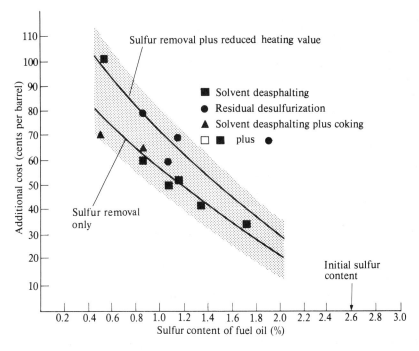

Figure 6.7 Desulfurization costs for fuel oil in 1965. (*From Engineer's Reference Library, by Editors of Power Magazine, McGraw-Hill, Inc., 1969.*)

desulfurization but only with limited success at this time. The Russians have developed an economical process for removing inorganic sulfur from coals containing a lot of iron pyrite but this system is not applicable to most of the coals found in the United States.

Most of the sulfur-emission limiting systems used in this country employ some form of flue-gas desulfurization (FGD) system in which the sulfur dioxide in the flue gas is removed by a wet-scrubbing process. There are two basic types of sulfur dioxide removal systems—the recovery or regenerative systems and the throwaway or nonregenerative systems. In the recovery systems, the reactant that is used to absorb the sulfur dioxide from the flue gas is recovered and reused. The final product in these systems is either sulfuric acid (H_2SO_4) or elemental sulfur (S). In the throwaway systems, the reactants are not recovered and the final products are sulfur salts of calcium and magnesium $(CaSO_3, CaSO_4, MgSO_3,$ and $MgSO_4)$.

There are many different proposed regenerative FGD systems. Among these are the Wellman-Lord process, the FW-Bergbau Forshung process, the wet-magnesium oxide process, the catalytic/IFP ammonia process, the aqueous-carbonate process, and the Foster-Wheeler dry adsorption process, among many others. The Wellman-Lord process and the FW-Bergbau Forshung process are shown in Figs. 6.8 and 6.9.

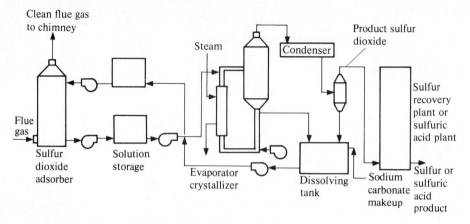

Figure 6.8 The Wellman-Lord absorption FGD regenerative system. *(From Heeney, 1976.)*

Some of these processes, including the Wellman-Lord process, are fully commercial and have been used in Japan and other countries but they are not widely used in the electrical power generation systems in the United States. In general, these systems remove the sulfur dioxide by absorbing it with an alkali (sodium, calcium, and/or magnesium) solution or by adsorbing it with char, charcoal, or alumina pellets. The resulting sulfur salt or concentrated sulfur adsorber is then collected and the sulfur dioxide is regenerated in much higher concentrations, freeing the absorbing or adsorbing material for reuse. The concentrated sulfur dioxide gas is then either oxidized to sulfur trioxide and converted to sulfuric acid or the sulfur dioxide gas is reduced to elemental sulfur. The resulting product can usually be sold to fertilizer companies and the resulting income helps defray the operating cost of the system.

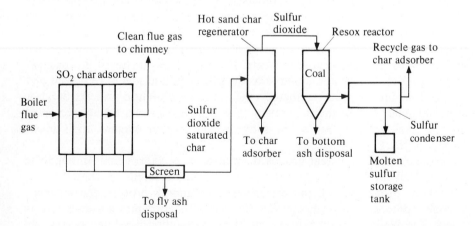

Figure 6.9 The FW-Bergbau Forshung adsorption FGD regenerative system. *(From Heeney, 1976.)*

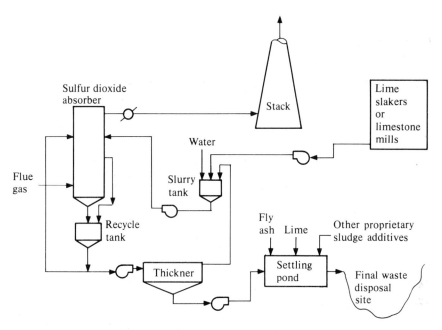

Figure 6.10 The lime-limestone FGD nonregenerative system. *(From Heeney, 1976.)*

Most of the FGD systems used in this country are the throwaway or non-regenerative systems using either lime or limestone as the principal reactant. A schematic diagram of a typical lime/limestone throwaway system is shown in Fig. 6.10. The lime or limestone can be used directly in the scrubber solution as in the process of Fig. 6.10, or it can be used to regenerate sodium carbonate (Na_2CO_3) in the double-alkaline process shown schematically in Fig. 6.11.

In the double-alkali process, the sodium carbonate solution is used to scrub the flue gas because it is more reactive than the calcium or magnesium solutions and it is nonscaling. The final products of the throwaway systems are calcium and magnesium sulfites ($CaSO_3$ and $MgSO_3$) and calcium and magnesium sulfates ($CaSO_4$ or $MgSO_4$). Limestone is usually less expensive and more abundant than lime but the lime system requires less fan and/or pump power and, moreover, experience with these systems indicates that the lime systems have better performance records. In 1976, the estimated capital cost of these systems is around 70 dollars/kW_e or around 70 million dollars for a 1000-MW_e power plant. In addition, there is the added cost of the reactant (lime or limestone). (See Table 6.1.)

At the present time, most of the sludge from the throwaway systems is discharged to holding ponds. This is considered to be a temporary solution and the EPA is concerned because it may cause groundwater pollution and also because it degrades large areas of land. A number of companies are working on the possibility of chemical fixation of the sludge so that it can be used as land fill or as a road base material.

Table 6.1 Estimated costs of flue-gas desulfurization

	Process									
	Lime		Limestone		Double-alkali		Magnesium oxide		Wellman-Lord	
Model plant characteristics	Capital cost, $/kW$_e$	Added cost, mills/kW·h	Capital cost, $/kW$_e$	Added cost, mills/kW·h	Capital cost, $/kW$_e$	Added cost, mills/kW·h	Capital cost, $/kW$_e$	Added cost, mills/kW·h	Capital cost, $/kW$_e$	Added cost, mills/kW·h
EPA Study										
250 MW$_e$										
Retrofit	76.64	7.154	90.92	5.994	99.54	6.558	101.07	6.253	99.73	5.783
New	64.29	5.223	77.08	5.286	83.83	5.745	90.80	5.712	86.07	5.170
500 MW$_e$										
Retrofit	62.52	5.020	71.61	4.889	93.03	5.862	84.93	4.657	91.54	5.196
New	53.90	4.572	62.59	4.388	81.88	5.286	74.89	4.201	82.49	4.788
1000 MW$_e$										
Retrofit	53.08	4.429	70.11	4.736	84.13	5.423	76.31	4.607	85.47	4.802
New	48.94	4.215	59.85	4.203	76.09	5.106	66.06	3.880	78.42	4.484
TVA Study										
500 MW$_e$										
New	61.10	3.65	68.40	3.41	......	......	71.70	4.02	84.80	5.37

Sources: U.S. Environmental Protection Agency, and Tennessee Valley Authority, 1976.

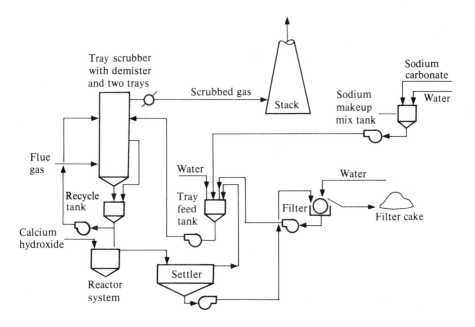

Figure 6.11 The double-alkali FGD nonregenerative system. *(From Heeney, 1976.)*

The emission of oxides of nitrogen (nitric oxide and nitrogen dioxide) are also limited by the Clean-Air Act of 1970. At the present time, the EPA is not actively enforcing the NO_x standards. The power industry, however, is taking steps to produce combustion systems that reduce the NO_x emissions. Since the formation of NO_x is strongly temperature dependent, the NO_x emissions can be reduced by lowering the combustion temperature and by eliminating the hot spots in the furnace. This is one of the major reasons that the cyclone furnace is no longer popular. The formation of nitrogen oxides can also be reduced by lowering the air-fuel ratio and/or by employing exhaust-gas recirculation. There are no exhaust-gas, NO_x removal systems in use in the United States, today, but there are some of these systems in use in Japan.

6.4 THERMAL POLLUTION

Another area of pollution associated with the generation of electrical energy is water pollution. The two general areas of water pollution associated with power generation are the chemical and solid pollution from mining, etc., and the thermal pollution at the power plant, itself. This study is primarily concerned with the discharge of thermal energy to the environment at the power plant, although the leaching of sulfuric acid from abandoned coal mines and slag piles is a major problem in some areas.

The discharge of thermal energy into natural waters is commonly called "thermal pollution" by environmentalists and antipower people. The addition of heat to water reduces the water's ability to hold dissolved gases, including the dissolved oxygen vital to aquatic life. If the water temperature exceeds 35°C (95°F), the dissolved oxygen content is too low to support life. At lower temperatures, however, the aquatic growth is usually enhanced by the warm water and the plants and fish grow at a faster rate. In fact, the warm-water discharge from a power plant is usually a very popular place to fish, particularly during cold weather.

In an effort to compare power generating systems with respect to the amount of thermal energy discharged to the environment, a term will be introduced here called the "thermal discharge index" or the TDI. The thermal discharge index of any power system is the number of thermal energy units discharged to the environment at the plant for each unit of electrical energy produced by the plant. The thermal discharge index may be expressed in terms of power units as

$$\text{TDI} = \frac{\text{thermal power to environment, MW}_{\text{th}}}{\text{electrical power output, MW}_e} \tag{6.3}$$

Needless to say, a low value of thermal discharge index is desirable but it cannot be zero or else the plant violates the second law of thermodynamics.

The thermal discharge index is strongly dependent on the thermal efficiency of the power plant η_{th}. If P_e is the electrical power output of the system in megawatts, then P_e/η_{th} is the thermal power input, in thermal megawatts. Subtracting these values gives the thermal energy discharged from the plant and this is equal to $P_e(1 - \eta_{\text{th}})/\eta_{\text{th}}$, in thermal megawatts. Thus, the thermal discharge index becomes

$$\text{TDI} = \frac{P_e(1 - \eta_{\text{th}})/\eta_{\text{th}}}{P_e} = \frac{1 - \eta_{\text{th}}}{\eta_{\text{th}}} \tag{6.4}$$

The newer fossil-fired power systems have thermal efficiencies up to 40 percent, which means that they have a thermal discharge index of 1.5. Thus, 1.5 MW·h of thermal energy is discharged to the environment at the plant for each MW·h of electrical energy produced. Light-water reactors, on the other hand, have thermal efficiencies of 32 to 33 percent. This means that the thermal discharge index of these systems is around 2.1. In the case of fossil-fired power systems, some of the thermal energy (the boiler losses) is discharged to the atmosphere from the stack while the balance of the waste heat is removed from the condenser. In the gas-turbine power plant, in fact, all of the waste heat is discharged in the exhaust. In the nuclear power systems, however, all of the waste heat is removed from the condenser of the steam plant. Thus, even if the thermal discharge index of a nuclear and a fossil-fired system are the same, the nuclear power plant requires more cooling water.

There are several ways to supply the condenser cooling water. These include once-through cooling from natural lakes and streams, once-through cooling from artificial cooling ponds or lakes, and the wet and dry cooling towers. The once-through cooling water system is one where the water from a river or lake is strained to remove the debris and aquatic life and then is pumped through the condenser of the power plant. This water normally undergoes a temperature rise of around 10 Celsius degrees as it passes through the condenser. This method of waste-heat removal is the least expensive method of discharging the waste heat and probably gives the highest thermal efficiency for the plant because it produces the lowest average sink temperature for the power system.

While the above system is the most desirable from the standpoint of the power engineer, it has met a lot of opposition from political and environmentalist groups. Some states have enacted laws that prohibit the introduction of water into a natural lake or river that is more than one Fahrenheit degree above the ambient water temperature. As a consequence, it is becoming increasingly difficult to use this method of waste-heat removal.

Some power companies are still using once-through cooling in their new plants but they are constructing large company-owned lakes to supply the water. If the cooling lake is too small, floating spray pumps are sometimes used on the lake to enhance the atmospheric cooling of the lake water. These lakes are usually popular fishing spots because the fish grow very fast and they are usually free of ice the year round. Some water makeup to the lake may be required to replace that lost by evaporation.

If it is not feasible to use cooling lakes, most power companies are opting for cooling towers to reject the waste heat to the atmosphere. There are two basic types of cooling towers—the wet cooling towers and the dry cooling towers. Most utilities use the wet cooling tower in which the hot condenser water is sprayed directly into the tower. The hot water makes direct contact with the air flowing up through the tower and heat transfer is accomplished as the result of sensible and latent (evaporative) heat transfer. The wet cooling tower requires continuous water makeup to compensate for the evaporative loss and the addition of large quantities of water vapor to the atmosphere can adversely affect the weather downwind of the tower, causing higher than normal precipitation.

The minimum sink temperature that can be achieved with a wet cooling tower approaches the wet-bulb temperature of the air. As a result, the overall plant efficiency is probably lower than that of a system using a once-through cooling system, particularly on a hot day when the demand is high.

Most of the large-capacity wet cooling towers are the natural-convection systems characterized by the very large hyperbolic structures that run to 500 ft in height and 500 ft in diameter at the base. A typical wet cooling tower is shown in Fig. 6.12. Air flow in a wet tower is established by the difference between the "cold" air entering the bottom of the tower and the density of the warm air and water vapor leaving the top of the tower. The estimated capital cost of these units is around 15 to 20 dollars/kW_e or up to 20 million dollars for a 1000-MW_e power system.

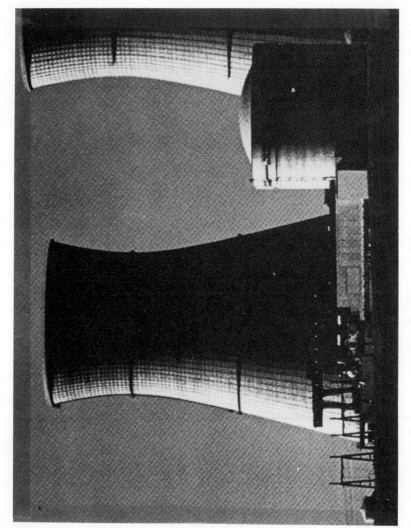

Figure 6.12 A nuclear power station with large natural-draft cooling towers. (*Courtesy of Babcock and Wilcox.*)

Smaller power systems may employ cooling towers with mechanical-draft systems using induced-draft or forced-draft systems. The use of mechanical-draft towers eliminates the need for height. These towers are somewhat less expensive than the natural-draft systems with an estimated capital cost of 10 to 15 dollars/kW$_e$. Although this system has somewhat lower capital costs, it does have higher maintenance and operating costs associated with it.

The dry cooling tower is a possible alternative to the wet cooling tower, particularly where makeup water is not available or where the large vapor plumes are not environmentally acceptable. In this system, the cooling fluid is contained in the tubes so that all the heat transfer for this system is sensible heat transfer and the minimum sink temperature is the dry-bulb temperature.

The dry cooling tower eliminates the need for a surface condenser at the turbine exhaust as the steam could theoretically be condensed directly in the tower. However, the steam pressure and density at the turbine exhaust is so low and the pipes in the tower would have to have such a tremendous volume that such a scheme is impractical. A possible dry cooling tower system is shown schematically in Fig. 6.13. In this system, the turbine steam is condensed in a direct-contact, spray condenser using part of the condensate that has been sub-cooled in the dry tower. The dry cooling tower is considerably more expensive that the wet tower. This system, like the wet tower, also affects the weather down-wind, except that, by superheating the air, it probably produces less rain than normal.

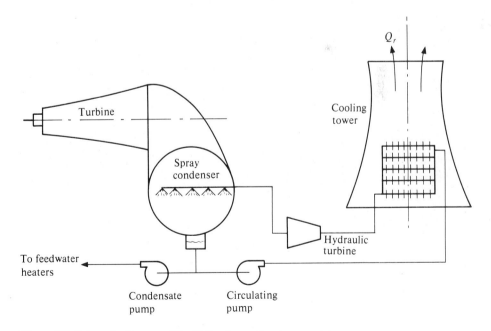

Figure 6.13 Schematic diagram of the Heller dry cooling tower system.

6.5 SOLID-WASTE POLLUTION

In addition to the particulate and gaseous material discharged to the atmosphere, the average coal-fired power plant produces a considerable amount of solid waste in the form of ash. A 500-MW_e coal-fired power plant burning coal with 10 percent ash will generate about 20 tons of ash per hour or around 165,000 tons of ash per year. Bottom ash or slag is relatively inert and can be used as landfill. The flyash collected at the electrostatic precipitator or baghouse is usually quite acidic, however, because of the sulfur dioxide in the flue gas, but it does find a use in the making of cement and concrete blocks.

If a fossil-fuel system (either coal or oil) uses a throwaway sulfur dioxide wet scrubber system, such as the lime/limestone scrubbers, massive amounts of calcium and magnesium salts will also be generated. The same 500-MW_e coal-fired power system, discussed in the preceding paragraph, burning a coal with a sulfur content of 2.72 percent, must collect about 2 percent of the sulfur and this will generate about 15 tons of calcium salts per hour or roughly 120,000 tons per year.

The solid waste from the nuclear power systems is the fission products and the transuranic isotopes from the fission reactor. These products are highly radioactive and are eventually separated from the unburned fuel during the chemical reprocessing of the fuel elements. Because of the massive amounts of radiation emitted by these products, it is imperative that they be isolated from the biosphere for a minimum of 1000 years when the activity will be the same as uranium ore.

There are several proposals for disposing of the fission products. Most of the storage schemes involve the fixation of the products in borosilicate glass and then storing this glass in leak-tight capsules. These capsules can then be stored in deep salt mines or in deep wells drilled in the stable ocean floor. The use of salt deposits for storage was suggested because the presence of the salt pockets indicates that there has been no groundwater in the vacinity for thousands of years. Other possible solutions for the disposition of the fission products include the separation and transmutation of the long-lived radioisotopes to short-lived or stable products following neutron absorption in a breeder or fusion reactor. Some thought has also been given to the possibility of firing these long-lived products into the sun or into a long-term stable orbit.

It is interesting to compare the amount of solid waste generated in a nuclear power plant with that produced in a coal-fired unit. The average person in the United States consumes about 10 $MW_e \cdot$h per year or approximately 800 $MW_e \cdot$h of energy during a lifetime. If this energy is produced in a coal-fired unit, it will lead to the production of about 27 tons of ash plus additional quantities of possible sulfur salts. If the same energy is produced from a light-water reactor, it will lead to the production of about 100 g or about $\frac{1}{4}$ lb of fission products. This is about one cubic inch of products, although the total volume including the glass matrix will be somewhat greater. A 1000-MW_e power system will generate about 300,000 tons of ash per year if it is coal fired and will generate about 1.1 tons of fission product wastes if it is nuclear powered.

Example 6.1 Estimate the condenser cooling water requirements for a 400-MW_e power plant if the water undergoes a 10 Celsius degree (18 Fahrenheit degrees) temperature rise and determine the solid waste generated, in short tons per year of full-power operation, if:

(a) The system is a coal-fired system burning a coal with a HHV of 30,200 kJ/kg, an ash fraction of 8.3%, and a sulfur fraction of 5.1%. Assume the system uses a throwaway FGD system using $CaCO_3$ (limestone) as the feed material. Assume the overall plant efficiency is 40% and the boiler efficiency is 80%.

(b) The system is a nuclear reactor with a thermal efficiency of 32%.

Since SO_2 emission requirements are expressed as 1.2 lbm $SO_2/10^6$ Btu, this problem will be worked using English units.

SOLUTION

(a)
$$(HHV)_{coal} = \frac{30{,}200 \text{ kJ/kg } (0.4536 \text{ kg/lbm})}{1.055 \text{ kJ/Btu}} = 12{,}985 \text{ Btu/lbm}$$

$$\text{Input thermal power} = \frac{P_e}{\eta_{th}} = \frac{400}{0.4} = 1000 \text{ MW}_{th}$$

$$\text{Coal rate} = \frac{1000 \text{ MW}_{th}(3.413 \times 10^6 \text{ Btu/MW} \cdot \text{h})}{12{,}985 \text{ Btu/lbm } (2000 \text{ lbm/ton})}$$

$$= 131.2 \text{ ton/h}$$

Thermal power to stack $= P_{in}(1 - \eta_b) = 1000(1 - 0.8) = 200 \text{ MW}_{th}$

Total thermal power to

$$\text{plant environment} = P_{in}(1 - \eta_{th}) = 1000(1 - 0.4)$$
$$= 600 \text{ MW}_{th}$$

Thermal power to condenser $= 600 - 200 = 400 \text{ MW}_{th}$

Condenser water rate:

$$q = \dot{m}c_p \, \Delta T = 400 \text{ MW}_{th}(3{,}413{,}000 \text{ Btu/MW} \cdot \text{h})$$
$$= \dot{m}(1 \text{ Btu/lbm} \cdot {}^\circ\text{F})(18{}^\circ\text{F})$$

$$\dot{m} = \frac{400(3{,}413{,}000)}{18} = 7.584 \times 10^7 \text{ lbm/h} = 1.264 \times 10^6 \text{ lbm/min}$$

$$\text{Water flow rate} = \frac{1.264 \times 10^6 \text{ lbm/min}}{8.33 \text{ lbm/gal}} = 151{,}700 \text{ gal/min}$$

$$\text{Total } SO_2 \text{ production rate} = (131.2 \text{ ton/h})(0.051 \text{ tons/ton})$$
$$\times (2 \text{ ton } SO_2/\text{ton S})$$
$$= 13.38 \text{ tons } SO_2/\text{h}$$

$$\text{Allowable } SO_2 \text{ release rate} = 1000 \text{ MW } (3{,}413{,}000 \text{ Btu/MW} \cdot \text{h})$$
$$\times (1.2/10^6) \text{ lbm/Btu}$$
$$= 4096 \text{ lbm/h} = 2.05 \text{ ton } SO_2/\text{h}$$

$$SO_2 \text{ collection rate} = 13.38 - 2.05 = 11.33 \text{ ton } SO_2/\text{h}$$

Chemical reaction: $SO_2 + CaCO_3 \longrightarrow CaSO_3 + CO_2$

$$\text{Mo. Wt. of } CaCO_3 = 40.08 + 12.01 + 3(16)$$
$$= 100.09 \text{ lbm/lbm} \cdot \text{mol}$$

$$\text{Mol. Wt. of } CaSO_3 = 40.08 + 32.06 + 3(16)$$
$$= 120.14 \text{ lbm/lbm} \cdot \text{mol}$$

$$\text{Mol. Wt. of } SO_2 = 32.06 + 2(16) = 64.06 \text{ lbm/lbm} \cdot \text{mol}$$

$$\text{Mass of } CaCO_3 \text{ required} = 11.33 \frac{100.09}{64.06} = 17.70 \text{ ton/h}$$

$$\text{Mass of } CaSO_3 \text{ produced} = 11.33 \frac{120.14}{64.06} = 21.25 \text{ ton/h}$$

$$= 186{,}300 \text{ ton/year}$$

$$\text{Ash rate} = 131.2(0.083)(8766) = 95{,}458 \text{ ton/year}$$

(b) Input power $= \dfrac{P_e}{\eta_{th}} = \dfrac{400 \text{ MW}}{0.32} = 1250 \text{ MW}_{th}$

Power to condenser $= 1250 - 400 = 850 \text{ MW}_{th}$

Condenser water rate:

$$\dot{m} = \frac{q}{c_p \, \Delta T} = \frac{850 \text{ MW } (3{,}413{,}000 \text{ Btu/MW} \cdot \text{h})}{(1 \text{ Btu/lbm} \cdot {}^\circ\text{F})(18{}^\circ\text{F})}$$

$$= 1.612 \times 10^8 \text{ lbm/h} = 2.686 \times 10^6 \text{ lbm/min}$$

$$\text{Flow rate} = \frac{2.686 \times 10^6 \text{ lbm/min}}{8.33 \text{ lbm/gal}} = 322{,}500 \text{ gal/min}$$

Number of U-235 atoms fissioned per year
$= 1250 \text{ MW}_{th} (3.1 \times 10^{16} \text{ fissions MW} \cdot \text{s})(8766 \text{ h/year})(3600 \text{ s/h})$
$(1 \times \text{atom/fission})$

$= N_{bu}$

$= 1.2229 \times 10^{27} \text{ atoms/year}$

Mass U-235 consumed per year $= N_{bu}(\text{Mol. Wt.})/(\text{Av No.})$

$$= \frac{1.2229 \times 10^{27}(235)}{6.023 \times 10^{26}}$$

$$= 477.1 \text{ kg/year} = 1051.9 \text{ lbm/year}$$

$$= 0.526 \text{ ton/year}$$

Fraction of fission mass that becomes fission products:

Initial mass of compound nucleus $= 236$ amu

Mass of fission neutrons per fission $= 2.5$ amu

Equivalent mass of fission energy $= \dfrac{200}{931}$

Fraction of U-235 that is fission products $= \dfrac{236 - 2.5 - 200/931}{235}$

$$= 0.9927$$

Mass of fission products produced per year $= 0.526(0.9927)$

$$= 0.522 \text{ ton/year}$$

PROBLEMS

6.1 If a given power plant with a 600-MW$_e$ unit and an overall heat rate of 9700 Btu/kW·h burns Sangamon County, Ill., coal, find the tons of flyash and the tons of sulfur dioxide discharged to the atmosphere per year. Assume that the plant has a load factor of 70%, that 70% of the ash is trapped in the cyclone furnaces, and that the electrostatic precipitator has a collection efficiency of 90%. If the boiler efficiency is 83% and the temperature rise across the condenser is 10 Celsius degrees, find the maximum flow rate of river water through the condenser in gallons per minute. Assume the moisture and ash fractions of the coal are 0.14 and 0.12, respectively.

6.2 A large fossil-fuel-fired power station has a total capacity of 3000 MW$_e$ and an overall thermal efficiency of 37.5%. The station burns Hopkins County, Ky, coal with moisture and ash fractions of 0.07 and 0.08, respectively. In the furnace, 70% of the ash is trapped and the balance is exhausted with the flue gas. The flue gas passes through a large electrostatic precipitator that removes 99.96% of the entrained ash. Calculate the mass flow rate per day of flyash and the sulfur dioxide from this plant. If a tax of 10 cents is levied on each pound of sulfur discharged to the environment, calculate the yearly tax for this facility.

6.3 A 2400-MW$_e$ coal-fired power plant burns Belmont County, Ohio, coal ($M = 5, A = 12$) with an overall heat rate of 9800 Btu/kW·h. The system is composed of dry-bottom pulverized-coal furnaces in which 20% of the ash is trapped inside the unit. An electrostatic precipitator removes 98% of the flyash in the flue gas and the boilers have a boiler efficiency of 87%. If the temperature rise of the cooling water through the condenser is 11 Celsius degrees, find:

  (a) The maximum flow rate of the condenser cooling water, in gallons per minute
  (b) The mass rate of ash, in pounds mass per hour, discharged to the atmosphere
  (c) The mass rate of SO$_2$, in pounds mass per hour, formed in the combustion chamber
  (d) The mass rate of lime (CaO), in pounds mass per hour, that must be added and the mass rate of CaSO$_3$ that must be collected to reduce the SO$_2$ discharge to the atmosphere to legal levels

6.4 Find the mass rate, in pounds mass per hour of fission products produced and the cooling-water flow rate, in gallons per minute, of a PWR system producing 1200 MW$_e$ at a thermal efficiency of 33%.

6.5 A 600-MW$_e$ coal-fired power plant has a thermal efficiency of 38%. The plant burns Polk County, Iowa, coal with a moisture and ash content of 14 and 8%, respectively. Seventy percent of the ash is trapped as molten slag at the bottom of the furnace and the remaining flyash is passed through an electrostatic precipitator that has a collection efficiency of 98.5%. The plant uses a wet lime-limestone process to reduce the sulfur dioxide emissions to the required level for the Clean-Air Act of 1970. For a 24-h period of full-power operation, determine: (a) the mass of ash discharged to the atmosphere, (b) the mass of SO$_2$ discharged to the atmosphere, (c) the mass of ash collected, (d) the mass of CaSO$_3$ generated, and (e) the total quantity of cooling water, in liters, required if the water enters at 16°C and leaves at 27°C and the boiler efficiency is 81%.

6.6 Consider two power plants, one coal and one nuclear, operating at a thermal power output of 3800 MW$_{th}$ for one year. Assume that both plants have an overall heat rate of 9225 Btu/kW·h. The coal-fired plant burns Randolph County, Mo., coal with $M = 12$ and $A = 12$. This plant has a boiler efficiency of 84% and it employs a wet lime-limestone (CaO) process to reduce SO$_2$ emissions to meet the levels specified by the Clean-Air Act of 1970. Find the mass of solid waste generated in each power plant, in tons per year, and find the cooling-water flow rate, in gallons per minute, to the condenser of each unit if the water temperature is increased 20 Fahrenheit degrees. If the pressure rise across the pumps supplying the water is 35 ft, find the pumping power required for each unit.

6.7 Compare a 1200-MW$_e$ nuclear power plant and a 1200-MW$_e$ coal-fired power system that are designed to operate for one year at a load factor of 75%. Assume that the coal-fired system burns Macoupin County, Ill., coal with a moisture and ash content of 12 and 15%, respectively. The coal-fired plant has an overall thermal efficiency of 38%, while that of the nuclear plant is 33%. Fifteen percent of the flyash is collected in the bottom of the boiler and an electrostatic precipitator removes 96% of the balance of the ash in the flue gas. The coal system uses a wet limestone (CaCO$_3$) scrubber to reduce SO$_2$ emissions to the level required by the Clean-Air Act of 1970.

For the coal-fired system, find:

(a) The mass of limestone (CaCO$_3$) required, in tons per year

(b) The minimum mass of CaSO$_3$ collected from the scrubber, in tons per year

(c) The mass of ash collected, in tons per year

(d) The mass of ash discharged to the atmosphere, in tons per year

(e) The mass of SO$_2$ discharged to the atmosphere, in tons per year

(f) The mass of CO$_2$ (assuming perfect combustion) discharged to the atmosphere, in tons per year

(g) The cooling-water flow rate through the condenser, in gallons per minute, if once-through cooling is used and the water temperature increases 10 Celsius degrees. Assume the boiler efficiency is 82%.

For the nuclear plant, find:

(h) The mass of U-235 consumed (fission and radiative capture), in tons per year

(i) The mass of fission products generated, in tons per year

(j) The condenser cooling-water flow rate, in gallons per minute, for a 10-Celsius-degree temperature rise.

REFERENCES

Aimone, R. J., B. J. Bourke, and J. J. Stuparich: Experience with Precipitators when Collecting Ash from Low-Sulfur Coals, *Proceedings of the American Power Conference*, 1974, vol. 36, pp. 615–620, Illinois Institute of Technology, Chicago, Ill.

Atkins, R. S., and D. H. Klipstein: Improved Precipitator Performance by SO$_3$ Gas Conditioning, *Proceedings of the American Power Conference*, 1975, vol. 37, pp. 693–700, Illinois Institute of Technology, Chicago, Ill.

Baruch, S. B.: Trends in Pollution Control at Coal-Burning Power Plants, *Proceedings of the American Power Conference*, 1976, vol. 38, pp. 817–824, Illinois Institute of Technology, Chicago, Ill.

Boexmann, J., and H. Heeren: Surface Condensers for Dry Cooling?, *Power*, vol. 110, no. 4, pp. 80–81, April, 1975.

EPA Picks Three Winners in SO_2 Removal Race, *Electric Light and Power*, Energy/Generation Edition, pp. 56–58, June, 1972.

Faith, W. L.: "Air Pollution Control," John Wiley and Sons, New York, 1959.

Foster, A. R., and R. L. Wright, Jr.: " Basic Nuclear Engineering," Allyn and Bacon, Inc., Boston, Mass., 1973.

Heeney, J. M.: "State of the Art of Flue Gas Desulfurization," Faculty Engineering Conference, Sargent and Lundy, Paper FEC-P 143, March 12, 1976, Chicago, Ill.

Helfritch, D. J., and G. H. Beach: Coal Fired Flyash Control by Fabric Filter Dust Collector, *Combustion*, vol. 48, no. 4, pp. 38–41, October, 1976.

Kinch, A. H.: Retrofitting Wet Scrubbing Systems to Industrial Boilers, *Proceedings of the American Power Conference*, 1974, vol. 36, pp. 636–640, Illinois Institute of Technology, Chicago, Ill.

McIlvaine, R. W.: Flue Gas Desulfurization—Controversial Aspects Examined, *Combustion*, vol. 48, no. 4, pp. 33–37, October, 1976.

Skrotski, B. J. A., and W. A. Vopat: " Power Station Engineering and Economy," McGraw-Hill Book Company, New York, 1960.

Tieman, J. W.: Available and Proposed Devices for the Control of Sulfur Oxide Emissions from Coal-Burning Boilers, *Proceedings of the American Power Conference*, 1973, vol. 35, pp. 495–503, Illinois Institute of Technology, Chicago, Ill.

SEVEN

PRODUCTION OF MECHANICAL ENERGY

7.1 INTRODUCTION

Mechanical energy is one of the more desirable energy forms in that it can be converted into thermal energy with an efficiency of 100 percent and it can also be converted into electrical energy with very high conversion efficiency. Unfortunately, the production of mechanical energy from thermal energy is not as easy. There are no large-scale processes or systems for the direct production of mechanical energy from nuclear energy, electromagnetic energy, and chemical energy. It can be argued that muscular contraction is actually a direct conversion of chemical energy into mechanical energy but this is not a large-scale source of mechanical energy.

Essentially all the mechanical energy is produced as the result of the conversion of thermal energy or from the direct conversion of electrical energy. The conversion of thermal energy into mechanical energy is normally accomplished in some sort of heat engine operating on a thermodynamic heat-engine cycle with limited conversion efficiency. The conversion of electrical energy into mechanical energy is usually accomplished as the result of the interacting magnetic fields in electric motors.

7.2 CONVERSION OF THERMAL ENERGY

7.2.1 Thermodynamic Power Cycles

Almost all of the mechanical energy produced today is produced from the conversion of thermal energy in some sort of a heat engine. The operation of all heat engines can usually be approximated by an ideal thermodynamic power cycle of some kind. A basic understanding of these cycles can often show the power engineer how to improve the operation and performance of the system. A brief review of the basic laws of thermodynamics, particularly as they apply to cycles, will be presented in this section.

Any thermodynamic cycle is composed of a series of thermodynamic processes that returns the working fluid to its initial state. During many of these processes, one property is commonly held constant. These include isothermal (constant temperature), isobaric (constant pressure), isometric (constant volume), isentropic (constant entropy), adiabatic (no heat transfer), and throttling (constant enthalpy) processes. Any reversible adiabatic process is also an isentropic process. It is common practice to plot the processes comprising the cycle on a graph of property coordinates, usually $P-v$ or $T-s$ plots. When plotting constant property lines on any property graph, two of the lines, a horizontal line for the ordinate property and a vertical line for the abscissa property, are easily established. The relative slope of the other constant property lines for the common working fluids can then be found from the horizontal and vertical property lines. Starting at the constant pressure line and going in a clockwise direction, the order of the other constant property lines are temperature, enthalpy, entropy, and volume, or P, T, h, s, v. For a phase change the isobaric and isothermal lines have the same slope, while for an ideal gas the isothermal and constant enthalpy lines have the same slope. The various constant property lines are plotted on the common property coordinates in Fig. 7.1.

For any thermodynamic cycle, the first law of thermodynamics states that the cyclic integral of work is equal to the cyclic integral of heat, or

$$\oint \delta W = \oint \delta Q \tag{7.1}$$

This means that the net work transferred from the cycle (designated as positive work) is equal to the net heat transferred to the cycle (designated as positive heat flow).

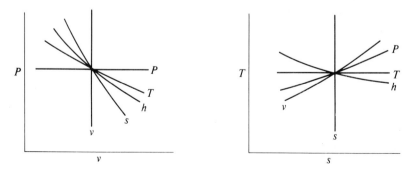

Figure 7.1 The relative slope of constant property lines.

The figure of merit of any thermodynamic cycle is defined as the ratio of the desired energy output to the energy that costs. For a thermodynamic heat-engine cycle, the figure of merit is called the thermal efficiency η_{th}. The desired energy output is the net work output of the cycle and the energy that costs is the heat added from the high-temperature heat source. Since the net work output according to the first law of thermodynamics is equal to the net heat addition, the thermal efficiency becomes

$$\text{Thermal efficiency} = \eta_{th} = \frac{\text{net work output}}{\text{heat added at high temperature}} = \begin{cases} \dfrac{\oint \delta W}{Q_a} & (7.2) \\[2ex] \dfrac{\oint \delta Q}{Q_a} & (7.3) \end{cases}$$

The thermal efficiency of any heat-engine cycle should be as high as possible, but according to the second law of thermodynamics the thermal efficiency must be less than 100 percent.

Another important parameter of any heat-engine cycle is the specific work w which is the net work output per pound of working fluid in the cycle. It is also equal to the areas enclosed by the cycle diagram when it is plotted on either a P–v or a T–s diagram, providing the mass flow of working fluid is the same throughout the cycle. A high value of specific work is desirable because a high value means that a lower flow rate of working fluid or a lower engine speed is required to produce a given power output. Thus, when evaluating a particular thermodynamic power cycle, a high value of thermal efficiency and a high value of specific work is desired.

7.2.2 Reversible Heat-Engine Cycles

One statement of the second law of thermodynamics states that it is impossible to build a device whose sole effect is to produce work while exchanging heat with a single thermal reservoir at constant temperature. Effectively, this means that it is impossible to construct a heat engine or develop a power cycle that has a thermal efficiency of 100 percent. This means that at least part of the thermal energy transferred to a power cycle must be transferred to a low-temperature sink.

In the search for the ideal or most efficient heat engine, it turns out that any power cycle that is completely reversible will give the maximum possible thermal efficiency of any power cycle for the given conditions of the heat source and sink. There are four phenomena that render any thermodynamic process irreversible. They are:

1. Friction
2. Unrestrained expansion
3. Mixing of different substances
4. The transfer of heat across a finite temperature difference

If a cycle is to be completely reversible, none of the processes can experience any of these four phenomena.

Of the four phenomena listed above, the one that has the greatest implica-, tion for completely reversible power cycles is the stipulation that any reversible heat transfer *must* take place across an infinitesimal temperature difference. Because of the second law, any heat-engine cycle must have two external heat-transfer processes—heat transfer to the working fluid from the high-temperature source and heat transfer from the working fluid to the low-temperature sink. Since it is illogical to expect that the source and sink temperatures will change at the same rate that the working fluid changes, these two processes *must* be reversible isothermal processes in order for the cycle to be completely reversible. This isothermal restriction on the heat-transfer processes applies only to those processes in which heat is transferred to external systems. This limitation does not apply to any heat-transfer process within the engine, providing the heat is transferred across an infinitesimal temperature difference in an internal cycle component.

There are three ideal power cycles that are completely reversible power cycles and these systems are called "externally reversible" power cycles. These three ideal cycles are the Carnot cycle, the Ericsson cycle, and the Stirling cycle. All of these cycles are four-process cycles and have two reversible isothermal processes for heat addition and heat rejection to the environment. The maximum possible thermal efficiency for these systems is

$$\eta_{th} = 1 - \frac{T_L}{T_H} \tag{7.4}$$

where T_H and T_L are the temperatures of the heat source and sink, respectively. The thermal efficiency for any heat-engine cycle operating between the fixed temperature limits, T_H and T_L, cannot exceed that given by Eq. (7.4).

The Carnot cycle is an externally reversible power cycle and is sometimes referred to as the optimum power cycle in thermodynamic textbooks. Actually, it is no better and, in fact, it is not as good as the other two externally reversible cycles. The Carnot cycle is composed of two reversible isothermal processes and two reversible adiabatic (isentropic) processes. This cycle is plotted on P–v and T–s coordinates in Fig. 7.2. The Carnot cycle is not a practical power cycle because it has so little specific work that any friction in the actual processes would probably "eat up" the net work output.

The Ericsson power cycle is another heat-engine cycle that is completely reversible or externally reversible. This cycle is composed of two reversible isothermal processes and two reversible isobaric processes as shown on the P–v and T–s coordinates in Fig. 7.3. Since there is heat transfer in all four processes, the heat transfer in the two isobaric processes must take place in an internal cycle component called a regenerator. A possible system, composed of a compressor, turbine, and counterflow heat exchanger, is also shown in Fig. 7.3. The counterflow heat exchanger serves as the regenerator in this system. The engine

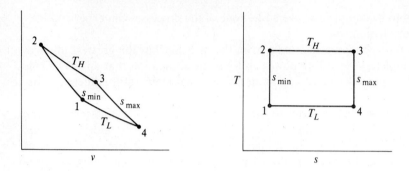

Figure 7.2 The Carnot power cycle.

shown in Fig. 7.3 is not a practical engine because it is difficult to add and extract heat from the gas as it passes through a compressor and/or turbine. The ideal thermal efficiency of the ideal Ericsson cycle is the same as that given by Eq. (7.4).

The Stirling cycle is also an externally reversible heat-engine cycle and is the

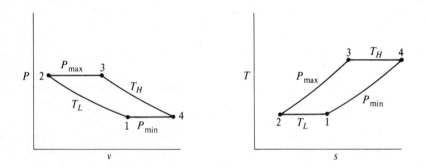

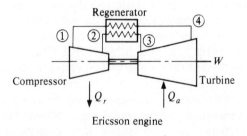

Figure 7.3 The Ericsson power cycle and engine.

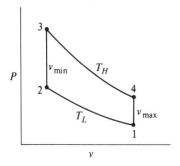 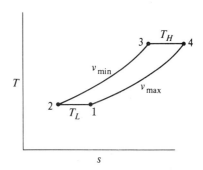

Figure 7.4 The Stirling power cycle.

only one of the three ideal power cycles that has seen considerable practical application. The Stirling cycle is composed of two reversible isothermal processes and two reversible isometric processes (constant volume), as is shown in Fig. 7.4. This cycle also employs a regenerator for reversible heat transfer to and from the working fluid during the two isometric processes. As was the case for the Ericsson and the Carnot cycles, the thermal efficiency of the Stirling cycle is equal to $(1 - T_L/T_H)$.

The N. V. Phillips Laboratory of the Netherlands has done a considerable amount of developmental research on Stirling-cycle engines. They have designed two basic Stirling engines—a single-acting, single-cylinder engine as shown in Fig. 7.5, and the double-acting, multiple-cylinder engine which is shown in Fig. 7.6.

The earliest Stirling engine was the single-acting engine of Fig. 7.5. This engine is an external-combustion system that has two pistons driven by concentric piston rods that are connected to a rhombic-drive system. The top piston is simply a displacer piston that drives the gas back and forth through the regenerator. The bottom piston is the power piston. This engine is perfectly balanced and is essentially vibrationless. In the early 1960s, General Motors Corporation, in cooperation with the Phillips Laboratory, conducted a considerable amount of research on this engine but General Motors is no longer interested in it.

In the last few years, the Phillips Laboratory has developed a double-acting multicylinder engine similar to that shown in Fig. 7.6. The Ford Motor Company is currently testing an experimental engine of this type in a Ford Torino for possible adaptation to automobile propulsion. This engine is an external-combustion system that uses hydrogen as the working fluid. The engine power is varied by controlling the hydrogen pressure which can run as high as 200 atmospheres (3000 lb/in^2). Power is transmitted from the pistons to the drive-shaft through a swash-plate drive system. The principal interest in this engine arises not because it is much more efficient than conventional engines but because of the promise of low pollution associated with external combustion and the possible utilization of low-grade fuels.

The single-acting Stirling engine—and how it works

Power piston and displacer piston have coaxial rods connected to different points on the rhomic drive. Rhomic drive is what keeps pistons in proper relation to each other. Each cylinder has its own burner. "Crankcase" is pressurized. At left, all gas is in cool space between pistons. Then power piston moves up, compressing cool gas at constant temperature. Next, displacer comes down, forcing the gas up through regenerator to the hot end, where the heated gas expands, driving the power piston, and also derives some push from the expanding gas. After the power stroke, the displacer returns to top position, returning hot gases via regenerator to cool space.

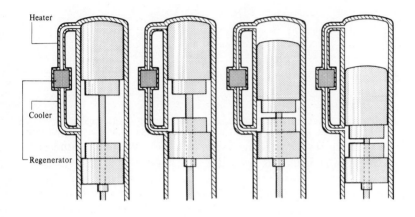

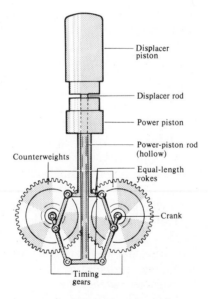

Figure 7.5 The single-acting, single-cylinder, Stirling-cycle engine. *(From Popular Science, vol. 202, no. 2, p. 74, February, 1973.)*

The new double-acting Stirling engine—and how it works

Single burner serves four cylinders grouped in a square. Hot end of one cylinder connects to cool end of neighbor via a regenerator that collects heat from hot gases on their way down, restores heat to cool gases on their way up. There's no displacer piston, and cool end is sealed off from crankcase, no longer pressurized. All four piston rods connect to a swashplate. Rods run in guides, do not bend or swing. They exert pressure against the angled plate, transmitting a wedging action causing swashplate to revolve and turn the output shaft. Four pistons, phased 90 degrees apart, deliver one power impulse for every quarter turn of the output shaft.

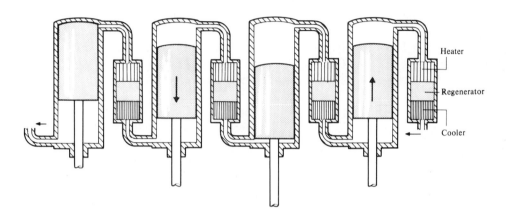

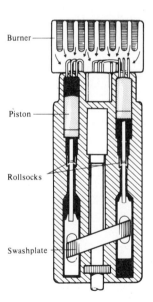

Figure 7.6 The double-acting, multiple-cylinder, Stirling-cycle engine. *(From Popular Science, vol. 202, no. 2, p. 75, February, 1973.)*

7.2.3 Internal-Combustion Engine Cycles

Internal-combustion engines cannot operate on an ideal externally reversible heat-engine cycle but they can be approximated by internally reversible cycles in which all the processes are reversible except the heat-addition and heat-rejection processes. In general, the internal-combustion (IC) engines are more polluting than external-combustion systems because of the formation of nitrogen oxides, carbon monoxide, and unburned hydrocarbons. Because of the cyclic nature of engine operation, reciprocating internal-combustion engines can operate at extremely high peak temperatures. Consequently, these systems can produce reasonably high values of thermal efficiency.

The Otto cycle is the basic thermodynamic power cycle for the spark-ignition (SI), internal-combustion (IC) engine. This cycle is a four-process cycle and is plotted on P–v and T–s coordinates in Fig. 7.7. Theoretically, an air-fuel mixture is compressed (process 1-2) reversibly and adiabatically (isentropically) to minimum volume (piston at top dead center); the mixture is then ignited (by spark plug) and the energy is added in a reversible isometric process (process 2-3, $v = v_{min}$); the hot working fluid then expands in a reversible adiabatic process (process 3-4, $s = s_{max}$); and the heat is rejected to the atmosphere (exhaust and intake strokes) in a reversible isometric process (process 4-1, $v = v_{max}$). This last process actually takes place in the atmosphere as the burned gases are expelled and are replaced by a fresh charge of fuel and air.

The important Otto-cycle parameter is the compression ratio r_v, which is the ratio of maximum-to-minimum volumes:

$$r_v = \frac{v_{max}}{v_{min}} = \frac{v_1}{v_2} = \frac{v_4}{v_3} \tag{7.5}$$

For the air-standard Otto cycle which is an ideal Otto cycle that uses cold air as the working fluid, the thermal efficiency can be shown to be equal to

$$\eta_{th} = \begin{cases} 1 + \dfrac{\text{heat rejected } (-)}{\text{heat added}} = 1 + \dfrac{Q_r}{Q_a} = 1 - \dfrac{c_v(T_4 - T_1)}{c_v(T_3 - T_2)} \\[2ex] 1 - \dfrac{T_1}{T_2} = 1 - r_v^{(1-k)} \end{cases} \tag{7.6}$$

where k is the ratio of specific heats, c_p/c_v.

If the Otto cycle was completely reversible, the thermal efficiency would be $(1 - T_1/T_3)$ instead of the lower value of $(1 - T_1/T_2)$. Since k is greater than unity, the thermal efficiency can be improved by increasing the compression ratio and/or by using a working fluid with a high value of k. Since this is an internal-combustion system and an air-fuel mixture must be used as the working fluid, k is essentially fixed at a value close to that for a diatomic gas ($k = 1.4$).

Effectively, the only way to increase the thermal efficiency of the ideal Otto cycle is to increase the compression ratio. There is a limit, however, to the maximum value of compression ratio that can be employed in an actual engine.

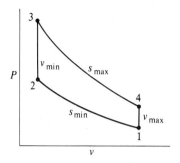
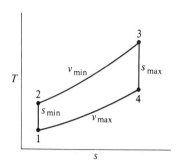

Figure 7.7 The Otto power cycle.

This limit is caused by the premature ignition of the fuel-air mixture before the flame front, initiated by the spark plug, reaches the unburned charge. The premature ignition of the unburned fuel charge produces a pressure imbalance that causes detonation or "ping." Detonation in a reciprocating engine can be very destructive to the engine.

Prior to 1972, some stock automobile engines had compression ratios as high as 12.5, but these engines had to use "premium" or high-octane gasoline. Since 1972, the increased concern about automobile emissions and pollution has brought about the development of systems in which the compression ratios have been lowered to less than nine.

The other principal internal-combustion heat-engine cycle is the Diesel cycle, which is the ideal thermodynamic heat-engine cycle for the compression-ignition (CI), internal-combustion engine—commonly called the diesel engine. In this system, atmospheric air is compressed to a very small volume, generating high pressures and temperatures. At or near top dead center (minimum volume), diesel fuel is injected into the hot air and this fuel, unlike the spark-ignition engine fuel, is designed to burn within a short time after it is injected into the cylinder. The combined effect of burning fuel and increasing volume makes the heat-addition process approach an isobaric process. At some point during the power stroke, called the fuel cutoff, the fuel addition is terminated and the piston continues the expansion process adiabatically until it reaches bottom dead center (maximum volume). At this point, the exhaust and intake strokes essentially produce a constant volume heat-rejection process that takes place in the atmosphere.

The ideal Diesel cycle is a four-process cycle that is composed of a reversible adiabatic $(s = s_{min})$ compression process (process 1-2); a reversible isobaric $(P = P_{max})$ heat-addition process (process 2-3); a reversible adiabatic $(s = s_{max})$ expansion process (process 3-4); and a reversible isometric $(v = v_{max})$ heat-rejection process (process 4-1). The Diesel cycle is plotted on P–v and T–s coordinates in Fig. 7.8.

The Diesel cycle has two important engine parameters. One of these parameters is the compression ratio r_v, which is the ratio of maximum-to-minimum

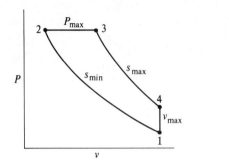
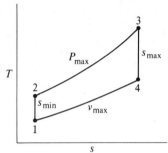

Figure 7.8 The Diesel power cycle.

volumes. The other parameter is the cutoff ratio r_{cf}, which is the ratio of the volume at the end of fuel addition (point 3) to the minimum volume (point 2):

$$r_{cf} = \frac{\text{volume at the end of fuel addition}}{\text{minimum volume}} \tag{7.7}$$

The thermal efficiency of the air-standard Diesel cycle is

$$\eta_{th} = \begin{cases} 1 + \dfrac{\text{heat rejected } (-)}{\text{heat added}} = 1 + \dfrac{Q_r}{Q_a} = 1 - \dfrac{c_v(T_4 - T_1)}{c_p(T_3 - T_2)} \\[4mm] 1 - \dfrac{(r_{cf}^k - 1)}{k r_v^{(k-1)}(r_{cf} - 1)} \end{cases} \tag{7.8}$$

Although there are three parameters in Eq. (7.8), that is, k, r_v, and r_{cf}, k is fixed at a value of about 1.4 as this system is only employed for internal-combustion systems. Effectively, only the compression and cutoff ratios can be altered in the engine design.

Since only air is compressed in the diesel engine (process 1-2), there is no problem with respect to detonation as in the Otto-cycle engine and these systems are operated at much higher compression ratios than the spark-ignition engines. Whereas spark-ignition engines are limited to compression ratios of 6 to 12, the compression-ignition engines have compression ratios of 18 to 25. For the same compression ratio, the Otto cycle is more efficient than the Diesel cycle, but the diesel engine operates at much higher compression ratios.

The cutoff ratio of a Diesel cycle must be greater than unity but it should be as low as possible to attain a high value of thermal efficiency. Although a lower cutoff ratio does improve the thermal efficiency, which is desirable, it also lowers the value of the specific work, which is undesirable. Consequently, the value of the cutoff ratio must be optimized to give a high thermal efficiency but also to give a reasonable value of specific work.

There are several other minor internal-combustion engine cycles and some are shown in Fig. 7.9. The limited-pressure cycle, sometimes called the combined

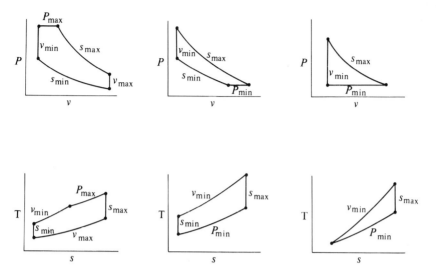

Figure 7.9 Secondary internal-combustion thermodynamic power cycles.

cycle, is a five-process cycle which is really a combination of the Otto and the Diesel cycles in that the heat-addition process is composed of an isometric and an isobaric process. If an actual Otto or Diesel engine cycle is to be approximated by an ideal thermodynamic power cycle, the limited-pressure cycle will probably give the best fit to the actual cycle.

The Atkinson cycle is an ideal internal-combustion engine cycle which is almost identical to the Otto cycle except that the heat-rejection process is a reversible isobaric process instead of an isometric process. This means that the hot exhaust gas must expand to atmospheric pressure in the prime mover before it is exhausted. Since the Atkinson cycle has a greater specific work (larger cycle diagram) than a comparable Otto-cycle engine with the same minimum volume, mass, and heat addition, it has a higher thermal efficiency. This cycle can be approximated by a series of free-piston compressors and a gas-turbine prime mover as shown in Fig. 7.10.

The Lenoir cycle is a three-process cycle that is the ideal thermodynamic power cycle for the oldest internal-combustion engine—the noncompression engine. In the noncompression engine, a charge of natural gas and air is drawn into the cylinder. Partway through the intake stroke, the intake valve is closed and the fuel-air mixture is ignited, producing essentially an isometric heat-addition process. The hot combustion products are then allowed to expand adiabatically to atmospheric pressure and the spent exhaust gases are then discharged to the atmosphere during the exhaust stroke. These engines were used to pump water out of the coal mines in the late nineteenth century.

The Lenoir cycle is also the ideal heat-engine cycle for the pulse-jet engine used by the Germans in the "V-I buzz bomb" of World War II. This engine

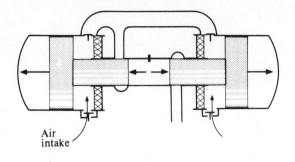

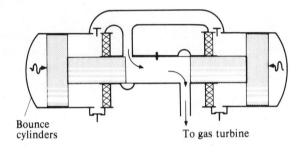

Air
intake

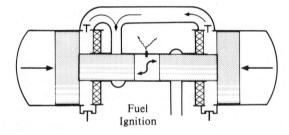

Bounce
cylinders

To gas turbine

Fuel
Ignition

Figure 7.10 A free-piston compressor and gas-turbine prime mover operating on the Atkinson power cycle. *(From Skrotzki and Vopat, 1960.)*

consists of a series of reed valves, a fuel-injection system, igniter, and expansion nozzle. This system is shown in Fig. 7.11. Although the Lenoir cycle is a very simple cycle, it does not find widespread application today because it has very low values of specific work and thermal efficiency.

7.2.4 Internal-Combustion Engines

Most of the internal-combustion engines are reciprocating engines with pistons and cylinders. There are several ways of classifying reciprocating engines but the two common classification systems are according to the order of firing or according to the cylinder arrangement. The two basic types of ignition or firing systems are the four-stroke-cycle engines, commonly called "four-cycle engines," and the two-stroke-cycle engines, commonly called "two-cycle engines."

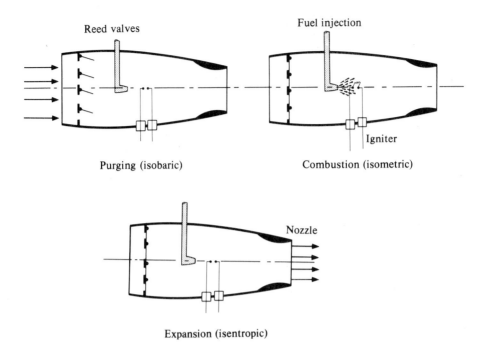

Reed valves

Fuel injection

Igniter

Purging (isobaric)

Combustion (isometric)

Nozzle

Expansion (isentropic)

Figure 7.11 The pulse-jet engine.

The four-stroke-cycle engine requires four full piston strokes to complete one complete cycle. The four strokes are shown in Fig. 7.12 and consist of the intake stroke (1) with the intake valve open, the compression stroke (2) with all valves closed, the power stroke (3) with all valves closed, and the exhaust stroke (4) with the exhaust valve open.

The two-stroke-cycle engine combines the four strokes required to complete a cycle in a four-cycle engine into just two strokes. A typical two-stroke-cycle engine is shown in Fig. 7.13. As the piston moves down the cylinder on the power stroke, the exhaust port is uncovered first and the pressurized exhaust gas starts to escape to the atmosphere. While the piston moves down the cylinder, it also compresses the air-gas-oil mixture in the crankcase. As the piston continues downward, it uncovers the intake port, allowing the compressed fuel-air mixture to enter the cylinder from the crankcase. On the compression (upward) stroke, the piston covers the intake port, then the exhaust port, and compresses the remaining fuel charge until the spark plug ignites the mixture slightly before top dead center. The piston is then forced down by the expanding combustion products in another power stroke. When the piston is near top dead center, the carburetor intake port is uncovered and an additional charge of air-fuel-oil is added to the crankcase. In this system, oil is mixed with the gasoline to provide lubrication for the crankshaft and bearings. Every cylinder in this engine fires every revolution.

1. Air intake	2. Compression	3. Expansion	4. Exhaust
In four-stroke diesel engine, descending piston draws air through open inlet valve, thus filling cylinder space.	Rising piston compresses air against closed valves. Injection of the fuel starts near the end of compression.	The burning mixture expands, forcing the piston down. This is the one stroke of the four that delivers power.	With the exhaust valve open, the rising piston forces products of combustion out of cylinder. Cycle repeats.

Figure 7.12 The four-stroke-cycle, compression-ignition, internal-combustion engine.

The four-stroke-cycle engine has a number of advantages over the usual two-stroke-cycle engine, including better fuel economy, better lubrication, and easier cooling. The two-stroke-cycle engine has a number of advantages, including fewer moving parts, lighter weight, and smoother operation. It should be pointed out, however, that there are some two-stroke-cycle engines that have valves and separate lubrication systems.

Figure 7.14 shows schematic diagrams of some of the different cylinder arrangements for reciprocating engines. The vertical in-line engine is commonly used today in four- and six-cylinder automotive engines. The V-engine is commonly employed in eight-cylinder automotive engines and are called V-8

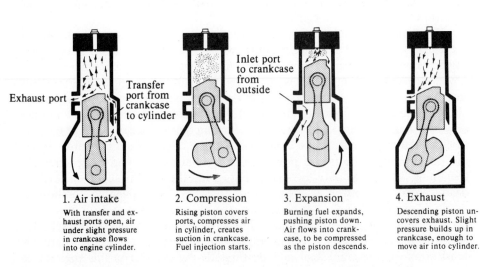

1. Air intake	2. Compression	3. Expansion	4. Exhaust
With transfer and exhaust ports open, air under slight pressure in crankcase flows into engine cylinder.	Rising piston covers ports, compresses air in cylinder, creates suction in crankcase. Fuel injection starts.	Burning fuel expands, pushing piston down. Air flows into crankcase, to be compressed as the piston descends.	Descending piston uncovers exhaust. Slight pressure builds up in crankcase, enough to move air into cylinder.

Figure 7.13 A typical two-stroke-cycle, compression-ignition, internal-combustion engine.

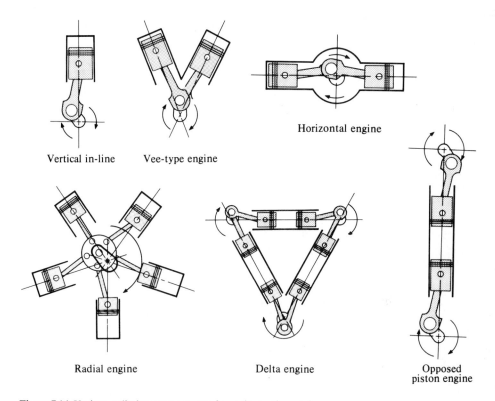

Figure 7.14 Various cylinder arrangements for reciprocating engines.

engines. The horizontal engine is essentially a V-engine with 180° between the opposed cylinders. This engine is used in the four-cylinder, air-cooled engine that powered the Volkswagon "bug."

The opposed-piston engine consists of two pistons, two crankshafts, and one cylinder. The two crankshafts are geared together to assure synchronization. These opposed-piston systems are often employed in large diesel systems. The delta engine is an engine that is composed of three opposed-piston engines connected in a delta arrangement. These systems find application in the petroleum industry.

The radial engine is composed of a ring of cylinders in one plane. One piston rod, the master rod, is connected to the single crank on the crankshaft and all the other piston rods, called articulated rods, are connected to the master rod. If the engine is a four-stroke-cycle system, each bank of cylinders must contain an odd number of cylinders to obtain balanced firing. These engines are commonly constructed with several banks of cylinders to increase the power output. The radial engine has a high power-to-weight ratio and these systems were commonly employed in large aircraft before the advent of the turbojet engine. In some of the early radial engines used to power World War I fighters the crankshafts were fastened to the aircraft, and the cylinder bank, with the propeller attached, revolved. Such engines were commonly called "rotary engines" and some interesting handling problems were posed for these airplanes because of the gyroscopic effects.

The term "rotary engine," when used today, implies something other than a radial engine with a stationary crank. The term "rotary" is essentially applied to any internal-combustion system, except a turbine, that is anything other than a reciprocating engine.

The most famous rotary engine at this time is the Wankel engine. This engine is shown in Fig. 7.15 and utilizes a triangular-shaped lobe as the rotating member. Power is transferred from the rotating lobe to the driveshaft by means of an internal gear. The angular velocity of the driveshaft is three times that of the rotating lobe. This system is inherently a high-speed engine with the driveshaft turning at 3000 to 8000 rev/min. While this engine is lighter, has fewer moving parts, and is cheaper to build than a comparable reciprocating engine, the sealing problems leading to poor fuel economy have reduced interest in the Wankel engine.

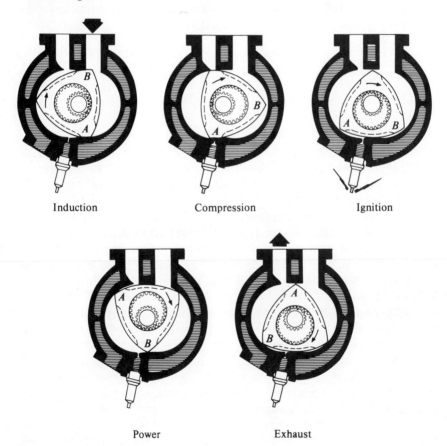

Figure 7.15 The Wankel rotary engine.

A number of different rotary engines have been proposed, and some of them, shown in Fig. 7.16, have been built and operated. The Tschudi engine has a torus-shaped cylinder with curved pistons. The Selwood engine is a rotary engine that also has curved cylinders in the plane of the crankshaft and curved pistons. The shaft of this engine is stationary and contains the carburetor while the outer housing rotates. The Mallory-vane engine operates like a sliding-vane air compressor. The Bricklin rotary engine is a V-engine with fixed floating pistons. There are many other kinds of rotary engines but, so far, none of them appear to be significantly better than the typical reciprocating engine.

7.2.5 Engine Performance

There are several performance factors that are common to all engines and prime movers. One of the main operating parameters of interest is the actual output of the engine. The brake horsepower (Bhp) is the power delivered to the driveshaft of the engine. This power is normally measured by means of some sort of dynamometer such as an electric dynamometer (generator or eddy-current), a water brake, or a friction (prony) brake. The brake horsepower is usually measured by determining the reaction by the dynamometer and using the following equation:

$$Bhp = \frac{2\pi WRN_d}{33,000} \tag{7.9}$$

where W is the net reaction force of the dynamometer, in pounds-force, R is the radius arm of the dynamometer, in feet, and N_d is the angular velocity of the dynamometer, in revolutions per minute. For an energy-absorbing machine, like a pump or compressor, the brake horsepower is the power actually supplied to the shaft of the machine.

Another power term associated with any fluid machinery is the power delivered to or by the working fluid. This power is called the indicated power because it used to be determined by the use of an indicator diagram. For a prime mover, the indicated horsepower (Ihp) is the power delivered to the piston or to the impeller by the working fluid. For a compressor or pump, the indicated horsepower is the power delivered to the working fluid by the piston or impeller. The difference between the indicated and the brake horsepower is the power dissipated in the machine and is called the friction horsepower (Fhp). For a prime mover,

$$Ihp = Bhp + Fhp \tag{7.10}$$

For a pump or a compressor, on the other hand,

$$Bhp = Ihp + Fhp \tag{7.11}$$

Tschudi engine

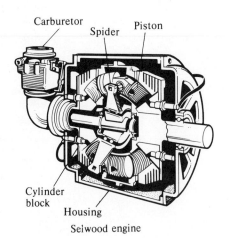

Carburetor Spider Piston

Cylinder
block
Housing

Selwood engine

Bricklin-Turner engine

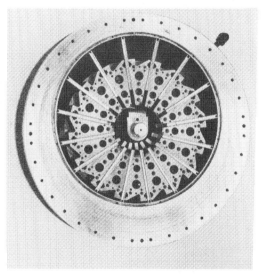

Mallory engine

Figure 7.16 Some unusual rotary engines. *(From Walton, 1968.)*

The mechanical efficiency η_m of an engine is the measure of the machine's ability to transfer mechanical energy. As is the case for most efficiencies, the mechanical efficiency is the ratio of the output power to the input power. For a prime mover, such as an engine or a turbine,

$$\eta_m = \frac{\text{Bhp}}{\text{Ihp}} \tag{7.12}$$

For a compressor or pump,

$$\eta_m = \frac{\text{Ihp}}{\text{Bhp}} \tag{7.13}$$

The mean effective pressure (mep) is a parameter that is applicable to only the reciprocating engines and is effectively the average gage pressure acting on the piston during the power stroke. Perhaps a better definition of the mean effective pressure is to define it as the net work output, in inch-pounds-force, per cubic inch of piston displacement. For a closed system, like a reciprocating engine during the power stroke, the reversible work is equal to $\int P \, dV$. If the effective average pressure or mean effective pressure is used, the work becomes equal to (mep) $\int dV$ or simply (mep)(V_{dsp}), where V_{dsp} is the piston displacement:

$$V_{dsp} = \frac{\pi(\text{bore})^2(\text{stroke})}{4} \tag{7.14}$$

Two mean effective pressures are commonly used—the brake mean effective pressure (bmep), which is obtained from the brake horsepower, and the indicated mean effective pressure (imep), which is obtained from the indicated horsepower. For a particular engine, the relationship between the mean effective pressures and the power is

$$\text{Horsepower (Bhp or Ihp)} = \frac{(\text{bmep or imep})(V_{dsp})N_p}{33,000} \tag{7.15}$$

where N_p is the number of power strokes per minute, or

$$N_p = \frac{CN_e}{a} \tag{7.16}$$

where C is the number of cylinders in the engine, N_e is the revolutions per minute of the engine, and a is equal to 1 for a two-stroke-cycle engine and 2 for a four-stroke-cycle engine.

The brake specific fuel consumption (bsfc) of an engine is a measure of the fuel economy of an engine and is normally expressed in units of mass of fuel consumed per unit energy output. In the United States, the common units of the brake specific fuel consumption are pounds-mass per brake horsepower per hour. The brake specific fuel consumption can be determined from the following equation:

$$\text{bsfc} = \frac{\text{fuel rate, lbm/h}}{\text{Bhp}} \tag{7.17}$$

The brake thermal efficiency of an engine η_{th}, unlike power plants, is based on the lower heating value (LHV) of the fuel. The relationship between efficiency and the brake specific fuel consumption is

$$\eta_{th} = \frac{2545}{(\text{bsfc})(\text{LHV})} \tag{7.18}$$

where the lower heating value is in units of British thermal units per pound-mass.

Another parameter associated with internal-combustion engines is the volumetric efficiency η_v of the engine. The volumetric efficiency is sometimes referred to as a measure of the ability of an engine to "breathe" and can be determined from the following equation:

$$\eta_v = \frac{\text{volume of air brought into the cylinder at ambient conditions}}{\text{piston displacement}} \tag{7.19}$$

The volumetric efficiency of an engine can exceed 100 percent if the system employs a supercharger.

7.2.6 External-Combustion Systems

External-combustion power systems have several advantages over internal-combustion systems. In general, the external-combustion systems are less polluting than internal-combustion engines. The primary pollutants from internal-combustion engines are unburned hydrocarbons, carbon monoxide, and oxides of nitrogen. In the external-combustion engines, the CH_x and CO can be drastically reduced by carrying out the combustion with excess air and the NO_x production can be markedly reduced by lowering the combustion temperature. By burning the fuel with excess air, more energy is released per pound of fuel.

Two other advantages of the external-combustion systems is that they can utilize less expensive fuels and different working fluids. All internal-combustion engines are limited to the use of air or air and fuel vapor and/or combustion products as the working fluid. The external-combustion engines can employ almost any fluid as the working fluid. While most internal-combustion engines are limited to the use of highly refined, expensive hydrocarbon fuels, the external-combustion systems can use coal, residual fuel oil, organic refuse, or any combustible fuel. These units can also utilize nuclear energy as the energy source.

There are essentially three general ideal external-combustion engine cycles. The Stirling cycle and the Brayton cycle are two ideal gas power cycles that can operate with external combustion. Vapor power cycles are cycles that can operate only with external combustion. The Stirling cycle is also a cycle that must operate with external combustion and this system was discussed earlier in the section on externally reversible power systems.

The Brayton power cycle is the ideal thermodynamic power cycle for the gas-turbine or turbojet engines. This is the only major power cycle that can

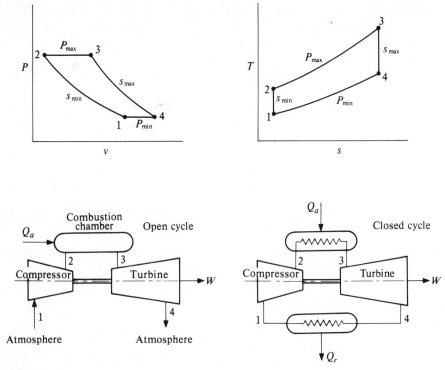

Figure 7.17 The Brayton power cycle and systems.

operate as either an internal- or as an external-combustion engine. The Brayton cycle is a four-process cycle and is shown in Fig. 7.17. In the ideal Brayton cycle, the working fluid is compressed reversibly and adiabatically (isentropically) in the compressor (process 1-2); heat is added in a reversible isobaric ($P = P_{max}$) process (process 2-3) in a combustion chamber or heat exchanger; the hot gas expands reversibly and adiabatically (isentropically, $s = s_{max}$) in the turbine (process 3-4); and then the heat is rejected in a reversible isobaric ($P = P_{min}$) process (process 4-1).

The Brayton cycle may be classed as either an open cycle or a closed cycle. In the open cycle, the working fluid is atmospheric air and the heat-rejection process occurs in the atmosphere as the turbine exhaust is discharged into the atmosphere. In the closed cycle, any working fluid can be employed and the heat-rejection process is accomplished in a heat exchanger. In this system, the working fluid is cycled continuously.

The turbojet engine, in which part of the expansion process occurs in a nozzle to produce thrust, must operate on the open cycle. All internal-combustion, Brayton-cycle engines must operate on the open cycle. It is also possible to use external-combustion systems using low-grade fuels or nuclear heat that operate on the open cycle, but the working fluid must be air.

Most external-combustion, Brayton-cycle systems are closed cycles in which the heat is added to the working fluid in a heat exchanger and the heat is rejected from the working fluid in another heat exchanger as shown in Fig. 7.17. The closed cycle has the advantage than it can employ any gas as the working fluid.

The important cycle parameter for the simple Brayton-cycle engine is the compressor pressure ratio r_p, which is the ratio of maximum-to-minimum system pressures. The thermal efficiency of the simple Brayton cycle is

$$\eta_{th} = 1 + \frac{Q_r}{Q_a} = 1 + \frac{mc_p(T_1 - T_4)}{mc_p(T_3 - T_2)} = 1 - \frac{T_1}{T_2} = 1 - r_p^{(1-k)/k} \qquad (7.20)$$

There are two parameters in this equation, k and r_p, and an increase in either of these parameters improves the efficiency of the simple Brayton cycle.

The so-called "noble" gases, such as helium, neon, argon, etc., are monatomic gases and consequently have the highest value of the ratio of specific heats k of any gaseous working fluid. The specific-heat ratio for these gases is five to three while that for a diatomic gas is seven to five and that for a triatomic gas is eight to six. The noble gases also make excellent working fluids since they produce no corrosion. Helium has the best thermal properties of any gas other than hydrogen.

Increasing the compressor pressure ratio r_p increases the thermal efficiency of the simple Brayton cycle and also increases the compressor discharge temperature. If the turbine-inlet temperature is fixed because of material limitations, an increase in the compressor pressure ratio may reduce the specific work of the cycle requiring a higher gas flow rate for the same power output. Forcing the compressor to operate over a wider range of pressures means that the mechanical efficiency of the compressor is reduced, and this can make the actual cycle less efficient.

There are several ways to improve the thermal efficiency of the simple cycle but they all involve the use of a heat-recovery device called a regenerator. The regenerator is essentially a counterflow heat exchanger in which heat is transferred from the hot exhaust gas to the compressor discharge gas. This cycle is shown schematically in Fig. 7.18 and is plotted on a T–s diagram. In the regenerator, the compressor discharge gas can be heated to some temperature, T_a, and the turbine discharge gas can be cooled to temperature T_b. Theoretically, T_a can approach the turbine exhaust-gas temperature T_4 and the value of T_b can approach the temperature of the compressor discharge temperature T_2. The performance parameter of the regenerator is the regenerator effectiveness ε_r, which is defined as the actual heat transfer divided by the maximum possible heat transfer in the device:

$$\varepsilon_r = \frac{T_4 - T_b}{T_4 - T_2} = \frac{T_a - T_2}{T_4 - T_2} \qquad (7.21)$$

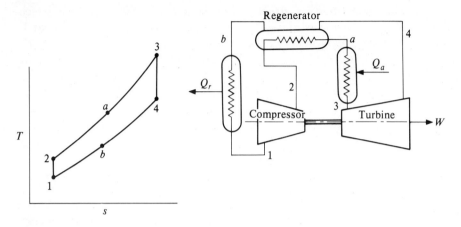

Figure 7.18 The regenerative Brayton power cycle and system.

The thermal efficiency of the regenerative Brayton cycle is given by the following equation:

$$\eta_{th} = 1 - \frac{T_b - T_1}{(T_3 - T_a)} = 1 - \frac{T_b - T_{min}}{T_{max} - T_a} \qquad (7.22)$$

If the regenerator effectiveness is 100 percent, T_a and T_b become equal to T_4 and T_2, respectively, and the above equation reduces to

$$\eta_{th} = 1 - \frac{T_{min}}{T_{max}} r_p^{(k-1)/k} \qquad (7.23)$$

The efficiency of the regenerative Brayton cycle can be increased by increasing the maximum temperature and/or by decreasing the minimum temperature, by decreasing the compressor pressure ratio r_p and/or by using a gas with a low value of k. These last two parameters, r_p and k, have the opposite effect on this cycle as they do for the simple cycle.

The Brayton-cycle efficiency can be increased by using multistage compression with intercooling, providing a regenerator is also used. The use of multistage compression with intercooling decreases the required compressor work and increases the specific work for this cycle. This system is shown schematically along with a plot on the T–s diagram in Fig. 7.19. The use of multistage compression with intercooling produces a lower compressor discharge temperature than that resulting from a single-stage compressor. Consequently, more heat is required to heat the gas to the turbine-inlet temperature and unless part of this energy is supplied by a regenerator, the ideal cycle efficiency will be lower than that of the comparable simple cycle. In the actual cycles, however, the use of multistage compression with intercooling may actually improve the cycle efficiency of the simple cycle because of the improved compressor efficiency over the smaller pressure range.

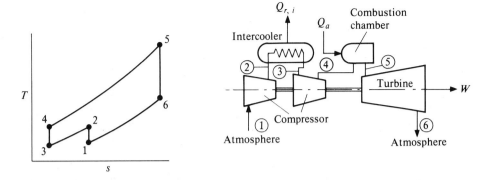

Figure 7.19 The Brayton power cycle and engine with multistage compression and intercooling.

For a two-stage compression process, there is some optimum interstage pressure that produces the minimum amount of compressor work. If the change in kinetic and potential energy is negligible, the ideal compressor work is equal to $-\int_1^2 V \, dP$. This means that an isothermal $(T = C)$ compression process gives a lower value of compressor work than an isentropic compression process because of the less-negative slope of the isothermal line on a P–v diagram, as shown in Fig. 7.20. The actual ideal compression process approaches a polytropic process where $Pv^n = \text{constant}$ and n ranges from a value of k for an isentropic process to 1.0 for an isothermal process. If the compression process is carried out in two stages with intercooling to the original temperature, as shown in Fig. 7.20, the compressor work is reduced by an amount indicated by the shaded area in the figure.

If j stages of compression are employed with intercooling to the original temperature between compression stages, the minimum compression work is realized when the same compression work is performed in each stage. As j approaches infinity, the compression process approaches an isothermal compression. While additional stages of compression with intercooling reduces the compression work, a point of diminishing returns limits the number of practical stages than can be used. For j compression stages with cooling to the initial

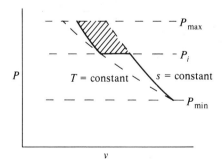

Figure 7.20 The effect of using multistage compression with intercooling.

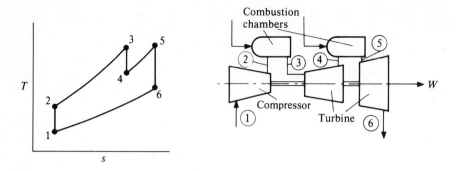

Figure 7.21 The reheat Brayton power cycle and system.

temperature, the optimum pressure ratio across each stage that minimizes the work should equal the jth root of the overall pressure ratio, or

$$\left(\frac{P_{out}}{P_{in}}\right)_{stage,\,opt} = \left(\frac{P_{max}}{P_{min}}\right)^{(1/j)} \tag{7.24}$$

In a two-stage compressor with intercooler, the optimum intercooler pressure P_i is found from Eq. (7.24):

$$\frac{P_i}{P_{min}} = \frac{P_{max}}{P_i} = \left(\frac{P_{max}}{P_{min}}\right)^{1/2}$$

or

$$P_i = (P_{min}\,P_{max})^{1/2} \tag{7.25}$$

The specific work of the Brayton cycle can also be increased by using the reheat Brayton cycle, although this cycle normally has a lower thermal efficiency than the simple Brayton cycle. The reheat Brayton-cycle engine is shown schematically in Fig. 7.21 along with a plot on the T–s diagram. Additional heat can be added to the gas by simply burning more fuel in the turbine gas. The air-fuel ratio is so high in the typical gas-turbine engine that there are no problems associated with the combustion of the additional fuel. The efficiency of the reheat Brayton cycle can be greatly improved with the use of a regenerator as the higher turbine-exit temperature makes the regenerator more effective.

When the open-cycle, turbojet engine is employed in military aircraft, reheat is often utilized under certain conditions to provide additional thrust. This reheat device is called an afterburner and is essentially a device that injects fuel into the turbine exhaust gas before it goes through the engine nozzle. While this device produces significantly higher thrust, it drastically reduces the engine efficiency and the range of the aircraft.

Example 7.1 The compression ratio of an Otto cycle is 8.8 to 1. The minimum pressure and temperature is 0.95 bar and 25°C. The amount of heat added to the air per cycle is 1400 kJ/kg. On the basis of the air-standard cycle, determine: (a) the pressures and temperatures at each point in the cycle, (b) the theoretical thermal efficiency, and (c) the specific work.

SOLUTION

(a) $T_1 = 25 + 273 = 298$ K

$T_2 = T_1(r_v)^{(k-1)} = 298(8.8)^{0.4} = 711.2$ K

$q = c_v(T_3 - T_2) = (0.7177 \text{ kJ/kg} \cdot \text{K})(T_3 - 711.2) \text{ K} = 1400 \text{ kJ/kg}$

$T_3 = T_2 + \dfrac{q}{c_v} = 711.2 + \dfrac{1400}{0.7177} = 2661.9$ K

$T_4 = T_3 \dfrac{T_1}{T_2} = 2661.9 \dfrac{298}{711.2} = 1115.4$ K

$P_2 = P_1(r_v)^k = 0.95(8.8)^{1.4} = 19.95$ bar

$P_3 = P_2 \dfrac{T_3}{T_2} = 19.95 \dfrac{2661.9}{711.2} = 74.68$ bar

$P_4 = P_3 \dfrac{P_1}{P_2} = 74.68 \dfrac{0.95}{19.95} = 3.556$ bar

(b) $\eta_{th} = 1 + \dfrac{q_r}{q_a} = 1 - \dfrac{T_4 - T_1}{T_3 - T_2}$

$= 1 - \dfrac{1115.4 - 298}{2661.9 - 711.2} = 58.10\%$

$= 1 - r_v^{(1-k)} = 1 - (8.8)^{-0.4} = 58.10\%$

(c) Specific work $= w_{net} = q_r + q_a = c_v(T_1 - T_4) + c_v(T_3 - T_2)$

$= c_v(T_1 + T_3 - T_2 - T_4)$

$= 0.7177(298 + 2661.9 - 711.2 - 1115.4)$

$= 813.4 \text{ kJ/kg}$

Example 7.2 A stationary gas-turbine power plant delivers 20,000 kW to an electric generator under the following conditions: air enters the compressor at 20°C and 0.98 bar and leaves at 4.2 bar, and the gas enters the turbine at 850°C. If the mechanical efficiency of the turbine and compressor are 80 percent, find: (a) the temperatures at each point in the cycle, (b) the specific work of the cycle, (c) the specific work of the turbine and the compressor, (d) the thermal efficiencies of the actual and ideal cycles, and (e) the required air flow rate, in kilograms per minute.

SOLUTION

(a) $T_1 = 20 + 273 = 293$ K

     $T_3 = 850 + 273 = 1123$ K

$$T_{2,s} = T_1 \left(\frac{P_2}{P_1}\right)^{(k-1)/k} = 293 \left(\frac{4.2}{0.98}\right)^{(0.4/1.4)} = 444.1 \text{ K}$$

$$T_{4,s} = T_3 \frac{T_1}{T_{2,s}} = 1123(293/444.1) = 740.9 \text{ K}$$

Actual temperatures:

$$T_{2,a} = T_1 + \frac{T_{2,s} - T_1}{\eta_c} = 293 + \frac{444.1 - 293}{0.8} = 481.9 \text{ K}$$

$$T_{4,a} = T_3 - \eta_t(T_3 - T_{4,s}) = 1123 - 0.8(1123 - 740.9) = 817.3 \text{ K}$$

(b) $q_a = c_p(T_3 - T_2) = 1.0048(1123 - 481.9) = 644.2 \text{ kJ/kg}$

          $q_r = c_p(T_1 - T_4) = 1.0048(293 - 817.3) = -526.8 \text{ kJ/kg}$

Net work $= w_{\text{net}} = q_a + q_r = 644.2 + (-526.8) = 117.4 \text{ kJ/kg}$

(c)   Specific work of turbine $= w_t = h_3 - h_4 = c_p(T_3 - T_{4,a})$

$$= 1.0048(1123 - 817.3)$$

$$= 307.2 \text{ kJ/kg}$$

Specific work of compressor $= w_c = h_1 - h_2 = c_p(T_1 - T_{2,a})$

$$= 1.0048(293 - 481.9)$$

$$= -189.8 \text{ kJ/kg}$$

Net work $= w_{\text{net}} = w_c + w_t = -189.8 + 307.2 = 117.4 \text{ kJ/kg}$

(d)   Actual thermal efficiency $= \dfrac{w_{\text{net}}}{q_a} = \dfrac{117.4}{644.2} = 18.22\%$

Theoretical thermal efficiency $= 1 - \dfrac{T_{4,s} - T_1}{T_3 - T_{2,s}}$

$$= 1 - \frac{740.9 - 293}{1123 - 444.1}$$

$$= 34.03\%$$

Theoretical thermal efficiency $= 1 - \left(\dfrac{P_1}{P_2}\right)^{(k-1)/k}$

$$= 1 - \left(\frac{0.98}{4.2}\right)^{(0.4/1.4)} = 34.02\%$$

(e)   Air flow rate $= \dfrac{P}{w} = (20{,}000 \text{ kW})(1 \text{ kJ/kW·s})$

$$\times \ (60 \text{ s/min})/(117.4 \text{ kJ/kg})$$

$$= 10{,}221.5 \text{ kg/min}$$

7.2.7 Vapor Power Cycles

All vapor power cycles must be classified as external-combustion systems and unlike the Brayton cycle they must employ a heat exchanger for the heat-addition process. Most of these systems also employ a heat exchanger (condenser) for the heat-rejection process although it is possible to operate a water system by discharging the steam directly to the atmosphere and supplying continuous water makeup. As was discussed earlier, the primary advantage of external-combustion systems is that they generally produce less pollution and they permit the use of low-grade inexpensive fuels as well as the utilization of nuclear energy.

The basic ideal thermodynamic vapor cycle is the Rankine cycle and the thermodynamic processes for this cycle are identical to the processes of the Brayton cycle—isentropic compression, isobaric heat addition, isentropic expansion, and finally isobaric heat rejection. The difference between these cycles is that the Rankine cycle employs a two-phase (liquid and vapor) working fluid, whereas the Brayton cycle is a gas power cycle.

Water is the common working fluid used in the vapor power cycles but a number of other fluids have been used or proposed for these cycles. Mercury was actually employed along with water in two high-temperature binary power systems constructed in the 1950s and has been proposed for use as the working fluid for some high-temperature space power systems. The binary mercury-water Rankine cycle is composed of a high-temperature mercury Rankine cycle in which the heat-rejection process in the mercury condenser is the heat-addition process in a conventional water cycle. The use of two Rankine cycle systems produces a higher overall thermal efficiency. Although mercury has been used as the high-temperature fluid in the past, mercury is very expensive, the vapor is chemically toxic, and it is very difficult to contain as it amalgamates most structural materials. Consequently, mercury is no longer commonly employed in vapor power systems.

Other proposed Rankine-cycle working fluids are potassium, sodium, rubidium, ammonia, and a number of chain and aromatic hydrocarbon compounds. Currently, there is some interest in developing a potassium-water binary vapor power system to achieve higher overall thermal efficiencies. Water is by far the most common fluid employed in these cycles and, in general, any further discussion of vapor power systems will be limited to those systems using water as the working fluid, unless specifically designated otherwise. The Rankine cycle for a water system is shown on a *T–s* diagram in Fig. 7.22 and the same cycle is shown for an organic working fluid in the same figure. There is a considerable

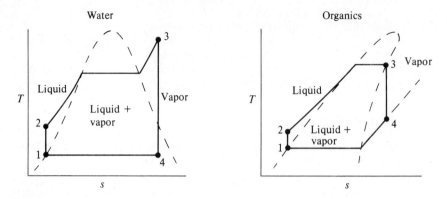

Figure 7.22 The Rankine vapor heat-engine cycle.

difference in the operation of these systems because of the different phase diagrams.

The simple steam power cycle is composed of the four components shown in Fig. 7.23 and one of the four thermodynamic processes occurs in each component. Process 1-2 is a reversible adiabatic (isentropic) compression process that takes place in the pump; process 2-3 is the reversible isobaric heat-addition process that occurs in the boiler; process 3-4 is the reversible adiabatic (isentropic) expansion in the turbine or steam engine; and process 4-1 is the reversible isobaric heat rejection in the condenser. If an energy balance is performed on each of these components, the steady-flow energy equation, neglecting the

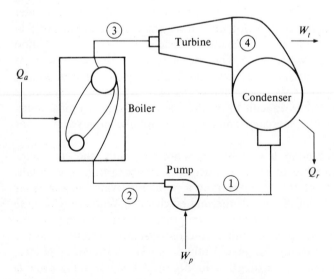

Figure 7.23 The Rankine-cycle components and system.

change in potential and kinetic energy, is

$$q = \Delta h + w \tag{7.26}$$

where q is the specific heat transfer in the component, in kilojoules per kilogram, Δh is the change in enthalpy across the component, in kilojoules per kilogram, and w is the thermal equivalent of the specific work done by the component, in kilojoules per kilogram.

Since the processes in the boiler and condenser are isobaric processes, the ideal work in these components is zero and the energy equation in (7.26) reduces to

$$q = \Delta h \tag{7.27}$$

Since the compression and expansion processes in the pump and turbine are adiabatic processes $(q = 0)$, Eq. (7.26) reduces to

$$w = -\Delta h \tag{7.28}$$

If the change in kinetic and potential energy in a steady-flow component is negligible, the reversible work produced by that component is equal to $-\int_a^b v \, dP$. Since the fluid in the pump is essentially incompressible, the ideal specific pump work is

$$_1 w_2 = -\Delta h = h_1 - h_2 = -\int_1^2 v \, dP \cong -v_1 \int_1^2 dP = -v_1(P_2 - P_1) \tag{7.29}$$

The negative value for this work term means that the work is done on the working fluid by the pump. Using the energy equation (7.28), the turbine work is

$$_3 w_4 = -\Delta h = h_3 - h_4 \tag{7.30}$$

The specific work for the ideal cycle is the sum of the specific works of the pump and the turbine, or

$$(w)_{net} = {}_1 w_2 + {}_3 w_4 = h_1 - h_2 + h_3 - h_4 = h_3 - h_4 - v_1(P_2 - P_1) \tag{7.31}$$

The thermal efficiency of the simple cycle η_{th} becomes

$$\eta_{th} = \frac{w_{net}}{q_{add}} = \frac{h_1 - h_2 + h_3 - h_4}{h_3 - h_2} = 1 - \frac{h_4 - h_1}{h_3 - h_2} \tag{7.32}$$

An important operating limitation for any vapor cycle using water as the working fluid is the moisture content of the steam at the turbine exhaust. The turbine-outlet steam is usually at a very, very low pressure which means that the density is also very low. This low steam density produces very high velocity in the low-pressure part of the turbine. If there are any moisture droplets in this steam, they can actually erode the surface of the low-pressure turbine blades like raindrops erode the ground. This condition is likely to be severe if the moisture content of the steam at the turbine exhaust exceeds 10 percent.

The thermal efficiency of the simple steam or Rankine cycle can be improved by increasing the maximum steam temperature (the vapor superheat), by increasing the maximum system pressure P_{max}, and/or by increasing the condenser vacuum (decreasing the minimum system pressure P_{min}). Increasing the superheat of the steam also increases the specific work and improves (lowers) the moisture content of the steam at the turbine exhaust. Increasing the maximum system pressure increases the efficiency and may increase the specific work if the same maximum temperature is maintained, but it always increases the moisture content at the turbine exhaust. Lowering the minimum steam pressure increases the thermal efficiency and the specific work of the cycle, which is desirable, but it also increases the moisture at the turbine exhaust, which is undesirable. The moisture, thermal efficiency, and the specific work of the basic Rankine cycle are plotted in Fig. 7.24 as a function of the maximum temperature and the maximum and minimum pressures for the ideal cycle.

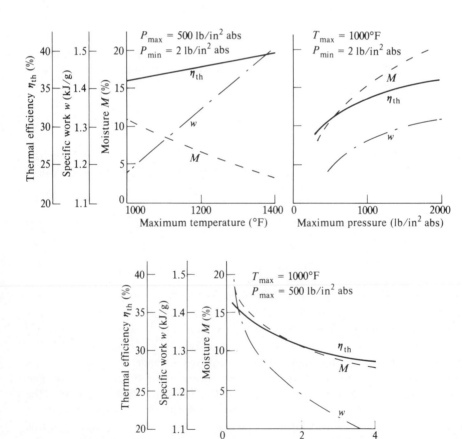

Figure 7.24 The variation of Rankine-cycle performance with cycle parameters.

The maximum steam temperature is normally limited by the materials employed in the superheater section of the steam generator and particularly by the materials in the inlet of the high-pressure turbine. Most modern steam systems are limited to a maximum operating temperature of 538 to 595°C (1000 to 1100°F). The maximum steam pressure is limited by the operating conditions in the steam generator and particularly by the moisture content at the turbine exhaust. The minimum vapor pressure in the condenser is limited by the temperature of the heat sink. In modern steam power plants, the minimum pressure ranges from about 1 in of mercury, absolute (79°F) to around 2 lb/in^2 (126°F).

The thermal efficiency of the ideal Rankine cycle is strongly affected by the compression and expansion efficiencies of the pump and turbine. The inefficiencies of these systems irreversibly converts some of the mechanical energy into thermal energy, thereby increasing the enthalpy of the fluid leaving the component. This effect, in a steam turbine, is sometimes referred to as "stage reheat." If the turbine efficiency is η_t, the actual enthalpy of the steam leaving the turbine $h_{4,a}$ is

$$h_{4,a} = h_3 - \eta_t(h_3 - h_{4,s}) = h_{4,s} + (1 - \eta_t)(h_3 - h_{4,s}) \tag{7.33}$$

where h_3 is the enthalpy of the turbine inlet steam and $h_{4,s}$ is the turbine outlet value for an isentropic expansion. If the pump efficiency is η_p, the actual enthalpy of the water leaving the pump $h_{2,a}$ is

$$h_{2,a} = h_1 + \frac{h_{2,s} - h_1}{\eta_p} = h_1 + \frac{v_1(P_2 - P_1)}{\eta_p} \tag{7.34}$$

Since $h_{4,a}$ is increased by the inefficiencies in the turbine, this improves (decreases) the moisture content at the turbine exhaust. Although this last effect is desirable, the lower turbine efficiency adversely effects the overall cycle performance.

The reheat Rankine cycle is a modification of the simple Rankine cycle in which all of the steam is removed from the turbine after partial expansion and returned to the boiler where additional heat is supplied to the steam at constant pressure in the reheat section of the boiler. The reheat vapor power cycle is shown plotted on a T–s diagram in Fig. 7.25 along with a schematic diagram of the system. This cycle permits the use of very high steam pressures without excessive moisture at the condenser inlet. The use of higher pressures increases the thermal efficiency of the cycle, if the cycle is a simple cycle. The efficiency of the reheat cycle may or may not be as efficient as a simple Rankine cycle operating between the same pressures and maximum temperature. As long as the reheat pressure is more than 80 percent of the maximum system pressure, the thermal efficiency will normally exceed that for the simple cycle.

Using the cycle notation as given in Fig. 7.25, the specific work for the cycle becomes

$$w = {}_1w_2 + {}_3w_4 + {}_5w_6 = h_1 + h_3 + h_5 - h_2 - h_4 - h_6$$

$$= h_3 + h_5 - h_4 - h_6 - \frac{v_1(P_2 - P_1)}{\eta_p} \tag{7.35}$$

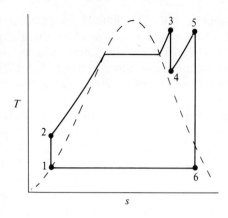

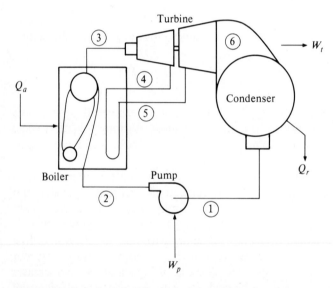

Figure 7.25 The reheat Rankine heat-engine cycle and system.

and the thermal efficiency for the cycle becomes

$$\eta_{th} = \frac{w}{q_{add}} = \frac{h_1 + h_3 + h_5 - h_2 - h_4 - h_6}{h_3 + h_5 - h_2 - h_4}$$

$$= 1 - \frac{h_6 - h_1}{h_3 + h_5 - h_2 - h_4} \tag{7.36}$$

The regenerative vapor power cycle is a modification of the basic Rankine cycle that improves the thermal efficiency of the basic simple cycle. In the regenerative vapor cycle, some steam is bled from the turbine before it reaches the condenser and this steam is used to heat the compressed boiler feed before it reenters the steam generator. The low-efficiency part of the Rankine cycle is that part of the cycle where heat is added to the feedwater. Thus, the thermal efficiency of the basic cycle can be improved if some of this heat addition is accomplished with the use of internal energy from the cycle. Extraction of underexpanded steam from the turbine for feedwater heating always reduces the specific work for the cycle and this means that higher steam flow rates are required for the same power output.

There are two basic types of feedwater heaters—open heaters and closed heaters (Fig. 7.26). In the open heater, the steam is mixed directly with the feedwater, whereas the closed heater is essentially a tube-and-shell heat exchanger with no direct contact between the fluids. The positioning of the heaters with respect to extraction pressures is important in order to obtain the maximum increase in the thermal efficiency for a given number of heaters.

If $h_{f,H}$ is the specific enthalpy of saturated liquid at the maximum pressure and $h_{f,L}$ is the specific enthalpy of saturated liquid at the minimum pressure, the increase in the specific enthalpy of the feedwater Δh_H across each of n heaters for maximum thermal efficiency is

$$\Delta h_H = \frac{h_{f,H} - h_{f,L}}{n+1} \tag{7.37}$$

The value of the specific enthalpy $h_{f,i}$ of the saturated water at the desired extraction pressure for the ith heater is

$$h_{f,i} = h_{f,L} + i(\Delta h_H) \tag{7.38}$$

where the first heater ($i = 1$) is the low-pressure heater. The optimum extraction pressure for the ith heater can then be found by determining the saturation pressure corresponding to $h_{f,i}$ from the steam tables.

The open feedwater heater has some advantages over the closed heater. The open heater is essentially a tank and consequently is less expensive than the closed heater. The heat transfer in the open heater is excellent since the water and steam can be mixed directly. However, this means that it is impossible to increase the temperature of the exit water above the saturation temperature of the extraction steam. In a properly designed closed heater, the feedwater exit temperature may exceed the saturation temperature of the extraction steam if this steam is superheated.

Each open heater must be accompanied by a feedwater pump and care must be taken to assure that there is no way that feedwater can be pumped up the extraction line into the turbine. Every regenerative power cycle normally contains at least one open feedwater heater that operates at a pressure slightly higher than atmospheric pressure. This heater, commonly called the deaerating

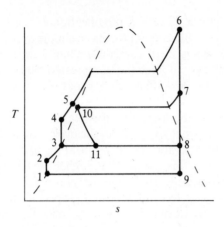

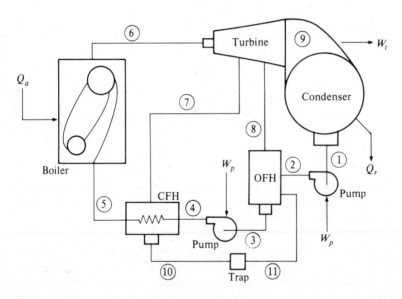

Figure 7.26 The regenerative vapor power cycle and system with open and closed heaters.

heater, is used to remove any dissolved gases, particularly oxygen, from the boiler feed. A typical deaerating heater is shown in Fig. 7.27.

The closed feedwater heater is essentially a tube-and-shell heat exchanger with the high-pressure water flowing in the tubes and the condensing steam in the shell. The condensate produced in the shell of the exchanger can either be pumped to higher pressure or it can be "trapped" to lower pressure. A "steam

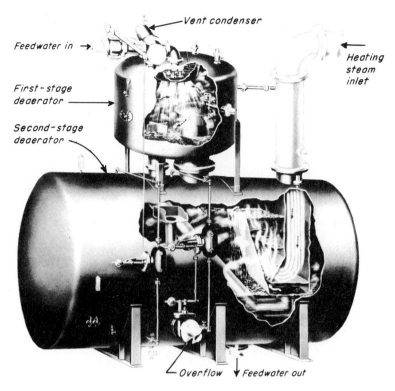

Figure 7.27 A spray-type deaerating heater. *(From " Power Generation Systems," 1967.)*

trap " is a float valve and when the water level reaches a given point, the float opens the valve and throttles the water to lower pressure. The use of a pump produces a slightly higher cycle efficiency but pumps are more expensive and require more maintenance than steam traps. Consequently, most closed heaters use a steam trap to throttle the condensate to the next lower heater or to the condenser.

The important performance parameter for feedwater heaters is the terminal temperature difference. The terminal temperature difference is equal to the saturation temperature of the extraction steam T_{sat} minus the outlet feedwater temperature $T_{w, out}$:

$$\text{Terminal temperature difference} = \text{TTD} = T_{sat} - T_{w, out} \tag{7.39}$$

This quantity should be as low as possible, preferably negative, but the limit value for an open heater is zero. If the extraction steam is superheated, proper design of a counterflow closed heater can produce a negative value of the terminal temperature difference.

The required fractional steam mass flow rate m_j to a feedwater heater is determined by performing an energy balance on the heater. The feedwater heater is a steady-flow system in which there is no work $(w = 0)$ and no external heat transfer $(q = 0)$. This means that the total enthalpy entering the heater must equal the total enthalpy leaving the heater:

$$H_{in} = \sum_j (m_j h_j)_{in} = H_{out} = \sum_j (m_j h_j)_{out} \tag{7.40}$$

It is not possible to develop a more specific equation for a given heater because the specific enthalpy equation depends on the way that the system components are connected and the type of heaters and pumps or traps used in the cycle. If more than one feedwater heater is employed in the cycle, start with the highest-pressure heater and work down through the lower-pressure heaters sequentially.

Example 7.3 Find the thermal efficiency and specific work of a simple Rankine cycle if the maximum temperature and pressure are 540°C and 70 bar and the minimum pressure is 0.1 bar. Assume the turbine and pump efficiencies are both 85 percent. Also find the specific work, the thermal efficiency, and the optimum turbine bleed pressure if one open feedwater heater is employed in the cycle. Assume that the terminal temperature difference is zero.

All steam and water properties are taken from steam tables by Keenan, Keys, Hill, and Moore (metric units).

SOLUTION

(a) The simple Rankine cycle is shown on the next page.

$$h_1 = h_f(\text{at } 0.1 \text{ bar}) = 191.8 \text{ kJ/kg}$$

$$h_2 = h_1 + v_1(P_2 - P_1)\eta_p = 191.8$$

$$+ \frac{(0.10102 \times 10^{-5} \text{ m}^3/\text{g})(70 - 0.1) \text{ bar } (10^5 \text{ N/m}^2 \cdot \text{bar})}{0.85}$$

$$h_2 = 200.1 \text{ kJ/kg}$$

$$h_3 = h(\text{at } 540°C, 70 \text{ bar}) = 3506.9 \text{ kJ/kg}$$

$$s_3 = s_{4,s} = 6.9193 \text{ kJ/kg·K}$$

$$M_{4,s} = \frac{s_g - s_{4,s}}{s_{fg}} = \frac{8.1502 - 6.9193}{7.5009} = 16.41\%$$

$$h_{4,s} = h_g - M_{4,s} h_{fg} = 2584.7 - 0.1641(2392.8) = 2192.0 \text{ kJ/kg}$$

$$h_{4,a} = h_{4,s} + (1 - \eta_t)(h_3 - h_{4,s})$$

$$= 2192 + (0.15)(3506.9 - 2192) = 2389.2 \text{ kJ/kg}$$

Net work = specific work = $w_{net} = h_1 + h_3 - h_{2,\text{act}} - h_{4,\text{act}}$

$$= 191.8 + 3506.9 - 200.1 - 2389.2 = 1109.4 \text{ kJ/kg}$$

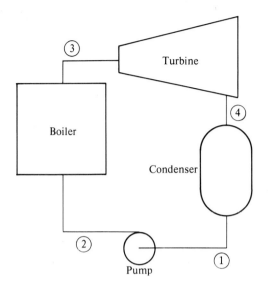

Thermal efficiency $= \eta_{th} = 1 + \dfrac{q_r}{q_a} = 1 + \dfrac{h_1 - h_{4,a}}{h_3 - h_{2,a}}$

$$= 1 + \frac{191.8 - 2389.2}{3506.9 - 200.1} = 33.55\%$$

(b) The regenerative Rankine cycle is shown on the next page.

$h_{f,L} = h_f(\text{at } 0.1 \text{ bar}) = 191.8 \text{ kJ/kg}$

$h_{f,H} = h_f(\text{at } 70 \text{ bar}) = 1267.0 \text{ kJ/kg}$

$n = 1$ heater

$$\Delta h = \frac{h_{f,H} - h_{f,L}}{n+1} = \frac{1267 - 191.8}{2} = 537.6 \text{ kJ/kg}$$

$h_{f,i} = h_{f,L} + \Delta h = 191.8 + 537.6 = 729.4 \text{ kJ/kg}$

Turbine bleed

pressure $= P_6 = P(\text{at } h_f = 729.4 \text{ kJ/kg}) = 8.4 \text{ bar}$

$h_1 = 191.8 \text{ kJ/kg}$

$$h_2 = 191.8 + \frac{0.10102(8.4 - 0.1)}{0.85} = 192.8 \text{ kJ/kg}$$

$h_3 = h_f(\text{at } 8.4 \text{ bar}) = 730.0 \text{ kJ/kg}$

$$h_{4,a} = 730.0 + \frac{0.11174(70 - 8.4)}{0.85} = 738.1 \text{ kJ/kg}$$

$h_5 = 3506.9 \text{ kJ/kg}$

$s_5 = s_6 = s_7 = 6.9193 \text{ kJ/kg}\cdot\text{K}$

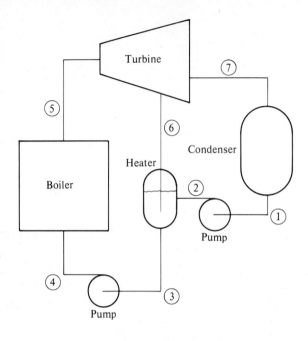

$h_{6,s}$(by interpolation from the steam tables) $= 2900.2$ kJ/kg

$$h_{6,a} = h_{6,s} + (1 - \eta_t)(h_5 - h_{6,s}) = 2900.2 + (0.15)(3506.9 - 2900.2)$$

$$= 2991.2 \text{ kJ/kg}$$

h_7[from part (a)] $= 2389.2$ kJ/kg

Steam extraction fraction $= m$

$$mh_{6,a} + (1 - m)h_{2,a} = h_3$$

$$m = \frac{h_3 - h_{2,a}}{h_{6,a} - h_{2,a}} = \frac{730.0 - 192.8}{2991.2 - 192.8} = 0.1920$$

Thermal efficiency $= \eta_{th} = 1 + \frac{q_r}{q_a} = 1 + \frac{(1 - m)(h_1 - h_7)}{h_5 - h_4}$

$$= 1 + \frac{(1 - 0.192)(191.8 - 2389.2)}{3506.9 - 738.1}$$

$$= 35.87\%$$

$$\text{Specific work} = w_{\text{net}} = q_a\eta_{\text{th}} = (3506.9 - 738.1)(0.3587)$$
$$= 993.2 \text{ kJ/kg}$$
$$= h_3 - h_4 + (1 - m)(h_1 - h_2) + h_5 - h_6$$
$$+ (1 - m)(h_6 - h_7)$$
$$= -8.1 + (0.808)(-1) + 3506.9$$
$$- 2991.2 + 0.808(2991.2 - 2389.2)$$
$$= 993.2 \text{ kJ/kg}$$

Two engineers at Dupont have developed and tested a novel rotary engine that operates on the Rankine cycle. This system operates with an organic working fluid and is novel in that it has both a rotating boiler and condenser. A test engine has been built and operated at a rated power of 20 hp. The schematic diagram of this engine is shown in Fig. 7.28.

—Du Pont's 20 hp laboratory test engine. Boiler, nozzle ring, and condenser are an integral unit—rotated by external electric motor (not shown) at 2500 rev/min. The expanding gas rotates the turbine in the opposite direction at 27,500 rev/min.

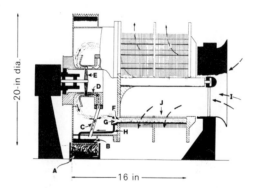

Figure 7.28 A rotary Rankine-cycle engine. External burner (A) boils fluid in annular chamber where liquid (B) is held on outer surface by centrifugal force. Hot vapor travels through tubes (C) to nozzle ring (D), expands through ring and drives turbine wheel (E). Engine power from turbine shaft drives external load (not shown). Vapor goes into condenser tubes (F). Condensed fluid flows back to channel (G). Centrifugal action forces liquid from channel through tubes (H) back to boiler (B). Cooling air is pulled into hollow center of condenser cylinder (I), then between condenser fins (J) by viscous drag of rotating fin assembly. (*From Doerner et al., 1972.*)

7.2.8 Combined Cycles

Earlier, it was pointed out that mercury and water have been employed in a binary vapor power cycle. When two power cycles, A and B, with respective thermal efficiencies of η_A and η_B are connected in series so that q_A is the specific

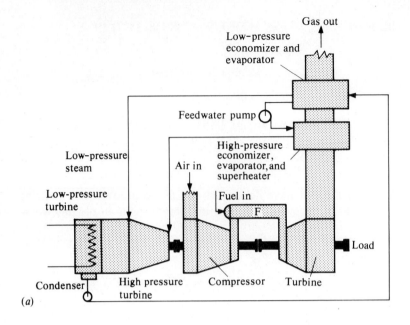

(a)

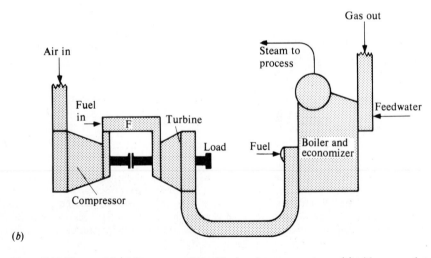

(b)

Figure 7.29 The combined (Brayton and Rankine)-cycle power systems, (a) with no supplementary firing and (b) with supplementary firing. (*From Skrotzki, 1963.*)

heat transfer to cycle A and the heat rejection from cycle A plus some additional quantity of heat q_B is the heat added to cycle B, the overall thermal efficiency of the combination is

$$\eta_{th} = \frac{q_A \eta_A + (1 - \eta_A) q_A \eta_B + q_B \eta_B}{q_A + q_B} = \eta_B + \frac{q_A \eta_A (1 - \eta_B)}{q_A + q_B} \qquad (7.41)$$

In recent years, considerable interest has been given to the possibility of combining the Brayton or gas-turbine cycle with the Rankine or vapor power cycle in the so-called combined-cycle power system. In these systems, a heat-recovery steam generator (HRSG) is connected directly to the exhaust of a gas turbine. The steam generator may operate with the turbine exhaust alone $(q_B = 0)$ or supplemental heat may be added to the turbine exhaust before it enters the steam generator $(q_B > 0)$. Supplemental heat is easily added to the turbine exhaust by simply burning more fuel in the exhaust because the typical gas turbine operates with very high excess air. The schematic diagrams of the unfired combined cycle (a) and also for the supplementary-fired combined cycle (b) are shown in Fig. 7.29. A more detailed diagram of the HRSG is shown in Fig. 7.30.

Since 1971, a number of companies have developed combined-cycle power systems in package form. Westinghouse has a system called PACE (an acronym for *power at combined efficiencies*). The General Electric system is called STAG (an acronym for *steam and gas* turbines) and Stone and Webster has a combined-cycle system called FAST.

The combined-cycle systems appear to be very promising for the so-called intermediate-load service between the peaking units (gas turbines and diesels) and the base-load units (large fossil-fuel or nuclear steam power plants). The combined-cycle plants are attractive for capacity factors ranging from 15 to 45 percent (1300 to 4000 h/year). For capacity factors of less than 15 percent (0 to 1300 h/year), the gas-turbine and diesel units are the best because of the low capital cost. For capacity factors greater than 45 percent (> 4000 h/year), the large coal-fired and nuclear power plants are the best because of the lower fuel costs.

7.3 TURBINES

7.3.1 Steam Turbines

General The turbine is a device that converts the stored mechanical energy in a fluid into rotational mechanical energy. There are several different types of turbines, including steam turbines, gas turbines, water turbines, and wind turbines or windmills. The material presented in this section is applicable to only steam or gas turbines, although some of the general principles are applicable to the other systems as well.

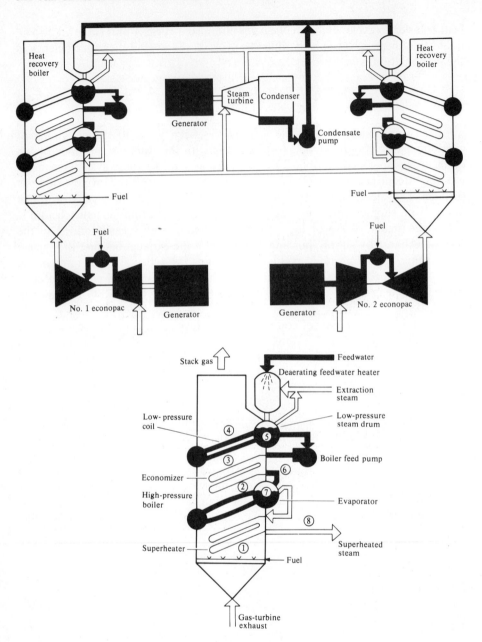

Figure 7.30 The combined cycle and the heat-recovery steam generator (HRSG). (*From Berman and Lebonette, 1971.*)

There are several different ways to classify steam turbines. One way of classifying steam turbines is with respect to the purpose of the turbine. This system includes the central-station units which are used to drive electrical generators at synchronous speed (usually 3600 or 1800 rev/min) and have power capabilities of 16 to 1500 MW_e.

Superposed or topping steam turbines are high-pressure turbines that are installed in older, low-pressure steam systems to improve the overall thermal efficiency of the power plant. The relatively high-pressure turbine exhaust is fed directly to the existing low-pressure turbine or to a manufacturing process.

Mechanical-drive turbines are used to power large draft fans, pumps, compressors, and other large rotating machinery. These systems normally operate between 900 and 10,000 rev/min and range in capacity from 0.5 to 10 MW. These systems have a number of advantages over electrical-drive systems, including better utilization of thermal energy, easy speed control, and fast startup; in addition, no sparks are generated during normal operation, they are amenable to hot, damp environments, and the low-pressure exhaust steam can be used for other purposes.

Steam turbines can also be classified according to the back or exhaust pressure of the unit. Under this classification system, turbines are classed as either condensing or noncondensing turbines. In the noncondensing turbine, the turbine-exhaust pressure is above or equal to atmospheric pressure and the system can operate with or without a condenser. This system may require a continuous water makeup. A condensing turbine normally discharges steam to a condenser at a very high vacuum and this improves the thermal efficiency of the cycle. There is no need for continuous makeup to this system.

Steam turbines may be classified according to the method of steam injection or extraction from the turbine. Bleeder or extraction turbines are used where turbine steam is removed partway through the turbine for process use or for feedwater heating. Reheat turbines are used in the reheat vapor power cycles. Extraction-induction turbines have ports for both the extraction and induction of steam at intermediate points in the turbine. The various turbine arrangements, including tandem and cross-compounded units, are shown in Fig. 7.31. The different types of steam turbines and their connections are given in Fig. 7.32.

Turbine blading Although there are a few radial-flow steam turbines, most of the turbines are axial-flow machines; a typical unit of this type is shown in Fig. 7.33. There are two basic types of turbine blading employed in common steam turbines—impulse blading and reaction blading. Two different types of impulse blades and one set of reaction blades are pictured in Fig. 7.34.

In the impulse turbine stage, all of the steam first passes through a nozzle block where the kinetic energy of the steam is increased. Flow through these nozzles approaches a reversible adiabatic (isentropic) process. If the inlet velocity is negligible, the exit steam velocity $\bar{V}_1$ becomes

$$V_1 = (-2g_c\,\Delta h)^{1/2} = 223.73(-\Delta h)^{1/2} \qquad \text{ft/s} \qquad (7.42)$$

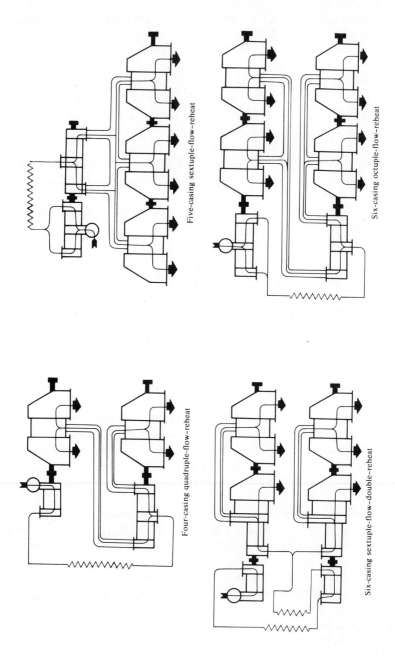

Five-casing sextuple-flow-reheat

Six-casing octuple-flow-reheat

Four-casing quadruple-flow-reheat

Six-casing sextuple-flow-double-reheat

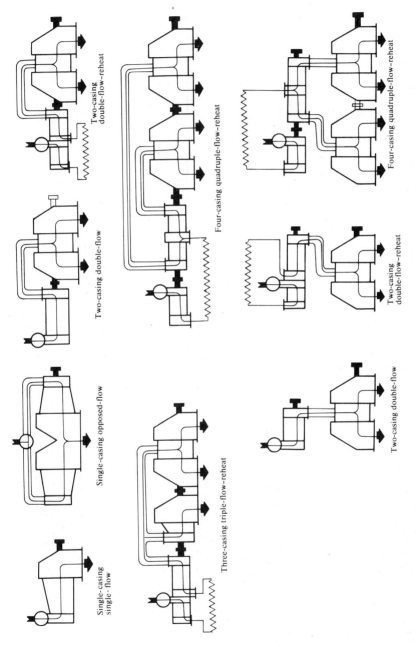

Figure 7.31 Casing and shaft arrangements for steam turbines, which depend on capacity and steam conditions. (*From Skrotzki, 1962.*)

Single-casing
single-flow

Single-casing opposed-flow

Two-casing double-flow

Two-casing
double-flow–reheat

Three-casing triple-flow–reheat

Four-casing quadruple-flow–reheat

Two-casing double-flow

Two-casing
double-flow–reheat

Four-casing quadruple-flow–reheat

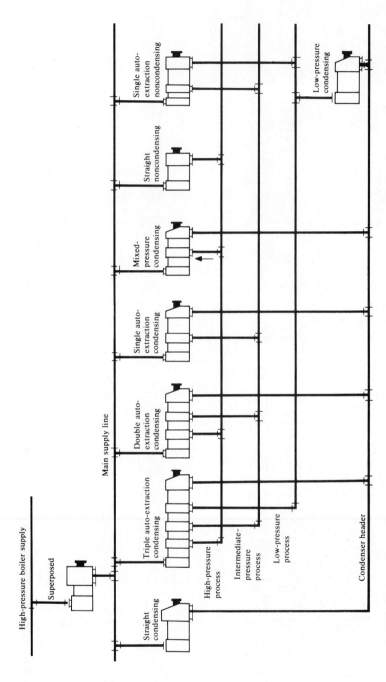

Figure 7.32 The different types of steam turbines and their connections. Heat-balance arrangements are almost endless. Turbines fit almost any heat balance as long as there is sufficient steam-pressure drop. Most efficient hookup occurs when process-steam flow and shaft-power demands coincide. Shaft-power needs in excess of that generated by process steam call for condensing units. Throttling valves meet any steam needs over and above what is required for shaft power. (*From Skrotzki, 1962.*)

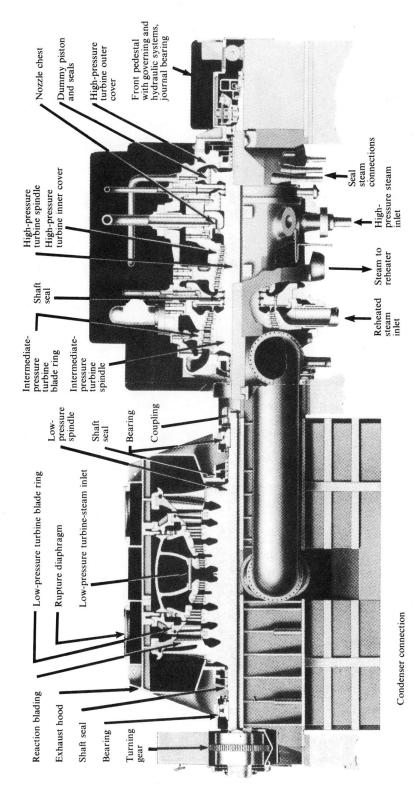

Reaction blading

Low-pressure turbine blade ring

Rupture diaphragm

Low-pressure turbine-steam inlet

Shaft seal

Bearing

Turning gear

Low-pressure spindle

Shaft seal

Bearing

Coupling

Intermediate-pressure turbine blade ring

Intermediate-pressure turbine spindle

Shaft seal

High-pressure turbine spindle

High-pressure turbine inner cover

Nozzle chest

Dummy piston and seals

High-pressure turbine outer cover

Front pedestal with governing and hydraulic systems, journal bearing

Seal steam connections

High-pressure steam inlet

Steam to reheater

Reheated steam inlet

Condenser connection

Figure 7.33 A typical steam-turbine installation. (*From Skrotzki, 1962.*)

349

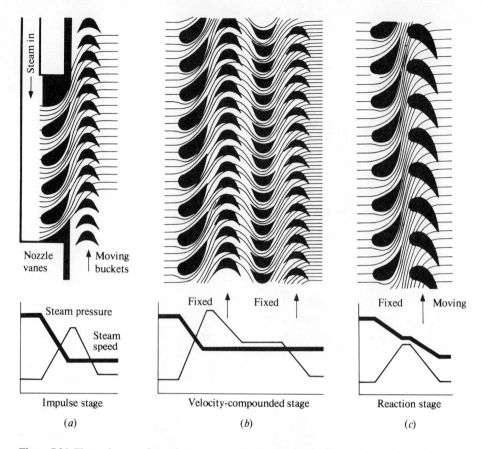

Figure 7.34 The various configurations, pressure distributions, velocity profiles, and specific-volume distributions for different types of turbine blading: (a) two Rateau impulse stages, (b) a Curtis impulse stage, and (c) two reaction stages. (*From Skrotzki, 1962.*)

where Δh is the enthalpy drop across the nozzle, in British thermal units per pound-mass. The maximum pressure ratio across each nozzle block is around 0.53 for superheated and about 0.58 for saturated steam.

The high-velocity steam jet leaving the nozzle is then directed toward the moving blades where part of the kinetic energy of the steam is then transferred to the blades. The velocity vectors in the tangential and axial directions of the turbine rotor are shown in Fig. 7.35. The force on the moving blades F_b is equal to ma_t/g_c, which can be written as $(m/g_c)(d\bar{V}_t/dt)$, or, letting $\dot{m}$ represent the steam flow rate through the blades,

$$F_b = \frac{\dot{m}}{g_c}(\bar{V}_{1,t} - \bar{V}_{2,t}) \qquad (7.43)$$

The energy transferred to the moving blades is equal to the product of the force on the blade times the distance moved by the blade, while the power

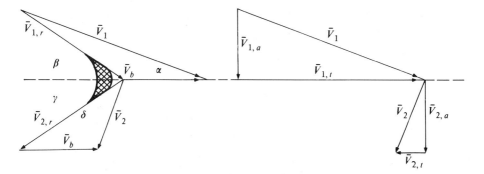

Figure 7.35 Steam velocity vectors for impulse turbine blading.

absorbed by the blade P_b is equal to the energy rate to the blade or the product of the blade force F_b and the blade velocity $\bar{V}_b$:

$$P_b = F_b \bar{V}_b = \frac{\dot{m}}{g_c}(\bar{V}_{1,t} - \bar{V}_{2,t})\bar{V}_b = \frac{\dot{m}}{g_c}(\bar{V}_1 \cos \alpha + \bar{V}_2 \cos \delta)\bar{V}_b \qquad (7.44)$$

Using geometry, it can be shown that Eq. (7.44) reduces to

$$P_b = \frac{\dot{m}}{2g_c}[(\bar{V}_1^2 - \bar{V}_2^2) - (\bar{V}_{1,r}^2 - \bar{V}_{2,r}^2)] \qquad (7.45)$$

If the steam flow through the blades is frictionless, the steam velocity, relative to the blade, is unchanged and $\bar{V}_{1,r} = \bar{V}_{2,r}$. In the actual case, however, there is friction and the relative steam velocity decreases as it flows through the blades, or $\bar{V}_{1,r} > \bar{V}_{2,r}$. For the case of frictionless flow, it can be seen from Eq. (7.44) that the effective rate of energy transfer to the blade is equal to the rate of change of the kinetic energy of the steam. For the case of flow with friction, the blade power is reduced by the second quantity of the equation, $(\dot{m}/2g_c)(\bar{V}_{1,r} - \bar{V}_{2,r})$. This last term represents the mechanical power that is converted into thermal power by friction.

The performance of a given turbine blade is given by the blade efficiency, which is defined as the fraction of the kinetic energy of the inlet steam transferred to the blade, or

$$\text{Blade efficiency} = \frac{2(\bar{V}_{1,t} - \bar{V}_{2,t})\bar{V}_b}{\bar{V}_1^2}$$

$$= \frac{(\bar{V}_1^2 - \bar{V}_2^2) - (\bar{V}_{1,r}^2 - \bar{V}_{2,r}^2)}{\bar{V}_1^2} \qquad (7.46)$$

For frictionless flow, the above equation reduces to

$$\text{Blade efficiency} = 1 - \left(\frac{\bar{V}_2}{\bar{V}_1}\right)^2 \qquad (7.47)$$

The blade efficiency and the blade power are a maximum when $\bar{V}_2$ is a minimum and this occurs when $\bar{V}_{2,t}$ is zero and $\bar{V}_2$ is equal to $\bar{V}_{2,a}$. When there is no friction, this condition is attained when the tangential component of the inlet steam velocity is equal to twice the blade velocity, or

$$\text{Optimum blade velocity} = \bar{V}_{b,\text{opt}} = \frac{\bar{V}_{1,t}}{2} = \frac{\bar{V}_1}{2} \cos \alpha \qquad (7.48)$$

The actual steam flow through the turbine blades does occur with friction and in this case $\bar{V}_2$ can be minimized by reducing the blade exit angle γ. This produces an asymmetric impulse turbine blade. Both the reduction in the blade exit angle and the presence of friction creates an axial load or thrust on the blade. Friction also increases the value of the exit steam enthalpy; this phenomenon is called "stage reheat."

As indicated earlier, there are two kinds of impulse blading. A Rateau stage is composed of a nozzle block followed by a single row of moving blades, as shown in Fig. 7.34a. A Curtis stage is composed of the nozzle block followed by a set of moving blades, followed by a set of stationary turning blades, which are followed, in turn, by another set of moving blades as shown in Fig. 7.34b. When two or more Rateau stages are put in series, they are said to be "pressure compounded" and if two or more Curtis stages are used in series, they are said to be "velocity compounded." The Rateau stage normally has a higher blade efficiency than the Curtis stage. The Curtis stage is commonly used as the first high-pressure stage in the turbine because it can accommodate a large pressure and enthalpy drop across the stage. Only a small fraction of the total stage power is generated in the second set of moving blades.

The other basic type of steam-turbine blading is the reaction blading and is shown in Fig. 7.34c. Reaction blades are shaped like airfoils and this produces a steam flow passage that is shaped like a converging-diverging nozzle. The stationary and moving blades have approximately the same shape. Thus, there is a pressure drop across both sets of blades. This is different from impulse blading which, barring friction, produces no pressure drop across the moving blades. The expansion through the moving blades gives an added thrust or reaction to the moving blades. If the enthalpy drop across both the stationary and the moving blades are the same, the reaction stage is said to be a "50 percent reaction stage." This condition results in $\bar{V}_1 = \bar{V}_{2,r}$.

The impulse stage produces the maximum power per stage and while the Rateau staging is more efficient, the Curtis staging can undergo a larger pressure drop. Impulse staging is commonly employed in the high-pressure part of the turbine because there is no pressure drop across the moving blades except that caused by friction. This produces minimal leakage around the ends of the moving blades and this is important in the high-pressure portion of the turbine where the blades are very short.

Reaction blading is commonly used in the intermediate and low-pressure parts of the turbine. Although end leakage is greater with reaction blading because of the higher pressure drop, the fractional loss is less because of the longer blades. Reaction blading is used in most of the turbine because it gives higher stage efficiencies than impulse staging.

7.3.2 Hydraulic Turbines

Hydraulic turbines convert the potential energy of the water into mechanical work. These systems provide approximately 17 percent of the electrical energy in the United States although this percentage will continue to drop as the electrical generation capacity is increased. The hydraulic units range in size from 2 to 615 MW for one of the units at the Grand Coulee power complex. In fact, when the Grand Coulee system is complete, it will have a total capacity of 7000 MW.

Hydraulic turbines may be classified as reaction, impulse, or mixed-flow turbines or they may be classified as radial, axial, or combined-flow turbines. There are three basic types of hydraulic turbines; they are shown in Fig. 7.36 along with a plot of their relative conversion efficiencies.

Propeller or axial turbines are used for low heads (10 to 100 ft). Fixed-bladed propeller turbines are inexpensive and have a high conversion efficiency at the design conditions but relatively poor conversion efficiency at lower flow rates. The Kaplan turbine is a propeller turbine with variable-pitch blades and these blades can be adjusted to give high efficiency for off-load conditions. Blade pitch is not achieved automatically but must be adjusted manually while the turbine is shut down.

The Francis turbine is one of the more common hydraulic turbines and is a radial, mixed-flow turbine that is employed for intermediate heads (15 to 1500 ft). These units have excellent conversion efficiency at the rated load but they have relatively poor efficiency at off-load conditions.

The Pelton wheel is an impulse hydraulic turbine which is normally employed at heads above 150 ft. These units require a relatively low water flow rate and the water velocity is controlled by a nozzle. Each "bucket" on the wheel has a "splitter" that deflects half of the flow to each side of the wheel to reduce the axial thrust. The Pelton wheel has a high conversion efficiency over a wide range of water flow.

7.3.3 Wind Turbines

Wind turbines or windmills convert the kinetic energy of the wind into mechanical work. For the production of alternating-current (ac) electricity, these systems must be designed to operate at a constant angular velocity over a wide range of wind speeds in order to produce a constant frequency. There are many different kinds of windmills, including propeller and other axial-flow turbines, as well as radial systems mounted on vertical axes. Some of the various wind-powered systems are shown in Fig. 7.37.

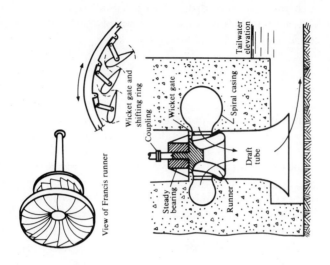

View of Francis runner

Wicket gate and shifting ring

Coupling

Wicket gate

Spiral casing

Steady bearing

Runner

Draft tube

Tailwater elevation

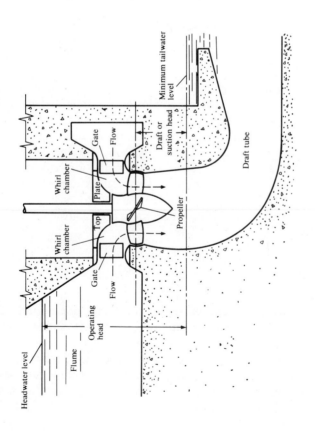

Headwater level

Flume

Operating head

Whirl chamber

Top

Plate

Gate

Flow

Whirl chamber

Gate

Flow

Propeller

Draft or suction head

Minimum tailwater level

Draft tube

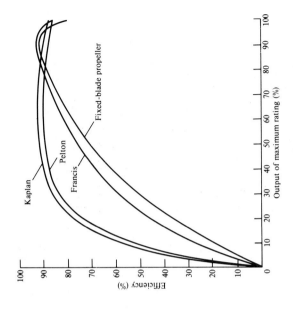

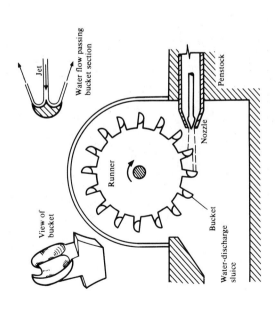

Figure 7.36 Hydraulic turbines and their performance. (*From Skrotzki and Vopat, 1960, and courtesy of V. A. Thiemann.*)

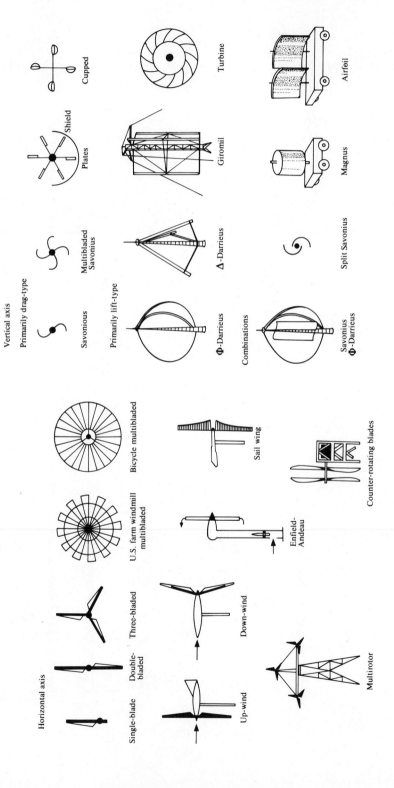

Cupped

Turbine

Airfoil

Shield

Plates

Giromil

Magnus

Vertical axis

Primarily drag-type

Savonious

Multibladed Savonius

Primarily lift-type

Δ-Darrieus

Split Savonius

Φ-Darrieus

Combinations

Savonius Φ-Darrieus

Bicycle multibladed

Sail wing

Counter-rotating blades

U.S. farm windmill multibladed

Enfield-Andeau

Horizontal axis

Three-bladed

Down-wind

Double-bladed

Multirotor

Single-blade

Up-wind

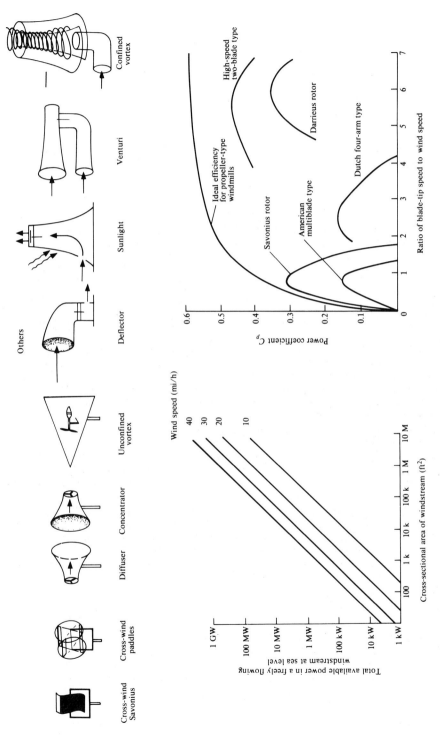

Figure 7.37 Basic wind-turbine systems and their performance. (*From Hirschfeld, 1977.*)

Wind-powered systems have been widely used since the tenth century for pumping water, grinding grain, and other low-power applications. There have been several attempts to build large-scale wind-powered systems to generate electricity. In 1931, the Russians built a large windmill with a 100-ft-diameter blade, but it had a very low conversion efficiency and was abandoned. In 1945, a Vermont utility built a large wind-powered generator to produce electricity. This system cost 1,250,000 dollars and had an electrical power output of 1250 kW_e. This unit operated for 23 days before one of the blades failed due to fatigue, and the project was abandoned.

NASA, in conjunction with ERDA, is currently testing a number of large wind-powered generators, including one at Sandusky, Ohio. This unit has a 125-ft-diameter blade, a capacity of 100 kW_e, and was built at a cost of 985,000 dollars. A diagram and picture of this unit is shown in Fig. 7.38.

In the design of windmills, it is important to keep the power-to-weight ratio as low as possible. This reduces the stresses caused by centrifugal forces in the blades. The theoretical maximum power that can be extracted from the wind by a ducted wind turbine is approximately 59 percent of the available wind power. An open windmill will normally attain 50 to 75 percent of this value due to spillage and other effects. The power produced by a wind turbine is directly proportional to the area swept by the blades and the wind power per unit area normal to the wind's velocity varies as the wind velocity cubed.

The Flettner rotor is a novel wind-powered system that is composed of a cylinder that is turned by an external power source. As the cylinder spins, it establishes a high pressure on that portion of the cylinder turning against the wind due to the stagnated flow in the boundary layer. This gives rise to a net force acting on the cylinder, as shown in Fig. 7.39, and this effect is called the Magnus effect or Magnus force. A wind-powered sailing vessel sailed across the Atlantic in the early 1930s using this mode of propulsion. A rather novel wind-powered electric generation system was proposed by J. D. Madaras. This system has a circular train track with a continuous train of flatcars. Each car has a 90-ft cylinder mounted on the car and each cylinder is spun at 120 rev/min. As the wind drives the cars around the circular track, the angular velocity of the cylinders is reversed. Electrical energy is produced by generators powered by the car wheels.

The design and successful operation of large-scale, wind-powered generators face a number of formidable problems. If the system is designed to produce ac power, a constant angular velocity and force is desirable. Unfortunately, the wind velocity is neither constant in magnitude nor direction and also varies from the bottom to the top of a large rotor. This imposes severe cyclic loads on the turbine blades, creating fatigue problems.

There are also times when the wind velocity is too low to produce significant power and this means that some sort of energy-storage system must be utilized with these systems. While these and other problems are difficult to overcome, the research and development of wind-powered systems continue because of the environmentally clean energy that is produced.

Figure 7.38 The 100-kW$_e$ wind-powered NASA-ERDA system under test at Sandusky, Ohio. (*From Kidd and Garr, 1972, and Lindsley, 1974.*)

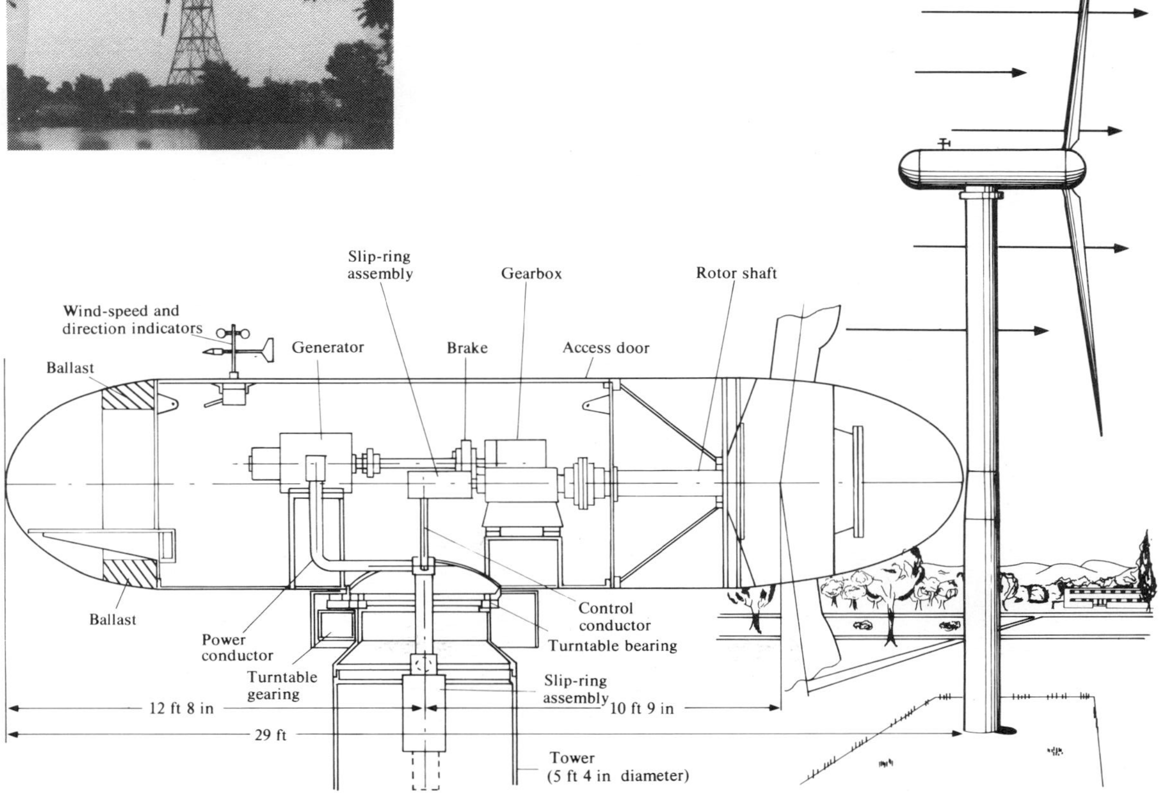

Rotor shaft

Access door

Gearbox

Brake

Slip-ring assembly

Generator

Wind-speed and direction indicators

Ballast

Control conductor

Turntable bearing

Slip-ring assembly 10 ft 9 in

Tower (5 ft 4 in diameter)

Power conductor

Turntable gearing

Ballast

29 ft

12 ft 8 in

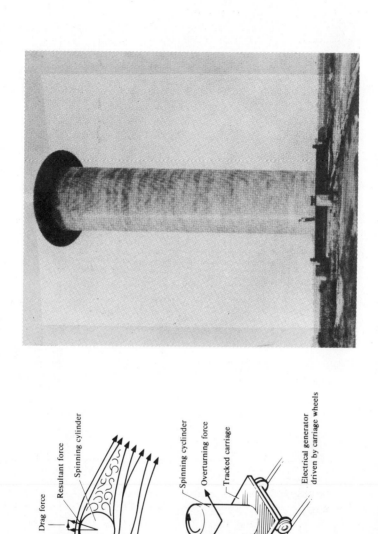

Figure 7.39 The proposed wind-powered Madaras electrical generating system and a picture of the pilot model of the Madaras plant that was erected in Burlington, N.J., in 1933 and that subsequently blew down. (*From Hirschfeld, 1977.*)

7.4 ELECTROMECHANICAL CONVERSION

7.4.1 General

Most of the systems that convert electrical energy into mechanical energy depend on the interaction of magnetic fields established by the flow of electrons (electric current) through conductors. When a direct current is passed through a conductor, a magnetic field is produced around the conductor, as shown in Fig. 7.40a. The "right-hand rule" states that if the thumb points in the direction of current flow, the fingers wind around the conductor in the direction of the magnetic field lines. The lines of force between the two poles of a magnet conventionally flow from the north pole to the south pole, as indicated in Fig. 7.40b.

If the conductor with the current flowing through it is placed between the two poles of the magnet, the magnetic field lines will have a high density on one side of the conductor and a low density on the other side. This produces a net force on the conductor in the direction of low flux density or perpendicular to the magnetic field. This phenomenon is demonstrated in Fig. 7.40c. The force on the conductor is proportional to the cross-vector product of the electric current through the conductor and the magnetic field strength. The force on the conductor, in dynes, is equal to

$$\text{Force} = \frac{L}{10}(\bar{I} \times \bar{B}) \tag{7.49}$$

where $\bar{I}$ is the current flowing through the conductor, in amperes, $\bar{B}$ is the magnetic flux density, in gauss, and L is the length of the conductor, in centimeters.

7.4.2 Electric Motors

The operation of an electric motor depends upon the interaction between the magnetic field of the stator or field winding and the magnetic field established in the rotor or armature winding. The magnetic field of the stator is established by a series of stationary coils wound around the field poles, as shown in Fig. 7.41.

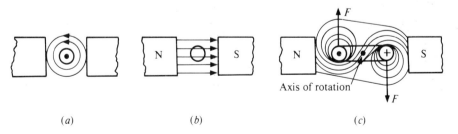

(a) (b) (c)

Figure 7.40 The interaction of magnetic fields around conductors.

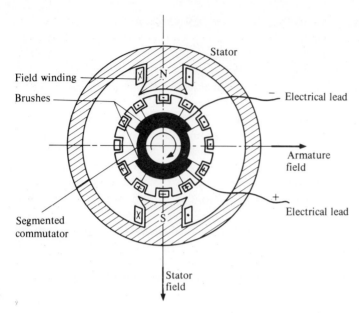

Figure 7.41 The magnetic fields in an electric motor.

The stator magnetic field is stationary if direct current is employed. If alternating current is passed through the field coils, however, the field-pole polarity reverses with the current reversal. Relative to the armature, it appears that the magnetic field of the stator revolves at synchronous speed. If 60-Hz current is put through the stator windings, the synchronous speed of the field is $7200/N$ rev/min, where N is the number of poles in the field.

In a dc motor, current is only supplied to those armature conductors that are directly opposite the stator poles, similar to that pictured in Fig. 7.41. This means that the magnetic field of the armature is 90° out of phase with the magnetic field of the stator, producing maximum torque. This condition is accomplished with the use of a segmented commutator and brushes.

In an ac motor, the magnetic field of the armature can be established by either introducing a direct current in the armature winding with a set of slip rings and brushes or by inducing or generating a current in the armature with the rotating field of the stator winding. When direct current is supplied to the armature of an ac motor, it effectively makes the rotor a permanent magnet and if the system is not overloaded the armature field will "lock" in with the rotating field of the stator and turn at synchronous speed. These motors are called synchronous motors and run at constant speed.

If the current flow through the rotor conductors is produced by the lines of force in the rotating field cutting the armature conductors, the motor is called an induction motor. The armature of an induction motor always runs at a slower angular velocity than the synchronous speed of the field and this speed differential is called "slip" and is usually reported in revolutions per minute. The slip increases as the load is increased. The rotor of an induction motor requires no electrical connections although the wound-rotor motor commonly uses a rheostat or variable resistor connected to the armature circuit by slip rings and brushes.

Direct-current motors have a number of advantages over the ac systems. The dc motors are more efficient than the ac systems because they do not have the hysteresis and eddy-current losses associated with ac windings. Consequently, it is common practice to employ dc motors for high-power applications if there is a source of direct current available. Direct-current motors are also best for variable-speed applications.

There are some problems associated with the use of dc motors. They have a higher initial cost and greater maintenance requirements than the ac motors. The use of segmented commutators and brushes greatly increases the required maintenance for these systems. The low availability of large-scale dc power also limits the application of these motors.

7.4.3 Types of Motors

There are three basic types of dc motors; they include the series-wound motor, the shunt-wound motor, and the compound-wound motor. In the series-wound motor, the field and armature windings are connected in series. This system has excellent starting torque but the speed varies widely with the load. In fact, this motor "runs away" if it is unloaded while running.

In the shunt-wound motor, the stator and armature windings are connected in parallel. These motors run at a fairly constant speed. These systems must have a variable resistor, called a starter or starting box, in the armature circuit. This resistor is put into the armature circuit and is removed as operating speed is attained because the back electromotive force (emf) generated in the armature then limits the armature current. The compound-wound motor has part of the field connected in series and part in parallel with the armature circuit. Some compound-wound motors have a centrifugal switch in the parallel-field winding so that the motor starts as a series-wound motor and runs as a compound-wound motor. The schematic diagrams for these motors along with their speed-torque characteristics and uses are shown in Fig. 7.42.

The two basic types of ac motors are the synchronous and induction motors. High-powered ac systems usually require polyphase ac power, usually three-phase ac power. Large synchronous motors require a dc power supply to the motor armature and this is provided through brushes and slip rings. These motors are started like a large induction motor.

Polyphase induction motors can be divided into either squirrel-cage motors or wound-rotor motors. Squirrel-cage motors have a rotor that literally looks like a cage and there is no electrical connection to the armature. The wound-rotor motor has no power to the armature but it does have a rheostat or variable resistor in the rotor circuit to produce better speed-torque characteristics. Although the induction motors are less expensive than the synchronous motor, and they require no power to the armature, they do produce a low, lagging power factor. This means that there is poor utilization of the ac energy because of the energy that is required to establish the magnetic fields in the

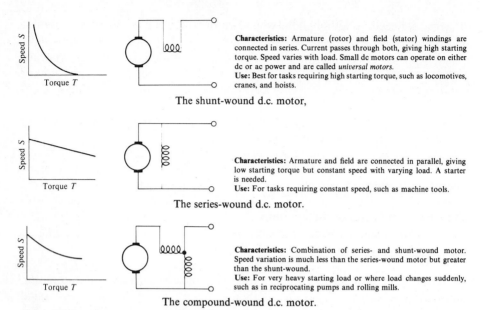

Characteristics: Armature (rotor) and field (stator) windings are connected in series. Current passes through both, giving high starting torque. Speed varies with load. Small dc motors can operate on either dc or ac power and are called *universal motors.*
Use: Best for tasks requiring high starting torque, such as locomotives, cranes, and hoists.

The shunt-wound d.c. motor,

Characteristics: Armature and field are connected in parallel, giving low starting torque but constant speed with varying load. A starter is needed.
Use: For tasks requiring constant speed, such as machine tools.

The series-wound d.c. motor.

Characteristics: Combination of series- and shunt-wound motor. Speed variation is much less than the series-wound motor but greater than the shunt-wound.
Use: For very heavy starting load or where load changes suddenly, such as in reciprocating pumps and rolling mills.

The compound-wound d.c. motor.

Figure 7.42 Schematics, characteristics, and uses of dc electrical motors. *(From Collins, 1969.)*

stator windings. The schematic diagrams of the various polyphase motors, along with the speed-torque characteristics and uses, are given in Fig. 7.43.

Almost all single-phase, squirrel-cage motors are limited to a maximum power of $7\frac{1}{2}$ hp. These systems normally require some method of phase displacement in part of the field winding so that they actually start as a two-phase motor in order to produce sufficient starting torque. This is usually accomplished by using an auxilliary starting-field winding with some sort of a reactance in the circuit to change the phase of the field current. This starting circuit commonly contains a centrifugal switch that opens slightly below the rated speed so that the motor starts as a two-phase motor but runs as a single-phase motor. The schematic diagrams of a number of single-phase, squirrel-cage motors are presented in Fig. 7.44 along with their speed-torque characteristics and their applications.

The single-phase, wound-rotor motors have an independent rotor circuit connected to a rheostat through slip rings. This system has excellent starting torque but the speed varies widely with load. Squirrel-cage rotors are often used with the wound-rotor system to produce a relatively constant angular velocity. The schematic diagrams for several of the single-phase, wound-rotor induction motors along with their speed-torque characteristics and their applications are given in Fig. 7.45.

Single-phase, synchronous motors are usually low-powered systems that depend upon reluctance and hysteresis effects to establish rotor polarity and there are no electrical connections to the armature. The reluctance motor has a dumbbell-shaped rotor that "locks" into the rotating magnetic field to give minimum magnetic reluctance. The schematic diagrams of some of the single-phase, synchronous motors are shown in Fig. 7.46 along with the speed-torque characteristics and the applications.

The National Electrical Manufacturers Association (NEMA) has specified normal service conditions for electrical motors. These conditions are as follows:

1. The ambient air temperature should not exceed 40°C (104°F).
2. The input voltage should be held within ±10 percent of the nameplate rating.
3. The frequency variation must be within ±5 percent of the nameplate rating.
4. The combined voltage and frequency swings should be within ±10 percent of the rated values but the frequency must not vary more than ±5 percent.
5. The motor should not be used at elevations more than 3300 ft above sea level.
6. Dust, moisture, dirt, and fumes must not seriously interfere with the motor ventilation.
7. All belt, chain, and gear drives must be solidly mounted.

It should be noted that if a standard motor is used in Denver, Colorado, it violates condition 5 because of the high altitude.

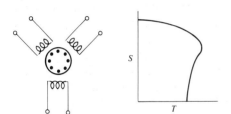

Characteristics: The rotor has copper conductor bars set into the periphery and connected to form a "squirrel cage." The simple design offers rugged service. Slip (difference between synchronous speed and actual speed of the rotor) and starting torque vary between the various types. Squirrel-cage motors are preferable to dc motors for constant-speed work, because the initial cost is less and the absence of a commutator reduces maintenance. Speed is determined by the number of poles.
Use: General-purpose, heavy applications such as blowers and drill presses. Absence of electrical contacts makes these motors ideal for operation in inflammable atmospheres.

Polyphase, squirrel-cage induction motor.

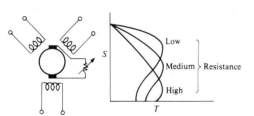

Characteristics: Wound-rotor induction motors are commutator machines very much like dc motors and have slip rings to provide external connections to the armature windings. Adjustable speeds are obtained through insertion of resistors in the armature circuit. The speed of maximum torque depends on the rotor resistance.
Use: Particularly adaptable wherever high starting torque and low starting current are required, where heavy or delicate loads must be accelerated gradually and smoothly, and where adjustable speed is desired.

Polyphase, wound-rotor induction motor.

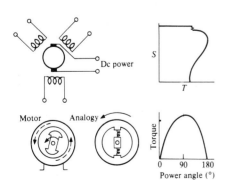

Characteristics: In contrast to the induction motor, the rotor of the synchronous motor rotates in exact step with line frequency (thus the name). The rotor is connected to a dc supply which provides a strong field in the rotor that locks in step with the stator's rotating ac field. However, small motors have what is effectively a permanently magnetized rotor. Speed is determined by the number of pole pairs and is always a ratio of line frequency. Since synchronous motors are not self-starting, a built-in squirrel cage in the large motors makes them start as an induction motor.
Use: Widely used to power electrical timing devices and machines which must operate in a synchronized manner with each other. Should not be used where torque fluctuations are violent. They are smaller in size than induction motors of the same horsepower and, therefore, cost less initially.

Polyphase, alternating-current, synchronous motor.

Figure 7.43 Schematics, characteristics, and uses of polyphase, ac motors. *(From Collins, 1969.)*

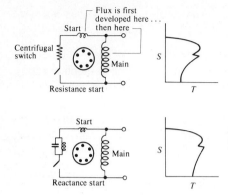

Characteristics: A split-phase motor has two windings—the main winding and start winding. Since the large main winding is highly inductive, its current lags applied voltage by nearly 90 degrees. All that remains is to provide a winding with no inductance, one in which the current and voltage are in phase, and a virtual two-phase motor is produced. This is done by a resistance circuit (called *resistance start*) or a reactance circuit (*reactance start*) which drops out after the motor reaches operating speed.

Use: These are low or moderate-starting-torque motors that are inexpensive. Since power is limited to about $\frac{1}{3}$ hp, these motors are used mainly for small appliances and office machines.

Split-phase, single-phase, squirrel-cage a.c. motors.

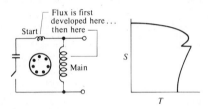

Characteristics: This motor has a capacitor connected to the stator to cause a momentary lead of current over voltage in that winding. The capacitor-start motor has fewer losses and a greater phase displacement than the split-phase motor, but is more expensive.

Use: Higher starting torque than the split-phase motor, hence useful for applications requiring more torque, such as household refrigerators. Available in up to 3 hp.

The capacitor-start, single-phase squirrel-cage a.c. motor.

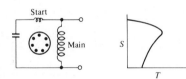

Characteristics: Also called capacitor split phase. A capacitor permanently connected in the start winding starts and runs the motor like a two-phase motor. This motor has the least starting torque of the three capacitor types, but it has a high maximum torque and requires the least maintenance of all single-phase motors.

Use: Due to high maximum torque, these motors are useful where momentary overloads are experienced, such as with grinders and sanders. Available in up to about $\frac{1}{3}$ hp.

The permanent-split-capacitor, single-phase, squirrel-cage a.c. motor.

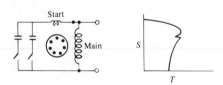

Characteristics: Capacitance in the rotor auxiliary circuit remains connected, but it is reduced in value after starting. Larger sizes are more expensive than three-phase motors, but three-phase power is usually not available.

Use: Probably the best of the single-phase motors. Exceptionally quiet. Available in fractional horsepower to 20 hp.

The two-value-capacitor, single-phase, squirrel-cage a.c. motor.

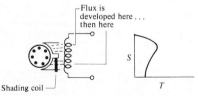

Characteristics: One pole piece in each pole is modified by placing a shorted copper turn or "shading coil" around it to split the component of the magnetic field. High slip, torque very slight.

Use: Low-torque applications, such as in small fans. Available in subfractional to fractional horsepower.

The shaded-pole, single-phase, squirrel-cage a.c. motor.

Figure 7.44 Schematics, characteristics, and uses of single-phase, squirrel-cage, ac motors. *(From Collins, 1969.)*

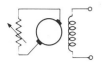

Characteristics: Like all wound-rotor motors, this type has an independent rotor circuit. However, unlike the polyphase motor, the commutator is placed so that the rotor's magnetic axis is offset from that of the stator to give the desired phase shift. Starting characteristics are very good, but the motor runs with varying speed.
Use: Like all repulsion types, this motor has good starting characteristics and is used for heavy loads.

Repulsion, single-phase, wound-rotor a.c. motor.

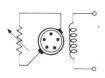

Characteristics: This motor has two windings on the rotor—a repulsion winding for starting and an inner squirrel-cage winding for running smoothly. However, electrical and mechanical noise, and extra maintenance are disadvantages.
Use: Applications requiring starting under full load, such as in conveyors and stokers.

The repulsion-induction, single-phase, wound-rotor a.c. motor.

Characteristics: This motor, also with two windings, starts as a repulsion motor with high torque and runs as a squirrel-cage motor with good constant-speed characteristics. Brush wear is minimized by using commutator circuit only at start.
Use: Most popular commutator, single-phase motor for industrial applications.

The repulsion-start, single-phase, wound-rotor a.c. motor.

Figure 7.45 Schematics, characteristics, and uses of single-phase, wound-rotor, induction, ac motors. *(From Collins, 1969.)*

Characteristics: The reluctance (resistance to establishing a magnetic field) rotor tends to align itself in the magnetic field so that magnetic reluctance is minimum. Roughly dumbbell shaped, the poles tend to align themselves along the flux field. Current alternating off and on continually pulls the poles around in exact step. The motor starts as an induction motor with split-phase or capacitor windings.
Use: This motor is inexpensive, simple, and suitable for light loads.

The single-phase, reluctance, synchronous motor.

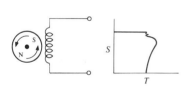

Characteristics: If the discrete poles of a reluctance motor are eliminated, the rotor still becomes magnetized, but the flux is not channeled as in the reluctance motor, and thus it is not as efficient. The rotor's tendency to maintain its magnetized condition in one position is called *hysteresis*. The hysteresis rotor uses this principle to run as the induced rotor poles are pulled around by the alternating field. However, the poles can move over the rotor surface slowly, and thus the rotor can slip while still providing torque. The hysteresis motor starts the same as the reluctance type.
Uses: Very steady speed makes this motor desirable for applications such as record-player drives. Having less tendency to hum and vibrate around synchronous speed, this motor is used more widely than reluctance motors.

The single-phase, hysteresis, synchronous motor.

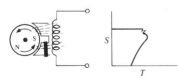

Characteristics: A special type of hysteresis motor, this type operates the same as a squirrel-cage shaded-pole motor, but at synchronous speed.
Use: Of all hysteresis motors, this type has the widest use in applications such as clocks, timing mechanisms, and advertising displays.

The single-phase, shaded-pole, synchronous motor.

Figure 7.46 Schematics, characteristics, and uses of single-phase, synchronous motors. *(From Collins, 1969.)*

7.4.4 Electromagnetic Pump

The electromagnetic pump converts electrical energy directly into kinetic energy of a conducting fluid. This system is sometimes used to pump liquid metals and other fluids with high electrical conductivities. In this pump, a low-voltage, high-amperage current is passed through the liquid metal by means of electrodes mounted on each side of the duct. The high-flux magnetic field, perpendicular to the current flow, produces a force on the conductor, which, in this case, is the fluid.

The electromagnetic pump has no moving parts and this can be important in the operation of a leak-tight, liquid-metal system. Unfortunately, this system has a conversion efficiency of around 30 to 40 percent, which is approximately half that of a conventional centrifugal pump. Consequently, electromagnetic pumps are used only where the expected flow rates are low and a leak-tight system is important. A schematic diagram of this pump is shown in Fig. 7.47.

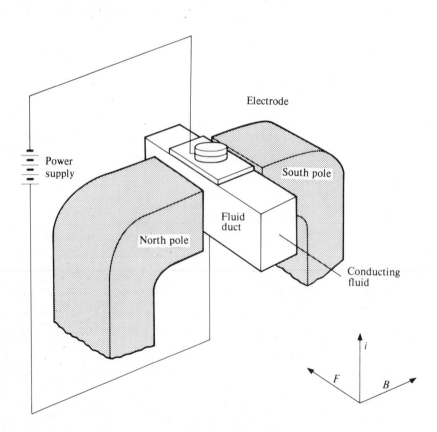

Figure 7.47 Schematic diagram of an electromagnetic pump.

PROBLEMS

7.1 A four-stroke, compression-ignition engine takes in ambient air at 10°C and 1 bar with a volumetric efficiency of 85 percent at 2000 rev/min. The effective fuel-air ratio is 0.058 and the LHV of the fuel is 44,200 kJ/kg. The six cylinders have a 0.15-m bore and a 0.18-m stroke. For an assumed thermal efficiency of 28 percent, estimate the brake horsepower, the brake specific fuel consumption, and the brake mean effective pressure.

7.2 Determine the approximate cylinder dimensions of a four-cylinder, four-stroke gasoline engine to deliver 70 Bhp at 2400 rev/min if the diameter-stroke ratio is 0.8. From past experience, it is anticipated that the bmep will be 120 lb/in² abs, the mechanical efficiency is about 80%, the bsfc is around 0.55 lbm/Bhp·h, and the LHV of the fuel is about 19,000 Btu/lbm. Assume the compression ratio is 7.0 and use an average value of k for the combustion products of 1.32. Calculate the expected and theoretical thermal efficiencies.

7.3 The air-standard Otto cycle has a compression ratio of 8.8. At the start of the compression process, the temperature is 27°C and the pressure is 0.95 bar. If the maximum temperature is 1650°C, determine the specific work of the cycle, the specific heat transfer to the cycle, and the thermal efficiency of the cycle.

7.4 A six-cylinder, four-stroke-cycle, single-acting automobile engine is required to develop 82 Bhp at 3600 rev/min. The probable friction horsepower, including friction losses, is 38 hp and the probable bmep is 5.85 bar. For $L/D = 1.35$, what is the bore and stroke? What are the probable brake torque, mechanical efficiency, and the indicated mean effective pressure?

7.5 A certain IC engine connected to an electric dynamometer when operated at full load gives these data: 1500 rev/min, 64.5 lbf on the dynamometer scale at the end of a 2-ft brake arm. At 1500 rev/min and no load except the dynamometer, the scale is adjusted to read zero so there is no tare. Fuel has a LHV of 16,280 kJ/kg and is used by this engine at the rate of 0.11 kg/min. Friction and windage losses are 2.5 hp. Determine the horsepower delivered to the dynamometer, the mechanical efficiency of the engine, and the thermal efficiency of the engine.

7.6 A four-stroke-cycle engine operates on the air-standard Diesel cycle. The engine has six cylinders with a 0.105-m bore and a 0.124-m stroke. The engine speed is 2000 rev/min. At the beginning of compression, the air is at 0.93 bar and 24°C and the maximum temperature is 1950°C. If the clearance volume is 6 percent of the displacement volume, compute the compression ratio, the pressure and temperature at the end of the compression process, the thermal efficiency, and the horsepower output of the engine. Also determine the efficiency and the power output if the cutoff ratio is halved.

7.7 The pressure and temperature before compression in an air-standard Diesel cycle are 0.94 bar and 22°C, respectively, and the conditions before and after the heat-addition process are 430 bar and 1850°C. Compute the compression ratio, the cutoff ratio, and the thermal efficiency of the cycle.

7.8 An eight-cylinder, four-stroke-cycle, single-acting automobile engine is required to develop 150 Bhp at 4000 rev/min. The friction torque, including fluid losses, is 68 ft·lbf at 4000 rev/min and the probable bmep is 6.26 bar. For an L/D ratio of 1.2, what are the bore and stroke and what are the probable torque and mechanical efficiency?

7.9 A certain generating plant requires 1 MW output. Two types of Diesel-engine installations are available. One unit is a low-speed unit that can operate on a low-grade fuel at 5.08 bar imep. The other two high-speed units can operate at 6.8 bar imep but require better fuel. If the isfc is 0.1015 kg/kW·h, if the fuel for the low-speed engine cost 288 dollars per ton and that for the high-speed engines is 320 dollars per ton, and if the friction mep is 1.22 bar in both cases, what is the fuel cost of the two installations in mils per kilowatthour?

7.10 An air-standard Brayton cycle operates with a compressor-pressure ratio of 4.2, the actual expansion and compression efficiencies of the gas processes are 0.88 and 0.82, respectively, and the maximum and minimum temperatures are 800 and 16°C, respectively. Compute the compression work, the expansion work, the ratio of expansion work to compression work, and the thermal efficiency, actual and theoretical. If the power output of the installation is 8 MW, determine the gas flow rate in kilograms per minute.

7.11 An open Brayton-cycle engine operates with a compressor-pressure ratio of 6, an inlet temperature of 20°C, and a turbine-inlet temperature of 950°C. The engine drives an electrical generator that produces 25 MW_e with a generator efficiency of 90 percent. For the following systems find the thermal efficiency, the specific work, and the air mass flow rate, if the compressor and turbine efficiencies are 80 percent.

  (*a*) A simple Brayton cycle

  (*b*) A regenerative Brayton cycle with a regenerator effectiveness of 80 percent

  (*c*) A Brayton cycle using two stages of compression with intercooling to 20°C

  (*d*) A Brayton cycle that employs both the regenerator of (*b*) and the intercooler of (*c*)

7.12 An open-cycle, regenerative, gas-turbine power plant receives air at 0.92 bar and 20°C. The air is compressed to 5.1 bar, is heated to 425°C in the regenerator, and then reaches a maximum temperature of 870°C in the combustion chamber. Assuming an air-standard cycle, compute the thermal efficiency of the plant if the mechanical efficiency of the compressor and turbine are 82 and 87 percent, respectively. Also determine the effectiveness of the regenerator.

7.13 A high-temperature, gas-cooled reactor (HTGR) is to be coupled to a gas turbine in a direct cycle, as shown below. The helium flow rate is 145,000 kg/h. Gas enters the turbine at 730°C and 25 bar, and is discharged at 450°C and 9.8 bar. The regenerator effectiveness is 88 percent. At the inlet

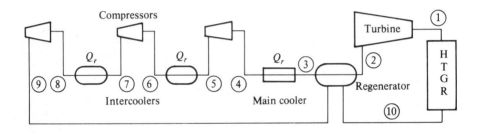

to the first stage of the three-stage compressor, the pressure is 9.52 bar and the temperature at all three inlets is 16°C. The outlet temperatures are each 66°C and the discharge pressure from the third stage is 25.5 bar. Find:

  (*a*) The pressure and temperature at each stage in the cycle

  (*b*) The reactor power in thermal megawatts

  (*c*) The fuel cost in mills per electrical kilowatthour if the fuel cost is 25 cents per 10^6 Btu

  (*d*) The thermal efficiency and the heat rate of the power cycle

7.14 An open Brayton cycle operates with a compressor-pressure ratio of 5, an inlet temperature of 80°F, and a turbine-inlet temperature of 1800°F, and produces a total of 50 MW of mechanical power. For the following systems, find the thermal efficiency, the specific work, the air mass flow rate, and the total compressor and turbine powers, in horsepower, assuming that the compressor and turbine efficiencies are 85 percent.

  (*a*) A simple Brayton cycle

  (*b*) A regenerative Brayton cycle with a regenerator effectiveness of 80 percent

  (*c*) A Brayton cycle using two stages of compression with one intercooler which cools the gas back to 80°F

  (*d*) A Brayton cycle that employs both the regenerator of (*b*) and the intercooler of (*c*)

7.15 Determine the thermal efficiency, the required steam flow rate, and the moisture at the turbine exhaust for a reheat-regenerative cycle which is to produce 200 MW at the turbine coupling if the throttle conditions are 160 bar and 540°C; reheat is at 100 bar and 590°C; one closed heater is at 34 bar; an open feedwater heater is at 1.7 bar; and the condenser pressure is 0.13 bar. The turbine and pump efficiencies are 88 percent. The terminal temperature difference from both heaters is 3 Celsius degrees, and the drain from the closed heater is trapped to the open heater. Also sketch the cycle on a T–s diagram.

7.16 Calculate the optimum heater pressures for Prob. 7.15 and recalculate the thermal efficiency for this cycle.

7.17 Determine the thermal efficiency, the required steam flow rate, and the moisture at the turbine exhaust for a reheat-regenerative cycle which is to produce 325 MW at the turbine coupling if the throttle conditions are 82 bar and 565°C; reheat is at 20.5 bar and 540°C; one closed heater is at 20.5 bar (reheater inlet); an open feedwater heater is at 1.4 bar; and the condenser pressure is 0.07 bar. The turbine and pump efficiencies are 88 and 82 percent, respectively. The terminal temperature difference from both heaters is 2 Celsius degrees and the drain from the closed heater is trapped to the open heater. Also sketch the cycle on a T–s diagram.

7.18 A 3800-MW$_{th}$ PWR is cooled with 153-bar water that enters the core at 300°C and leaves at 332°C. In the once-through steam generator, the high-pressure water is used to produce steam at 72 bar and 317°C (1). This steam expands to 6.8 bar (2) in the high-pressure turbine. Moisture (12) is separated and the saturated steam (3) is reheated with live steam (1) to 288°C (4). Before it enters the low-pressure turbine, the steam expands to 1.1 bar (5), where a fraction is bled to a closed feedwater

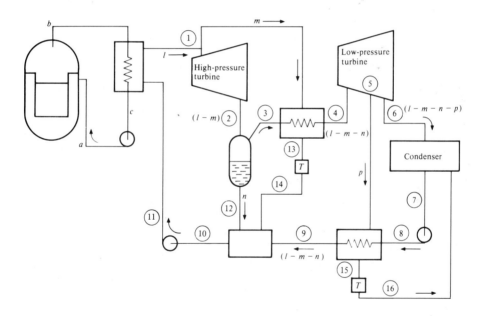

heater. Expansion continues to the condenser pressure of 0.1 bar (6). The separated moisture (12) is drained to an open feedwater heater and the reheater condensate (13) is "trapped" (14) to the same heater. The closed feedwater heater has a terminal temperature difference of 3 Celsius degrees. Each segment of the turbine expansion is 85 percent efficient and the pumps are 75 percent efficient.

(a) Sketch the cycle on the T–s plane.

(b) Find the cooling-water flow rate in the core, in liters per minute.

(c) Find the steam generation rate in kilograms per hour.

(d) Determine the power output of the system, in electrical megawatts, if the turbines drive a generator with an efficiency of 94 percent.

(e) What is the thermal efficiency of the cycle?

7.19 A reaction stage receives steam at 40 bar and 305°C and expands the steam through a pressure range of 3.4 bar. A flow of 13,600 kg/h occurs and at this rate the stage has an efficiency of 78 percent. Compute the blade work by the steam.

7.20 Steam at 85 bar and 510°C drops to 77 bar through an impulse stage. The nozzle makes an angle of 15° with the blade entrance and 92 percent of the ideal enthalpy drop through the nozzle is converted into kinetic energy. The ratio of blade to the inlet steam velocity is 0.45. Determine the steam work against the blade face per pound of steam flow.

REFERENCES

Berman, P. A., and F. A. Lebonette: Combined-Cycle Plant Serves Intermediate System Loads Economically, *Combustion*, vol. 42, no. 9, pp. 19–23, March, 1971.

Collins, G. W.: The Family Tree of Electric Motors, *Machine Design*, vol. 41, no. 1, pp. 152–156, January 9, 1969.

Dillio, C. C., and E. P. Nye: "Thermal Engineering," International Textbook Company, Scranton, Pa., 1963.

Doerner, W. A., R. J. Dietz, O. R. Van Buskirk, S. B. Levy, R. J. Rennolds, and M. F. Bechtold: "A Rankine Cycle Engine with Rotary Heat Exchangers," SAE publication 720053, 1972.

Hirschfeld, F.: Wind Power: Pipe Dream or Reality?, *Mechanical Engineering*, vol. 99, no. 9, pp. 20–28, September, 1977.

Kidd, S., and D. Garr: Electric Power from Windmills?, *Popular Science*, vol. 201, no. 5, pp. 70–72, November, 1972.

Krenz, J. H.: "Energy Conversion and Utilization," Allyn and Bacon, Inc., Boston, Mass., 1976.

Lindsley, E. P.: Wind Power: How New Technology is Harnessing an Age-Old Energy Source, *Popular Science*, vol. 205, no. 1, pp. 54–59, July, 1974.

"Power Generation Systems," by Editors of *Power Magazine*, McGraw-Hill Inc., New York, 1967.

Skrotzki, B. G. A. (associate ed.): "Steam Turbines," a *Power Magazine* special report, McGraw-Hill, Inc., New York, June, 1962.

Skrotzki, B. G. A.: "Gas Turbines," a *Power Magazine* special report, McGraw-Hill, Inc., New York, 1963.

Skrotzki, B. G. A., and W. A. Vopat: "Power Station Engineering and Economy," McGraw-Hill Book Company, New York, 1960.

Walton, H.: "The How and Why of Mechanical Movements," Popular Science Publishing Company, E. P. Dutton & Co., Inc., New York, 1968.

Wark, K.: "Thermodynamics," McGraw-Hill Book Company, New York, 1977.

Wood, B. D.: "Applications of Thermodynamics," Addison-Wesley Publishing Company, Inc., Reading, Mass., 1969.

PRODUCTION OF ELECTRICAL ENERGY

8.1 INTRODUCTION

Electrical energy is the energy associated with the flow or accumulation of electric charge. Electrical energy is a very useful form of energy as it can be easily converted into almost every other energy form with a high conversion efficiency. Moreover, electrical energy can be produced directly from the other energy forms without going through an intermediate form such as thermal or mechanical energy.

Some of the conversion systems used to produce electricity are often called direct-energy converters. Thermal energy can be converted directly to electrical energy in the thermoelectric converter and the thermionic converter, among others. These systems, however, are saddled with the maximum possible thermal efficiency of an externally reversible, heat-engine cycle, or $(1 - T_L/T_H)$. Chemical energy can be converted directly into electricity in the fuel cell and batteries. Electromagnetic energy can be converted directly into electricity in the photovoltaic or solar cell. Nuclear energy is converted directly into electricity in the nuclear battery. Mechanical energy can be converted into electricity in the conventional generator or alternator or in a fluiddynamic converter such as an EGD or MHD system.

In all of these conversion systems, except the ones involving the conversion of thermal energy, the conversion efficiency is not limited to the efficiency of an externally reversible, heat-engine cycle. As a result, there is considerable interest and research in some of these "direct converters" in order to produce a more efficient electrical generation system.

8.2 CONVERSION OF THERMAL ENERGY INTO ELECTRICITY

8.2.1 Thermoelectric Converters

As indicated earlier, there are a number of different systems that can be used to convert thermal energy into electricity but only the thermoelectric generator and the thermionic generator have found widespread application. All of these systems have a maximum thermal efficiency of less than $(1 - T_L/T_H)$.

The operation of the thermoelectric generator or converter depends on the Seebeck effect, the Peltier effect, and the Thomson effect. The Seebeck effect was discovered in 1822 by the German physicist, Thomas J. Seebeck. According to the Seebeck effect, a voltage is produced in a circuit of two dissimilar materials if the two junctions are maintained at different temperatures. The Seebeck coefficient S is a material property and gives the rate of change of thermoelectric potential E_s with temperature T, or

$$S = \frac{dE_s}{dT} \tag{8.1}$$

The Seebeck coefficients are strong functions of temperature.

The induced thermoelectric potential E_s produced in a circuit composed of two materials is obtained from the following equation:

$$E_s = \int_{T_L}^{T_H} (S_a - S_b)\, dT = \int_{T_L}^{T_H} S_{ab}\, dT \tag{8.2}$$

The combined Seebeck coefficient S_{ab} is defined as positive if the electrical current (the flow of positive charges) flows from material A to material B at the cold junction where the heat of recombination is released.

Some typical values of Seebeck coefficients are listed in Table 8.1. It will be noted that the Seebeck coefficients for metals and alloys are very low compared to those for semiconducting materials. The combined Seebeck coefficient for an iron-constantan circuit is 60.6 μV/K, while that for a combination of germanium

Table 8.1 Seebeck coefficients (at 100°C)

Material	S, V/K
Aluminum	-0.2×10^{-6}
Constantan	-47.0×10^{-6}
Copper	$+3.5 \times 10^{-6}$
Iron	$+13.6 \times 10^{-6}$
Platinum	-5.2×10^{-6}
Germanium	$+375.0 \times 10^{-6}$
Silicon	-455.0×10^{-6}

and silicon is 830 μV/K. The Seebeck coefficients for metals and alloys are too low for the efficient generation of electricity, although the dissimilar metal junctions are commonly used in thermocouple circuits to monitor temperatures.

The Seebeck coefficients of n-p semiconductors are also high and these are the materials that are commonly used in thermoelectric generators. Two common materials that are used are lead telluride or bismuth telluride. An n-type semiconductor is one in which impurity atoms are added to the lattice that have one more electron than is needed to satisfy the valence-bond requirements. Thus, the material has extra negative electrons in the lattice although the material has no net charge associated with these electrons. A p-type semiconductor is one in which impurity atoms with one less electron than is needed to satisfy the valence-bond requirements is added to the lattice. This introduces positive "holes" into the lattice although, again, the material is neutrally charged.

In a thermoelectric converter constructed of semiconductors, both the holes and the extra electrons migrate toward the cold junction where they "pile up" and combine. The combined Seebeck coefficient for these lattices is $S_{ab} = S_{pn} = -S_{np}$, and the thermoelectric potential becomes

$$E_s = \int_{T_L}^{T_H} (S_p - S_n)\, dT = \int_{T_L}^{T_H} S_{pn}\, dT \tag{8.3}$$

The Peltier effect was discovered in 1844 by the French physicist, J. C. A. Peltier. The Peltier effect states essentially that if a direct current is passed through a circuit of dissimilar materials one of the dissimilar-metal junctions will be heated and the other will be cooled. This is the reversed Seebeck effect and it is also reversible in that, if the direction of current flow is reversed, the junction that was formerly heated will be cooled and the formerly cooled junction will be heated.

The Peltier coefficient for a circuit composed of material A and material B is designated π_{ab}, and is defined as

$$\pi_{ab} = \frac{-Q}{i_{ab}} \tag{8.4}$$

where $-Q$ is the heat-transfer rate from the junction, in watts, and i_{ab} is the direct current flowing in the generator, in amperes.

Like the Seebeck coefficient, the Peltier coefficient is a strong function of temperature and is related to the Seebeck coefficient by the following relationship:

$$\pi_{ab} = T_{(L \text{ or } H)} S_{ab} = T_{(L \text{ or } H)} (S_a - S_b) = -\pi_{ba} \tag{8.5}$$

where $T_{(L \text{ or } H)}$ is either the absolute temperature of the cold junction T_L or the absolute temperature of the hot junction T_H. The Peltier coefficient π_{ab} is positive if heat is generated in the junction when the direct current flows from material A to material B.

The Thomson effect was discovered in 1854 by the English physicist, William Thomson (Lord Kelvin). This effect says that there is reversible absorption or liberation of heat in a homogeneous conductor exposed to a simultaneous temperature gradient and an electrical gradient. The Thomson coefficient is designated τ and is given by the following equation:

$$\tau = \frac{Q}{i}\,\Delta T \tag{8.6}$$

where Q is the heat-transfer rate, in watts, absorbed by the conductor when the current flows toward the higher temperature. The Thomson coefficient is related to the Seebeck coefficient by the following relationship:

$$\tau = T\frac{dS}{dT} \tag{8.7}$$

This coefficient is positive for p-type materials and negative for n-type materials. The Thomson effect is only of secondary importance in the operation of thermoelectric generators.

A typical p-n thermoelectric generator is shown in Fig. 8.1. As can be seen in the figure, the legs or elements of the generator are connected in series for the flow of electricity, and they are connected in parallel for the flow of heat. The total electrical resistance of the converter is R_g and for a series connection is equal to the sum of the resistances of each of the legs:

$$R_g = m(R_p + R_n) \tag{8.8}$$

where m is the number of pairs of p-n legs of the generator, and R_p and R_n are the resistances of the p-leg and n-leg, respectively:

$$R_p = \frac{\rho_p L_p}{A_p} \quad \text{and} \quad R_n = \frac{\rho_n L_n}{A_n} \tag{8.9}$$

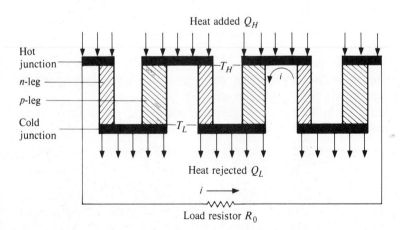

Figure 8.1 The typical n-p thermoelectric generator.

In the previous equations, ρ is the electrical resistivity of the materials, in ohm-meters, L is the length of the semiconductor legs, in meters, and A is the cross-sectional area of the legs, in square meters. It is normally assumed that the metallic connections between the semiconductor legs have negligible resistance.

The thermal conductance K_g of the generator is equal to the sum of the thermal conductances (the reciprocal of the thermal resistance) of the semiconductor legs, or

$$K_g = m(K_p + K_n) \tag{8.10}$$

where

$$K_p = \frac{k_p A_p}{L_p} \quad \text{and} \quad K_n = \frac{k_n A_n}{L_n} \tag{8.11}$$

and k is the thermal conductivity of the semiconductor materials in watts per meter per Celsius degree.

An energy balance on either the hot or cold node is composed of four energy terms. First, there is the heat-transfer rate to or from the junction from the environment, $\pm Q$. Second, there is the heat-transfer rate through the generator from the hot to the cold junction, $\pm K_g \Delta T$. Third, there is heat transfer due to the Peltier effect, $\pm \pi_{pn} i = \pm T_{(L \text{ or } H)} i \overline{S}_{pn}$. Fourth, there is the power dissipated in the device due to Joule heating and it can be shown that effectively half of the resistance heating is generated in each junction, $+i^2 R_g / 2$.

At the hot junction, the Peltier heat-transfer rate is $m\pi_{pn} i$ or $m\overline{S}_{pn} T_H i$, in watts, where

$$\overline{S}_{pn} = \int_{T_L}^{T_H} \frac{(S_p - S_n) \, dT}{T_H - T_L} \tag{8.12}$$

The energy or power entering the hot junction is equal to $i^2 R_g / 2$ plus Q_H, while the power leaving the hot junction is equal to the sum of $K_g \Delta T$ and $m\overline{S}_{pn} T_H i$. Combining these terms gives

$$Q_H = m\overline{S}_{pn} T_H i + K_g(T_H - T_L) - \frac{i^2 R_g}{2} \tag{8.13}$$

At the cold junction, the power transferred from the junction to the environment is equal to $-Q_L$, and all the other power terms are ultimately transferred to the cold junction:

$$-Q_L = m\overline{S}_{pn} T_L i + K_g(T_H - T_L) + \frac{i^2 R_g}{2} \tag{8.14}$$

The useful power produced by the device is equal to the power dissipated in the external load. Since this system generates direct current, the useful power generation is $i^2 R_0$, where R_0 is the resistance of the external load. The thermal efficiency of the thermoelectric generator is

$$\eta_{\text{th}} = \frac{i^2 R_0}{Q_H} = \frac{i^2 R_0}{m\overline{S}_{pn} T_H i + K_g \Delta T - i^2 R_g / 2} \tag{8.15}$$

Multiplying the numerator and the denominator by $\Delta T/R_g i^2$ and letting M equal the ratio of the external load resistance to the generator resistance R_0/R_g,

$$\eta_{\text{th}} = \frac{M\,\Delta T}{m\overline{S_{pn}}\,T_H\,\Delta T/iR_g + K_g\,\Delta T^2/i^2 R_g - \Delta T/2} \tag{8.16}$$

The current in the converter is equal to the total generated voltage divided by the total resistance of the circuit, or

$$i = \frac{v_t}{R_t} = \frac{m\overline{S_{pn}}\,\Delta T}{R_g + R_0} = \frac{m\overline{S_{pn}}(T_H - T_L)}{R_g(1 + M)} \tag{8.17}$$

Substituting this equation into Eq. (8.16) gives

$$\eta_{\text{th}} = \frac{M\,\Delta T}{(1 + M)T_H + (1 + M)^2/Z - \Delta T/2} \tag{8.18}$$

where Z is defined as the figure of merit of the generator and is equal to

$$Z = \frac{m^2 \overline{S_{pn}}^2}{K_g R_g} \tag{8.19}$$

The figure of merit of a generator is a function of the properties of the generator materials (S, k, and ρ) and the dimensions of the generator legs (A and L). To improve the thermal efficiency of the generator, the figure of merit Z should be as large as possible. Once the generator materials have been picked, the minimum product of $K_g R_g$ gives the maximum figure of merit Z_{max}:

$$K_g R_g = \left(\frac{k_p A_p}{L_p} + \frac{k_n A_n}{L_n}\right)\left(\frac{\rho_p L_p}{A_p} + \frac{\rho_n L_n}{A_n}\right)m^2$$

$$= m^2\left(k_p \rho_p + k_p \rho_n x + \frac{k_n \rho_p}{x} + k_n \rho_n\right) \tag{8.20}$$

where $x = A_p L_n/A_n L_p$. The optimum value of x that gives the minimum value of $K_g R_g$ or Z_{max} can be found by setting $[d(K_g R_g)/dx]$ equal to zero and solving for x. This gives

$$x_{\text{opt}} = \frac{A_p L_n}{A_n L_p} = \sqrt{\frac{\rho_p k_n}{\rho_n k_p}} \tag{8.21}$$

and this gives

$$Z_{\text{max}} = \frac{\overline{S_{pn}}^2}{(\sqrt{k_p \rho_p} + \sqrt{k_n \rho_n})^2} \tag{8.22}$$

The above equation is independent of the geometry of the system as long as the area and length of the generator elements are proportioned according to the electrical resistivity and the thermal conductivity of the materials, as given in (8.21).

The other variable in Eq. (8.18) that can be easily adjusted to improve the generator efficiency is the ratio of the external-load resistance to the generator resistance M. The optimum value of M that gives the maximum thermal efficiency can be determined by setting $(d\eta_{th}/dM)$ equal to zero and solving for M. This gives

$$M_{opt} = (R_0/R_g)_{opt} = \sqrt{1 + Z_{max}T_{ave}} \tag{8.23}$$

where T_{ave} is the average absolute temperature in the generator, $(T_H + T_L)/2$. Substituting this expression into the thermal efficiency equation yields

$$\eta_{th} = \frac{M_{opt}\,\Delta T}{T_H(1 + M_{opt}) + (1 + M_{opt})^2/Z_{max} - \Delta T/2} \tag{8.24}$$

While the above equation gives the condition for maximum thermal efficiency, there may be some times when the system will be operated at the condition for maximum power output rather than maximum efficiency. The output voltage of the generator is equal to the total generated voltage minus the internal voltage drop in the generator:

$$v_{out} = m\overline{S_{pn}}\,\Delta T - iR_g \tag{8.25}$$

and the output power is

$$P_{out} = iv_{out} = m\overline{S_{pn}}i\,\Delta T - i^2R_g \tag{8.26}$$

Differentiating Eq. (8.26) with respect to i and setting (dP_{out}/di) equal to zero gives the ideal current for maximum power output $i_{max\,P}$:

$$i_{max\,P} = \frac{m\overline{S_{pn}}\,\Delta T}{2R_g} \tag{8.27}$$

This gives a maximum power output for the generator equal to

$$P_{out,\,max} = \frac{m^2\overline{S_{pn}}^2\,\Delta T^2}{4R_g} \tag{8.28}$$

The output voltage for the condition of maximum power is found by substituting Eq. (8.27) into Eq. (8.25):

$$v_{out,\,max\,P} = i_{max\,P}R_0 = m\overline{S_{pn}}\,\Delta T - i_{max\,P}R_g = 2i_{max\,P}R_g - i_{max\,P}R_g$$

$$= i_{max\,P}R_g \tag{8.29}$$

Equation (8.29) shows that for the condition of maximum power $R_0 = R_g$, or

$$M_{max\,P} = \left(\frac{R_0}{R_g}\right)_{max\,P} = 1.0 \tag{8.30}$$

Comparing Eq. (8.23) with Eq. (8.30) indicates that, for the condition of maximum thermal efficiency, the load resistance R_0 should exceed the generator resistance R_g while, for the condition of maximum power, the two resistances must be equal (matched impedances).

8.2.2 Thermoelectric Refrigerators

Thermoelectric systems can also be used as a heat-pump or refrigeration system as well as electric generator. When used as a thermoelectric refrigerator, a direct current is passed through the unit and one junction is heated and the other is cooled. Reversing the flow of direct current reverses the mode of operation of the system. The energy relationships for the thermoelectric generator are essentially the same as those for the thermoelectric generator.

The figure of merit of any refrigeration system is the coefficient of performance β, which is defined as

$$\beta = \frac{\text{heat added at low temperature}}{\text{input energy}} = \frac{Q_A}{P_{\text{in}}} \tag{8.31}$$

The coefficient of performance should be as high as possible and may be greater or less than unity. An energy balance on the cold junction shows that the input power terms are the heat-transfer rate to the junction Q_A, the conduction heat-transfer rate from the hot junction $K_g \, \Delta T$, and the Joule heating in the junction $i^2 R_g/2$. The energy removal rate from the cold junction is equal to the Peltier heat flow rate $m\overline{S_{pn}} \, T_L i$. This gives the following equation for Q_A:

$$Q_A = m\overline{S_{pn}} T_L i - \frac{i^2 R_g}{2} - K_g \, \Delta T \tag{8.32}$$

The power input to the thermoelectric cooler is equal to iv_{in}, where the input voltage v_{in} is the sum of the Peltier voltage drop in each junction and the drop due to the internal resistance:

$$v_{\text{in}} = m\overline{\pi_{pn}} + iR_g = m\overline{S_{pn}} \, \Delta T + iR_g \tag{8.33}$$

and

$$P_{\text{in}} = m\overline{S_{pn}} \, \Delta T i + i^2 R_g \tag{8.34}$$

Substituting Eqs. (8.32) and (8.34) into Eq. (8.31) gives the following equation for the coefficient of performance:

$$\beta = \frac{m\overline{S_{pn}} T_L i - i^2 R_g/2 - K_g \, \Delta T}{m\overline{S_{pn}} \, \Delta T i + i^2 R_g} = \frac{N T_L - N^2/2 - \Delta T/Z}{\Delta T N + N^2} \tag{8.35}$$

where Z is the figure of merit of the generator $(m^2 \overline{S_{pn}}^2/K_g R_g)$ and N is equal to $(R_g i/m\overline{S_{pn}})$.

In order to achieve as high a value of the coefficient of performance as possible, the figure of merit Z should be a maximum, as given by Eq. (8.22). The optimum value of N for maximum β can be found by taking the derivative of Eq. (8.35) with respect to N and setting $(d\beta/dN)$ to zero and solving for N. The optimum value of N is

$$N_{\text{opt}} = \frac{R_g i_{\text{opt}}}{m\overline{S_{pn}}} = \frac{\Delta T}{\sqrt{1 + T_{\text{ave}} Z_{\text{max}}} - 1} \tag{8.36}$$

The optimum input current for the maximum coefficient of performance is

$$i_{\text{opt}} = \frac{m\overline{S}_{pn} N_{\text{opt}}}{R_g} \tag{8.37}$$

If the temperatures T_L and T_H are fixed across a given thermoelectric refrigerator, the heat-pumping rate from T_L to T_H is

$$Q_A = m\overline{S}_{pn} T_L i - i^2 R_g/2 - K_g \Delta T \tag{8.38}$$

The maximum heat-transfer rate $Q_{A,\text{ max}}$ can be found as a function of the input current. If the current is too high the Joule heating term predominates, but if it is too low the Peltier pumping rate is too low. The value of current that gives the maximum pumping rate is

$$i_{\text{max}\,Q} = \frac{m\overline{S}_{pn} T_L}{R_g} \tag{8.39}$$

If this current is maintained to a thermoelectric refrigerator and the heat-pumping rate is decreased, the temperature of the cold junction approaches some minimum value as Q_A approaches zero. This minimum temperature T'_L is

$$T'_L = \frac{T_H + Z_{\text{max}} T_L^2/2}{1 + Z_{\text{max}} T_L} \tag{8.40}$$

An even lower temperature can be achieved in this device if the current is lowered as the heat flow from the cold junction is decreased. According to Eq. (8.38) and (8.39), the lowest possible temperature is

$$T_{L,\text{ min}} = \frac{\sqrt{1 + Z_{\text{max}} T_H} - 1}{Z_{\text{max}}} \tag{8.41}$$

Example 8.1 A thermoelectric generator that will operate between 30 and 500°C is to be constructed of n-p semiconductors with the following properties:

Material	n-type	p-type
S, μV/K	-170	210
ρ, $\mu\Omega\cdot$m	14	18
k, W/m$\cdot$K	1.5	1.1

For this system that is designed to produce 500 W$_e$ at the highest possible thermal efficiency, find the number of element-pairs, the maximum possible thermal efficiency, and the maximum possible power output. Assume that the cross-sectional area and the length of the n-leg are 1 cm^2 and 1 cm, respectively, and that the length of the p-leg is 1 cm.

SOLUTION

Average Seebeck coefficient $= \overline{S_{pn}} = \overline{S_p} - \overline{S_n} = 210 - (-170) = 380 \ \mu V/K$

Maximum figure of merit $= Z_{max} = \dfrac{\overline{S_{pn}}^2}{[(\rho_n k_n)^{1/2} + (\rho_p k_p)^{1/2}]^2}$

$$= \dfrac{(380 \times 10^{-6} \ V/K)^2}{[(14 \times 10^{-6} \times 1.5)^{1/2} + (18 \times 10^{-6} \times 1.1)^{1/2}]^2 \ \Omega \cdot W/K}$$

$$= 0.00177/K$$

For minimum value of $K_g R_g$,

$$\dfrac{A_n L_p}{A_p L_n} = \left(\dfrac{\rho_n k_p}{\rho_p k_n}\right)^{1/2} = \left(\dfrac{14 \times 10^{-6} \times 1.1}{18 \times 10^{-6} \times 1.5}\right)^{1/2}$$

$$= 0.7552$$

$$A_p = \dfrac{A_n L_p}{(0.7552)L_n} = 1.3241 \ cm^2 = 0.00013241 \ m^2$$

$$K_g = m\left(\dfrac{k_p A_p}{L_p} + \dfrac{k_n A_n}{L_n}\right)$$

$$= m\left[\dfrac{1.1(0.0001341)}{0.01} + \dfrac{1.5(0.0001)}{0.01}\right]$$

$$= m(2.975 \times 10^{-2}) \quad W/K$$

$$R_g = m\left(\dfrac{\rho_p L_p}{A_p} + \dfrac{\rho_n L_n}{A_n}\right)$$

$$= m\left(\dfrac{18 \times 10^{-6}}{0.013241} + \dfrac{14 \times 10^{-6}}{0.01}\right)$$

$$= 0.002759m \quad \Omega$$

$$M_{opt} = (1 + Z_{max} T_{ave})^{1/2} = \left[1 + \dfrac{1.77 \times 10^{-3}(303 + 773)}{2}\right]^{1/2} = 1.3972$$

$$R_0 = M_{opt} R_g = 1.3972(0.002759m) = 0.003855m \quad \Omega$$

Total circuit resistance $= R_0 + R_g = \dfrac{\overline{S_{pn}} \Delta T}{R_T}$

$$= 0.002759m + 0.003855m$$

$$= 0.006614m \quad \Omega$$

$$\text{Current} = i = \dfrac{v_T}{R_T} = \dfrac{0.1786m}{0.006614m} = 27.00 \ A$$

Power output $= P_e = 500 \ W_e = iv_L = i^2 R_0$

$$= (27.00)^2(0.003855m) = 500$$

$$\text{Number of elements in the converter} = m = \frac{500}{(27)^2(0.003855)} = 177.9$$

Use 178 pairs of elements.

$$\text{Maximum thermal efficiency} = \frac{M_{opt}\,\Delta T}{T_H(1 + M_{opt}) + (1 + M_{opt})^2/Z_{max} - \Delta T/2}$$

$$\eta_{opt} = \frac{1.3972(470)}{773(2.3972) + (2.3972)^2/0.00177 - 470/2}$$

$$= 13.5\%$$

$$\text{Maximum output power} = \frac{m^2 \overline{S_{pn}}^2 \, \Delta T^2}{4R_g} = \frac{178(380 \times 10^{-6})^2(470)^2}{4(0.002759)}$$

$$= 514.5 \text{ W}$$

8.2.3 Thermoelectric Systems

As indicated earlier, n-p semiconductors are commonly used in thermoelectric systems. Since a low thermal conductivity or thermal conductance is desired, materials with high atomic or molecular weights are commonly employed because they have lower values of thermal conductivity. The thermal conductivity can be reduced further by using materials in the form of solid solutions. The product ZT of some of the proposed thermoelectric materials is shown in Fig. 8.2 for both n and p materials. Lead telluride (PbTe) is commonly employed in thermoelectric generators while bismuth telluride (Bi_2Te) is commonly employed in thermoelectric refrigerators.

The thermoelectric generator can use almost any source of thermal energy to produce electricity. Figure 8.3 pictures a kerosene-powered thermoelectric generator and Fig. 8.4 shows a radioisotope-powered, thermoelectric SNAP system similar to that used to power the experimental laboratories on the moon.

Thermoelectric generators have a number of advantages over the conventional thermodynamic heat-engine systems. The thermoelectric systems are compact, rugged, reliable, and have no moving parts. Unfortunately, the thermal efficiency of these systems is very low, normally 5 to 10 percent, and the components are quite expensive. Consequently, the thermoelectric generating systems are usually employed for very low-power applications.

8.2.4 Thermionic Converters

In the operation of a thermionic generator, the electrons are effectively "boiled off" the hot cathode and "condensed" on the cold anode. These electrons then flow back to the cathode through the external load resistor, producing useful energy. Thermionic emission was discovered in 1883 by Thomas A. Edison.

The valence electrons in orbit around a nucleus have an average energy equal to a quantity called the Fermi-level energy. The electrons vibrate about

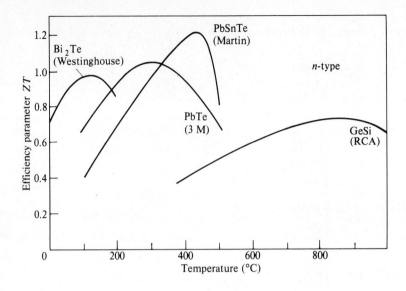

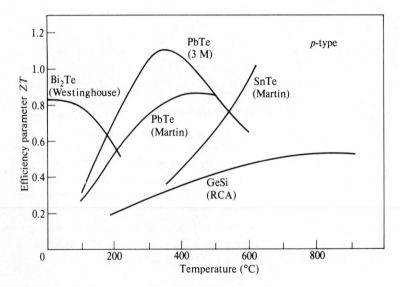

Figure 8.2 Properties of some typical thermoelectric materials.

this Fermi level with an amplitude that is proportional to the absolute temperature. The work-function energy ϕ is the amount of energy required to strip the valence electron from an atom. This energy must be overcome before the electron can leave the surface. The work-function and Fermi-level energies vary from material to material.

Figure 8.3 A kerosene-powered thermoelectric generator. *(Courtesy of Global Thermoelectric Power Systems, Ltd.)*

The rate of thermal emission of electrons from a given surface is given by the Richardson-Dushman equation. According to this relationship, the current density J_0, in amperes per square meter, is

$$J_0 = \xi T^2 \exp \frac{-e\phi}{kT} \tag{8.42}$$

where e is the charge of an electron, k is the Boltzmann constant (1.551×10^{-4} eV/K), T is the absolute temperature of the surface in kelvin, ϕ is the work function of the surface in electronvolts, and ξ is a constant with units of amperes per square meter per kelvin squared. ξ was supposed to be a universal constant with a value of 1.2×10^6 A/m$^2 \cdot$K^2; however, it has been found to vary from material to material. The thermionic emission properties of some typical materials used in these systems are presented in Table 8.2.

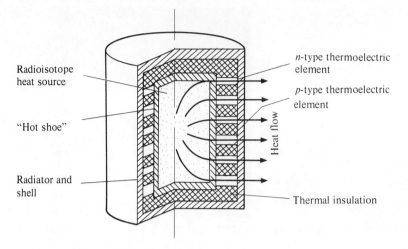

Figure 8.4 A radioisotope-powered thermoelectric generator.

A typical vacuum thermionic diode is shown in Fig. 8.5, along with the electron energies. The vacuum thermionic converter or generator is composed of the hot emitter or cathode and a cold collector electrode, the anode, separated by a vacuum gap. When electrons leave the cathode, the cathode is left with a positive charge and the free electrons are attracted back to the surface as well as being repelled by the anode surface. Other emitted electrons already in the interelectrode gap also tend to drive the emitted electrons back to the cathode surface. As a result, electrons emitted from the hot cathode tend to "pile up" in the interelectrode gap creating an additional energy barrier to the flow of electrons. This additional energy barrier is called the space–charge-barrier energy and is designated by ϕ_b.

Table 8.2 Thermionic-emission properties of some materials

Material	ϕ, V	ξ, A/m$^2 \cdot$K^2
Cs	1.89	0.50×10^6
Mo	4.2	0.55×10^6
Ni	4.61	0.30×10^6
Pt	5.32	0.32×10^6
Ta	4.19	0.55×10^6
W	4.52	0.60×10^6
W + Cs	1.5	0.03×10^6
W + Ba	1.6	0.015×10^6
W + Th	2.7	0.04×10^6
BaO	1.5	0.001×10^6
SrO	2.2	1.00×10^6

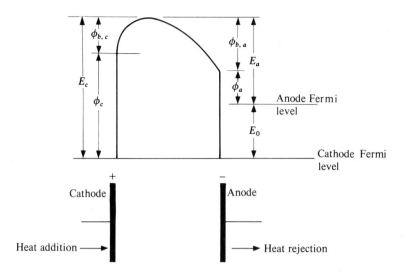

Figure 8.5 The thermionic-diode generator.

In order for an electron to reach the anode, it must have a total energy E_c equal to the sum of the cathode work function and the space–charge-barrier energy, or $E_c = \phi_c + \phi_b$. With the existance of the space–charge-barrier energy, the net current density from the cathode is

$$J_0 = \xi T^2 \exp \frac{-eE_c}{kT} \tag{8.43}$$

Using the energy diagram in Fig. 8.5, it can be seen that the output voltage E_0 from the generator is equal to the sum of the Fermi-level energy and the work-function energy of the cathode plus the space–charge-barrier energy minus the sum of the Fermi-level energy, the work-function energy, and the space–charge-barrier energy of the anode:

$$E_0 = E_c - E_a \tag{8.44}$$

For a high output voltage, the cathode should have a low Fermi level and a high work function. It should also have a high value of ξ because the high work function inhibits the flow of electrons. The anode should have a high Fermi level and a low work function, even though this does increase the electron emission from the anode, thereby reducing the net flow of electron emission from the cathode. For optimum performance, the approximate temperature–work-function relationship is

$$\frac{\phi_a}{T_a} = \frac{\phi_c}{T_c} \tag{8.45}$$

In order to keep the space–charge-barrier energy within reason for a vacuum converter, the interelectrode gap must be kept very small. In fact, for optimum thermal efficiency, this gap is only several microinches in width. It is extremely difficult to fabricate such a system and maintain the small gap spacing.

The space–charge-barrier energy is a difficult problem to overcome in a vacuum converter and for that reason the vacuum thermionic converter is seldom used. One solution to the barrier problem is to use a low-pressure, ionized gas or plasma in the interelectrode gap. The use of an ionized gas or vapor at an absolute pressure of around one inch of mercury significantly reduces the charge barrier, permitting the use of a wider electrode gap in the generator. The principal disadvantage of using an interelectrode gas is that it increases the convection heat transfer across the gap.

Most of the developmental work in the area of thermionic conversion has been centered on those systems employing cesium vapor between the two thermionic electrodes. Cesium has two major advantages over other gases. First, since cesium is a metallic vapor that ionizes easily, it effectively reduces the space–charge-barrier energy. Second, cesium has a low work function (1.89 eV) and if the cesium condenses on the cold anode it effectively lowers the anode work function. In order to achieve appreciable ionization of the cesium, the cathode or emitter temperature must exceed 1800 K (3240°R). This system may be operated in either the ignited (continuous discharge) mode or in the unignited mode.

Other research has been directed on systems using an ionized inert gas, such as helium or argon. The use of an inert gas minimizes the corrosion problems in the converter. In these systems, the inert gas is ionized by supplying a high voltage to a third (corona discharge) electrode. This permits operation at relatively low temperatures with the cathode operating in the neighborhood of 1500 K (2700°R).

The thermal efficiency of the thermionic conversion system is difficult to analyze explicitly because the losses are strongly dependent on the system geometry and the mode of operation. Since this system produces dc electricity, the output power of the system is simply equal to the product of the current i and the external load voltage v_L:

$$P_{out} = iv_L = J_0 A_c v_L \qquad (8.46)$$

A_c is the surface area of the cathode, in square meters, and v_L is the voltage drop across the external load resistor. This voltage drop is equal to the output voltage of the converter E_0 as given by Eq. (8.44) minus the voltage drop in the electrical leads connecting the cathode and the anode to the load resistor.

There are three major losses in the operation of the thermionic converter. One of these losses is the heat transferred between the cathode and the anode, either by radiation for the vacuum converter or by combined radiation and convection for the gas or vapor-filled converters. Another thermal loss is the conduction heat transfer along the electrical leads connected to the cathode and anode. The last major system loss in the device occurs in the anode as the work-function and the space–charge-barrier energies of the "condensing" electrons are converted into thermal energy.

The radiation heat-transfer rate or power P_r between the cathode and the anode can be estimated by using the conventional radiation heat-transfer equations. The spacing between the two electrodes is small enough that they may be approximated by infinite plane surfaces. The resulting equation is

$$P_r = \frac{\sigma A_c (T_c^4 - T_a^4)}{1/\varepsilon_c + 1/\varepsilon_a - 1} \tag{8.47}$$

where σ is the Stefan-Boltzmann constant, 5.67×10^{-8} W/m$^2\cdot$K^4, and ε_c and ε_a are the emissivities of the cathode and anode, respectively. If there is a low-pressure vapor between the electrodes, the heat-transfer rate will be increased because of convective heat transfer.

The biggest single power loss from the cathode is the energy carried away by the electrons P_{el}. The potential energy of an electron leaving the cathode must exceed the work-function energy plus the space–charge-barrier energy for the electrode, $(\phi_c + \phi_{b,c})$. In addition, each electron has an average total kinetic energy of $2kT_c$. For a net emitter current density of J_0, the energy rate from the emitter associated with electron flow is

$$P_{el} = J_0 A_c \left(\phi_c + \phi_{b,c} + \frac{2kT_c}{e} \right) \tag{8.48}$$

If the input lead wire, connected to the cathode, has an electrical resistivity of ρ_w, a length L_w, a thermal conductivity of k_w, and a cross-sectional area A_w, the conduction heat-transfer rate from the cathode combined with the Joule heating rate to the cathode is

$$P_w = \left(\frac{k_w A_w}{L_w} \right)(T_c - T_0) - \frac{(J_0 A_c)^2 (\rho_w L_w / A_w)}{2} \tag{8.49}$$

where $(T_c - T_0)$ is the linear temperature drop over the length L_w.

Since the sum of these three energy rates is the power loss from the cathode and must be made up by heat addition to the converter, the thermal efficiency of a thermionic converter can be estimated by the following equation:

$$\eta_{th} = \frac{P_{out}}{P_r + P_{el} + P_w} \tag{8.50}$$

Example 8.2 A thermionic converter is operating with a thoriated-tungsten (W + Th) emitter at 1900 K with a space–charge-barrier energy of 0.3 V and collector-barrier energy of 0.5 V. Find the emitter area needed to produce 100 W if the collector (anode) is constructed of barium oxide (BaO).

SOLUTION From Table 8.2:

$$\xi_c = 0.04 \times 10^6 \text{ A/m}^2 \cdot \text{K}^2 \qquad \phi_c = 2.7 \text{ V}$$

$$\xi_a = 0.001 \times 10^6 \text{ A/m}^2 \cdot \text{K}^2 \qquad \phi_a = 1.5 \text{ V}$$

Emitter current density $= J_c = \xi_c T_c^2 \exp \dfrac{-e(\phi_c + \phi_{b,c})}{kT}$

$$= 0.04 \times 10^6 (1900)^2$$

$$\times \exp - \frac{11{,}600(2.7 + 0.3)}{1900}$$

$$= 1603.7 \text{ A/m}^2$$

Output voltage $= v_L = \phi_c + \phi_{b,c} - \phi_a - \phi_{b,a}$

$$= 2.7 + 0.3 - 0.5 - 1.5 = 1.0 \text{ V}$$

Optimum anode temperature $= T_a = \dfrac{\phi_a T_c}{\phi_c} = \dfrac{1.5(1900)}{(2.7)} = 1056 \text{ K}$

Anode current density $= J_a = \xi_a T_a^2 \exp \left[\dfrac{-e(\phi_a + \phi_{b,a})}{kT_a} \right]$

$$= 0.001 \times 10^6 (1056)^2$$

$$\times \exp - \frac{11{,}600(1.5 + 0.5)}{1056}$$

$$= 0.3 \text{ A/m}^2$$

Converter area:

Output power $= P_{\text{out}} = 100 \text{ W}_e = A(J_c - J_a)v_L$

$$= A(1603.7 - 0.3)(1.0)$$

Converter area $= A = \dfrac{100}{1603.4} = 0.06236 \text{ m}^2 = 623.6 \text{ cm}^2$

Neglecting the radiation and conduction losses from the system, calculate the thermal efficiency, η_{th}.

$$\eta_{\text{th}} = \frac{P_{\text{out}}}{(\phi_c + \phi_{b,c} + 2kT_c/e)J_0 A}$$

$$= \frac{100}{[2.7 + 0.3 + 2(1900)/(11{,}600)](100)} = 30.05\%$$

The thermionic conversion systems require a high-temperature, high heat-flux source of thermal energy. Possible energy sources include radioisotopes, nuclear reactor fuel elements, concentrated solar energy, and combustion

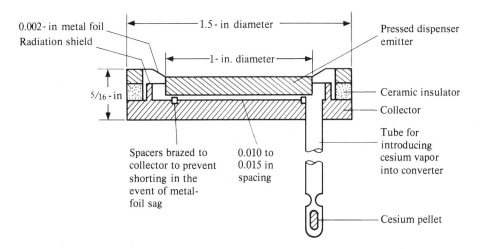

Figure 8.6 A cesium thermionic converter. *(From Angrist, 1976.)*

systems using heat pipes. The heat pipe is a device that permits the transfer of thermal energy fluxes across a relatively large distance with a very small temperature drop. In this system, heat is added to a working fluid in the pipe by boiling heat transfer and the heat is then transferred to the thermionic emitter by condensation. The resulting condensate in the pipe is returned to the hot end of the pipe by either gravity or the capillary action of a wick.

A typical thermionic converter, using cesium vapor to reduce the space charge barrier, is shown in Fig. 8.6. Like the thermoelectric conversion systems, these converters have no moving parts but these systems commonly have thermal efficiencies that range from 15 to 20 percent, considerably above that for the thermoelectric systems. Unfortunately, these systems are not as rugged as the thermoelectric systems, they are complicated to build, and they are very expensive. Thermionic converters also experience a problem due to the deterioration of the electrodes, particularly the hot cathode.

8.2.5 Other Thermal-Electric Conversion Systems

There are a number of other minor systems that convert thermal energy directly into electricity, including the ferroelectric converter, the thermomagnetic converter, the Nernst-effect generator, and others. Most of these systems are minor systems but some of them present some interesting concepts for the production of electricity.

One of the more unusual methods of thermal energy conversion takes place in the ferroelectric system shown in Fig. 8.7. This system essentially consists of a capacitor that uses barium titanate as the dielectric material. At low temperatures, barium titanate has a very high dielectric constant which means the capacitor has a very high value of electrical capacitance. This system is charged at

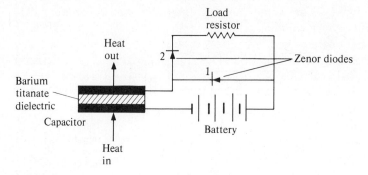

Figure 8.7 A schematic of the ferroelectric conversion system.

low voltage through zenor diode number 1. Once the capacitor is charged, heat is applied to the capacitor. When barium titanate is heated above 49°C (120°F), its Curie point, the value of the dielectric constant drops drastically. This means that the value of the capacitance drops significantly and, for the same charge on the plates, the voltage increases markedly. The high-voltage current then flows through the second zenor diode and flows through the external-load resistor back to the battery. The capacitor is then cooled and the cycle is repeated.

The thermomagnetic converter operates in somewhat the same manner as the ferroelectric converter except that it operates with inductive-field energy instead of electrostatic-field energy. In the thermomagnetic generator, a ferromagnetic material is placed in the core of an inductor and the core is heated above the Curie point making it paramagnetic. This results in a marked decrease in the inductance of the coil and the collapsing magnetic field produces a high-voltage pulse.

The Nernst-effect generator operates by passing a thermal heat flux through a semimetal or a semiconductor in the presence of a perpendicular magnetic field. This system, as well as the other systems discussed in this section, are not practical modes for the production of electrical energy at this time.

8.3 CONVERSION OF CHEMICAL ENERGY INTO ELECTRICITY

8.3.1 General Systems

The battery and the fuel cell are systems in which stored chemical energy of the system is converted directly to electrical energy. Since these systems do not go through the thermal energy regime, they are not limited by the efficiency of an externally reversible, heat-engine cycle, or $(1 - T_L/T_H)$. For this reason, considerable interest and research has been generated by these systems.

Batteries and fuel cells are very similar in operation with the major difference being that the battery contains a fixed quantity of fuel or chemical energy whereas the fuel cell operates with a continuous supply of fuel. Some batteries are reversible devices in that the products of the chemical reactions are separated back into the original reactants by supplying electricity to the battery during recharging. Fuel cells cannot be recharged as the products of the chemical reactions are thrown away.

Batteries are employed as energy-storage systems and can be divided into two major categories—primary batteries and secondary batteries. A primary battery, like the conventional "C" and "D" cells, cannot normally be recharged, while the secondary battery, like the lead-acid automobile battery, can be recharged many times.

The fuel cells and batteries are similar in composition in that both normally contain two electrodes separated by an electrolyte solution or matrix. In the fuel cell, the fuel reactant, usually hydrogen or carbon monoxide, is fed into one porous electrode and oxygen or air is fed into the other porous electrode.

The fuel-cell electrodes must accomplish three things. The electrode must be porous so that both the fuel and electrolyte can penetrate it to achieve proper contact. The pore size of the electrodes is very critical. If the pores are too large, the fuel gas will "bubble through" and be wasted. If the pores are too small, there will not be sufficient contact between the reactant and the electrolyte and the capacity of the cell is reduced. The electrode must contain a chemical catalyst that breaks the fuel compound into atoms so that they are more reactive. The most popular catalysts used today are platinum and sintered nickel. Finally, the electrodes must be able to conduct the electrons to the terminal.

The electrolyte solution must be highly permeable to either a H^+ or OH^- ion which is produced as an intermediate product at one of the electrodes. This same ion is transferred through the electrolyte to the other electrode where it combines with the other reactant. The electrons travel through the external circuit to the other electrode where the oxidation product is formed.

If the cell "burns" oxygen and hydrogen and has an acidic electrolyte, the intermediate ion is H^+ and the general reaction is

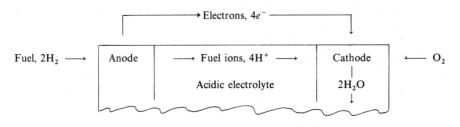

The anode reaction is: $2H_2 \longrightarrow 4e^- + 4H^+$

The cathode reaction is: $4e^- + 4H^+ + O_2 \longrightarrow 2H_2O$

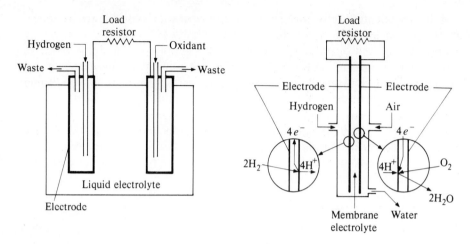

Figure 8.8 Schematic diagrams of some typical fuel cells.

In an oxygen-hydrogen fuel cell with an alkaline electrolyte (such as potassium hydroxide), the intermediate ion is OH^- and the general reaction is

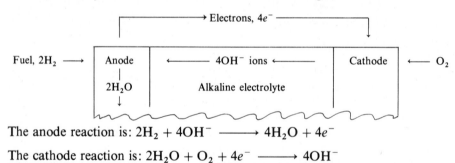

The anode reaction is: $2H_2 + 4OH^- \longrightarrow 4H_2O + 4e^-$

The cathode reaction is: $2H_2O + O_2 + 4e^- \longrightarrow 4OH^-$

Some fuel cells employ a solid electrolyte that is actually an ion-exchange membrane. One such membrane was composed of finely powdered sulphenated polystyrene resin held in an inert polymer. This membrane is flexible, has good mechanical strength, is chemically inert, and is impervious to the reactant gases. While the membrane does permit the passage of ions, it has a relatively high electrical resistance, even though the total thickness is about 3 mm. This type of converter was used as an auxilliary power source on some of the early Gemini two-man orbital space missions. A fuel cell of this type along with a typical fuel cell is shown in Fig. 8.8.

8.3.2 Fuel-Cell Performance

The total energy released in any chemical reaction is equal to the change in the enthalpy of formation ΔH. The change in the enthalpy of formation is equal to the sum of the enthalpy of formation for the reactants minus the sum of the enthalpy of formation of the products:

$$\Delta H = \sum (\Delta H)_{\text{reactants}} - \sum (\Delta H)_{\text{products}} \tag{8.51}$$

The change in the enthalpy of formation is equivalent to the higher heating value (HHV) of the fuels, discussed in Chap. 2.

Values of ΔH are given in Table 8.3 for the oxidation of various fuels at 25°C and 1 atm. Normally, all naturally occurring elements have a ΔH of zero. Consider the following combustion reaction:

$$C + O_2 \longrightarrow CO_2$$

The change in the enthalpy of formation H for this reaction is

$$\Delta H = \Delta H_C + \Delta H_{O_2} - \Delta H_{CO_2} = 0 + 0 - (-394 \times 10^6)$$

$$= 394 \times 10^6 \text{ J/kg·mol CO}_2$$

$$= 8.953 \times 10^6 \text{ J/kg CO}_2 = 32.81 \times 10^6 \text{ J/kg C}$$

$$= 14{,}107 \text{ Btu/lbm C}$$

Theoretically, all of the energy in the enthalpy of formation could be converted into electrical energy if none of it is converted into another form of energy. Unfortunately, in a reversible reaction some of the chemical energy is converted into thermal energy and only the rest can be theoretically converted into electricity. The minimum amount of thermal energy that is generated in any reversible reaction is equal to $\int T\,dS$. Since the fuel cell is essentially an isothermal device, $\int T\,dS = T \int dS = T\,\Delta S$. This means that the amount of electrical energy W_e produced in a fuel cell is

$$W_e \leq \Delta H - T\,\Delta S \tag{8.52}$$

Table 8.3 Enthalpy of formation ΔH^0 and Gibbs free energy ΔG^0 of compounds and ions (at 1 atm and 298°)

Compound or ion	Enthalpy of formation ΔH^0, J/kg·mol	Gibbs free energy ΔG^0, J/kg·mol
CO	-110.0×10^6	-137.5×10^6
CO_2	-394.0×10^6	-395.0×10^6
CH_4	-74.9×10^6	-50.8×10^6
Water	-286.0×10^6	-237.0×10^6
Steam	-241.0×10^6	-228.0×10^6
LiH	$+128.0 \times 10^6$	$+105.0 \times 10^6$
$NaCO_2$	-1122.0×10^6	-1042.0×10^6
CO_3^{--}	-675.0×10^6	-529.0×10^6
H^+	0.0	0.0
Li^+	-277.0×10^6	-293.0×10^6
OH^-	-230.0×10^6	-157.0×10^6

The electrical generation per kilogram-mole of fuel is equal to $(\Delta H - T\,\Delta S)$ for a reversible reaction but is less than this quantity if there are any irreversibilities in the cell.

Gibbs free energy G is a thermodynamic function that is defined as the enthalpy H minus the product of temperature T and entropy S:

$$G = H - TS \tag{8.53}$$

The derivative of this equation for an isothermal process $(dT = 0)$ is

$$dG = dH - T\,dS \quad \text{or} \quad \Delta G = \Delta H - T\,\Delta S \tag{8.54}$$

Thus, Eq. (8.52) reduces to

$$W_e \le \Delta G \tag{8.55}$$

where the change in Gibbs free energy for a given reaction, ΔG, is

$$\Delta G = \sum (\Delta G)_{\text{reactants}} - \sum (\Delta G)_{\text{products}} \tag{8.56}$$

Gibbs free energy is a function of the pressure and temperature at which the reaction takes place. The general thermodynamic property relationship is

$$T\,dS = dH - V\,dP \quad \text{or} \quad dH = T\,dS + V\,dP \tag{8.57}$$

Substituting Eq. (8.57) into Eq. (8.54) gives

$$dG = V\,dP \tag{8.58}$$

For each mole of gaseous constituent, $V = R_u T/P$, so that for an isothermal process

$$\int_{G^0}^{G} dG = \int_{P^0}^{P} \frac{R_u T}{P}\,dP = R_u T \int_{P^0}^{P} \frac{dP}{P} = G - G^0 = R_u T \ln \frac{P}{P^0} \tag{8.59}$$

Since G^0 is the value of Gibbs free energy for 1 atm, $P^0 = 1.0$, if P is the constituent pressure, in atmospheres, Eq. (8.59) reduces to

$$G = G^0 + R_u T \ln P \tag{8.60}$$

Consider the following chemical reaction for a fuel cell:

$$aA + bB \longrightarrow cC + dD$$

Assume that the reactants (A and B) and the products (C and D) are all gases at temperature T with partial pressures P_A, P_B, P_C, and P_D. The change in Gibbs free energy for this reaction becomes

$$\Delta G = \Delta G^0 + R_u T \ln \frac{P_A^a P_B^b}{P_C^c P_D^d} \tag{8.61}$$

In some cases, the reaction products may appear in the liquid phase, and in this case Eq. (8.61) can be used providing the partial pressure of component x, P_x, is replaced by a quantity θ_x, where θ_x is called the activity of component x and has a value equal to the partial pressure of component x only when the component obeys the ideal-gas laws.

Since both ΔG and ΔH are expressed in units of energy (in joules) per kilogram-mole, this means that W_e is the electrical energy associated with the passage of 1 kg·mol of electrons through the electrical circuit. One kg·mol of electrons is equal to Avagadro's number (6.023×10^{26}) and the charge of one electron is 1.602×10^{19} C. This means that the charge associated with 1 kg·mol of electrons is 9.65×10^7 C, and this quantity of charge is defined as one faraday, F_y. If n moles of electrons are released in the reaction and the internal cell voltage is E_g, the value of W_e is

$$W_e = nF_y E_g \leq \Delta G \tag{8.62}$$

The maximum value of the internal cell voltage is

$$E_g = E_g^0 - \frac{R_u T}{nF_y} \ln \frac{\theta_C^c \theta_D^d}{\theta_A^a \theta_B^b} \tag{8.63}$$

This equation is called the Nernst equation. It will be noted that increasing the fuel-cell temperature T decreases the value of the output voltage and hence the value of the electrical energy output.

As was discussed earlier, the fuel cell is not limited by the efficiency of an externally reversible, heat-engine cycle, but there is an upper limit to the conversion efficiency. This maximum conversion efficiency is equal to

$$\eta_{max} = \frac{W_{e,\,max}}{\Delta H} = \frac{\Delta G}{\Delta H} = 1 - \frac{T \, \Delta S}{\Delta H} \tag{8.64}$$

The actual conversion efficiency of a fuel cell is lower than that given by Eq. (8.64) and may be calculated as

$$\eta_{act} = \frac{W_{e,\,act}}{\Delta H} = \frac{nF_y v_L}{\Delta H} \tag{8.65}$$

where v_L is the output voltage of the cell. Efficiencies as high as 60 to 70 percent may be realized with most fuel cells without resorting to the high temperatures required for the heat-engine cycles. In fact, as mentioned earlier, an increase in operating temperature decreases the actual output and hence the conversion efficiency.

Example 8.3 Find the output voltage and theoretical conversion efficiency of an oxygen-hydrogen fuel cell operating at 600°C. Assume that the hydrogen is supplied at 1.1 atm and the oxygen is supplied from air at 1.2 atm and assume that the steam product is at 1 atm.

SOLUTION From Table 8.3:

$$\Delta G^0 \text{ (for steam)} = 228 \times 10^6 \text{ J/kg·mol } H_2O$$

$$\Delta H^0 \text{ (for steam)} = 241 \times 10^6 \text{ J/kg·mol } H_2O$$

$$\Delta H^0 \text{ (for water)} = 286 \times 10^6 \text{ J/kg·mol } H_2O$$

$$nF_y E_g^0 = \Delta G^0$$

For this reaction, $n = 2$ kg·mol electrons/kg·mol H_2O

$$F_y = 9.65 \times 10^7 \text{ C/kg·mol electrons}$$

$$E_g^0 = \frac{\Delta G^0}{nF_y} = \frac{228 \times 10^6 \text{ W·s/kg·mol } H_2O}{(2 \text{ mol } e^-/\text{mol } H_2O)(9.65 \times 10^7 \text{ C/mol } e^-)}$$

$$= 1.181 \text{ V}$$

Partial pressure of the O_2 in the reactants $= 1.2(0.21) = 0.252$ atm

Partial pressure of the H_2 in the reactants $= 1.1$ atm

Partial pressure of the H_2O in the products $= 1.0$ atm

$$E_g = E_g^0 + \frac{R_u T}{nF_y} \ln \frac{P_{H_2} P_{O_2}^{1/2}}{P_{H_2O}}$$

$$= 1.181 + \frac{8315(873)}{2(9.65 \times 10^7)} \ln \frac{1.1(0.252)^{1/2}}{1.0}$$

$$= 1.181 - 0.022 = 1.159 \text{ V}$$

Maximum conversion efficiency $= \eta = \dfrac{\Delta G}{\Delta H}$

$$\Delta G = \Delta G^0 + R_u T \ln \frac{P_{H_2} P_{O_2}^0}{P_{H_2O}} = 228 \times 10^6 + 8315(873) \ln [1.1(0.252)^{1/2}]$$

$$= 223.7 \times 10^6 \text{ J/kg·mol } H_2O$$

$$\eta = \frac{223.7 \times 10^6}{286 \times 10^6} = 0.7822 = 78.22\%$$

The performance of a fuel cell is a function of the operating characteristics of the cell. Like a battery, the output voltage of a fuel cell decreases as the electrical output of the cell increases. There are three major voltage losses associated with operation of the cell. The chemical-polarization or activation-polarization loss v_c is essentially the voltage required to cause the ions to break away from the electrode where they are formed. The internal-resistance loss is equal to the product of iR_i, where i is the current flow and R_i is the internal cell resistance. At very high loads, there is an additional loss called the concentration-polarization loss which is associated with electrostatic effects and concentration gradients associated with the localized depletion of ions in the vicinity of the electrodes at a high power output. The typical voltage-current characteristics of a fuel cell are shown in Fig. 8.9.

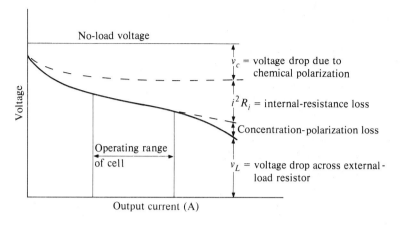

Figure 8.9 The current-voltage characteristics of fuel cells.

8.3.3 Types of Fuel Cells

Most of the operational fuel cells are low-temperature fuel cells employing hydrogen and oxygen as the reactants. Most of these cells operate at temperatures below 500 K, and while lowering the operating temperature improves the conversion efficiency the oxidation rate or power output of the cell can be increased by increasing the system pressure and/or temperature.

The redox fuel cell is a hydrogen-oxygen fuel cell that differs from the normal fuel cell in that the chemical reactions do not occur at the electrodes. A schematic diagram of the redox fuel cell is shown in Fig. 8.10. The redox cell has two electrolyte solutions separated by an ion-exchange membrane and the reactant gases are bubbled through each electrolyte. This cell is inherently less efficient than the conventional cell but it has lower resistance and polarization

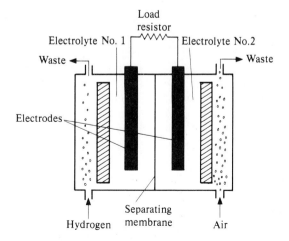

Figure 8.10 The redox fuel cell.

losses and potentially higher current densities. A primary advantage of this cell is that it can operate on relatively impure reactants.

A lot of research is being expended on the development of high-temperature fuel cells that can be used to operate with impure and inexpensive fuels, like hydrocarbons. In order to utilize a hydrocarbon fuel, it must first be "cracked" into hydrogen and carbon monoxide. The use of two fuel gases (carbon monoxide and hydrogen) in a fuel cell complicates the choice of an electrolyte in these systems. Moreover, since these systems operate at temperatures up to 1200 K, it precludes the use of aqueous and other electrolytes with low boiling points. This limits the choice of an electrolyte to either a solid electrolyte or to a molten salt. Two possible electrolytes for these systems are a mixed oxide or some sort of molten carbonate (CO_3^{--}) salt.

Any successful fuel cell should satisfy two major requirements—the invariance and the reactivity requirements. The invariance requirement specifies that the system must be designed to operate reliably for long periods of time. There must be no poisoning of the catalysts by impurities in the reactants, no clogging of the electrode pores, no "bubbling through" of the reactants, and no interdiffusion of the reactants.

The reactivity requirement is concerned with obtaining the maximum possible energy from the chemical reactions at relatively high reaction rates. Thus, it is important that all fuel atoms are completely oxidized during operation of the cell. The reaction rate can be increased by using porous electrodes that are very large to produce a large interaction interface between the gas and electrolyte. The reaction rate can also be increased by increasing the operating pressure and/or temperature of the cell. Unfortunately, the steps that are taken to increase the reaction rate are usually in conflict with the invariance requirement.

A number of fuel cells have been used successfully for special applications. These systems have been used extensively in the space program for relatively short-term applications such as manned orbital space flights and Apollo missions to the moon. Unfortunately, fuel cells have not progressed to the point where they can supply large quantities of economical electrical energy in an efficient manner for a long time period. In fact, some people working in this area have doubts that the fuel cell will ever satisfy that particular application.

8.4 CONVERSION OF ELECTROMAGNETIC ENERGY INTO ELECTRICITY

8.4.1 Operational Characteristics of Solar Cells

Electromagnetic energy can be converted directly to electrical energy in the photovoltaic cell, commonly called the solar cell. Like the fuel cell, the maximum conversion efficiency of this system is not limited by the efficiency of an externally reversible heat-engine cycle. Despite this, however, the conversion of solar energy into electrical energy is limited to relatively low conversion efficiencies.

The principle of operation of the photovoltaic cell was discovered by Adams and Day in 1876, using selenium. In 1919, Coblenz discovered that a voltage is induced between the illuminated and dark regions of semiconducting crystals. However, photoelectric conversion was essentially a laboratory phenomenon until 1941 when Ohl discovered the photovoltaic effect at a *p-n* junction of two semiconductors.

Primary interest in these systems concerns the possible conversion of the electromagnetic energy from the sun directly into electricity. Using the solar constant of 1395 W/m^2, it can be shown that the effective radiating temperature at the surface of the sun is around 6000 K (10800°R). According to the Wiens displacement law for thermal radiation, the most probable energy of the solar radiation is about 2.8 eV. While this value is very small compared to the energies encountered in nuclear reactions, it is more than sufficient to strip the valence electrons from many materials.

The successful operation of a solar cell relies on the action of the *p-n* junction. When a *p-n* junction is first formed, there is a transient charging process that establishes an electric field in the vicinity of the junction. Although both the *n*-type and the *p*-type semiconductors are neutrally charged by themselves, the electron concentration in the *n*-type material is so high that when it is combined with the *p*-type semiconductor some of the electrons from the *n* material "spill over" into the holes of the *p* material. This essentially makes the *n* material positively charged and the *p* material negatively charged in the vicinity of the junction. This charging process continues until the electric field or junction potential inhibits further net flow and the electron and hole flow is the same in both directions, as indicated in Fig. 8.11*a*.

If a forward-bias voltage v_L is applied across the junction, the junction potential is reduced by the amount of the bias voltage. The forward-bias voltage increases the flow of majority carriers (electrons to the *p* material and holes to the *n* material) across the junction as indicated in Fig. 8.11*b*. The net current density J across the junction is

$$J = J_0 \exp\left(\frac{ev_L}{kT} - 1\right) \tag{8.66}$$

where J_0 is the reverse-saturation current density. The reverse-saturation current density is the current that flows when a large reverse bias is applied across the junction and the current flow is due to only the minority carriers (electrons to the *n* material and holes to the *p* material).

The photons react with the valence electrons near a *p-n* junction to produce an effect similar to that produced by the forward-bias voltage. In this case, v_L is the external voltage that is generated by the photons. A typical solar-cell schematic is shown in Fig. 8.12. The nonreflected photons incident on the surface of the cell enter the thin outer layer of the semiconducting material and are either converted into heat or produce ion-pairs by stripping the valence electrons from the semiconductor atoms. In order to produce an ion-pair, the incoming photon must have an energy in excess of E_g, which is called the excitation

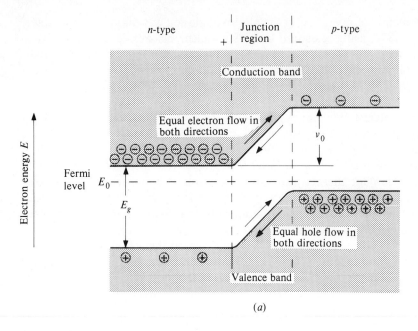

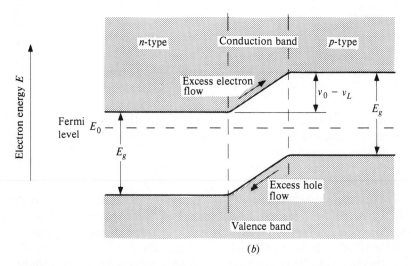

Figure 8.11 The charge distribution at an *n-p* junction of semiconductors (*a*) without and (*b*) with an applied voltage v_L. (*From Walsh, 1967.*)

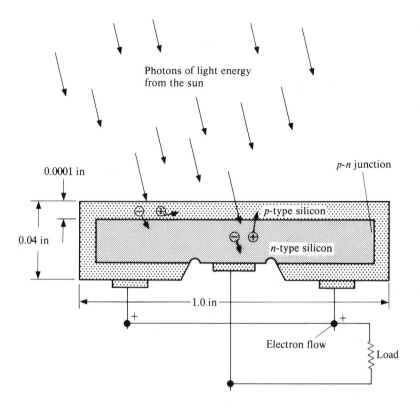

Figure 8.12 Schematic diagram of a typical solar cell. (*From " Encyclopedia of Energy," 1976.*)

energy. Some of these ions will be separated by the electric field of the junction. These ions reduce the electric field at the junction and this increases the flow of the majority carriers producing a current flow as shown in Fig. 8.13. The typical current-voltage characteristics of a solar cell are shown in Fig. 8.14.

The conversion efficiency of a solar cell is not limited like the thermal efficiency of a heat-engine cycle but there are some inherent losses that severely limit the cell performance. The two major losses are the junction loss and the spectrum loss. The junction loss is that loss due to the flow of minority carriers in the junction [the -1 in Eq. (8.66)]. Although this flow is usually small compared to the flow of majority carriers, it is not negligible. For silicon solar cells exposed to solar radiation, the junction loss reduces the conversion efficiency to about 50 percent. The junction loss decreases as the radiation intensity is increased as this effectively increases v_L in the current equation, (8.66). Care must be taken in increasing the radiation intensity because if the temperature T increases very much, it can negate the increase in v_L.

Another major solar-cell loss is the energy-spectrum loss. This loss is associated with the energy spectrum of the incident photons and the excitation

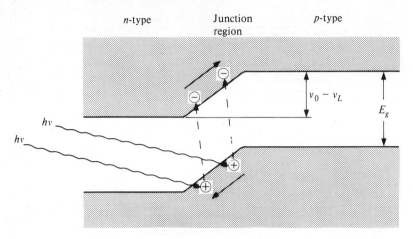

Figure 8.13 The charge-carrier flow due to irradiation of the solar cell. *(From Walsh, 1967.)*

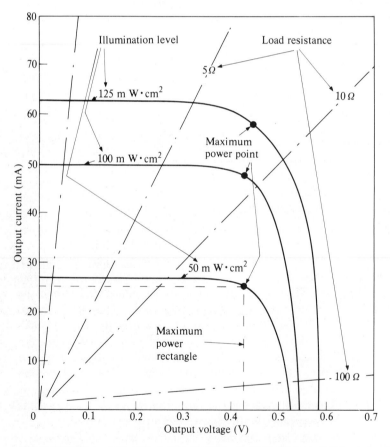

Figure 8.14 The current-voltage characteristics of a typical solar cell. *(From Angrist, 1976.)*

energy of the semiconductor material. Any photon with an incident energy less than the excitation energy E_g cannot produce an ion-pair and the photon energy is converted into thermal energy and lost. Those photons that have an energy in excess of the excitation energy will normally produce one ion-pair and the excess energy will be converted into thermal energy, although the excess energy may help prevent some recombination of the ion-pairs. For silicon solar cells, the excitation energy is 1.1 eV and the spectrum loss for solar radiation is about 50 percent.

There are a number of minor losses associated with operational solar cells, including the reflection of the photons, recombination of the ion-pairs before they reach the junction, and also the loss due to joule heating, particularly in the thin outer semiconductor layer. Taking into consideration all the losses, the maximum conversion efficiency of a silicon solar cell, which has one of the highest practical conversion efficiencies of any solar cell, is about 25 percent while the actual efficiency ranges between 15 and 20 percent.

Silicon is one material that is commonly used as the base material in solar cells because it gives one of the highest conversion efficiencies. The silicon is doped with phosphorus to produce the n-type semiconductor and it is doped with boron to produce the p-type semiconductor. A considerable amount of research has also been carried out on cadmium sulfide (CdS) cells, and while these systems are less expensive than the silicon cells they have lower conversion efficiencies. Research has been carried out and is continuing on the following semiconductor materials: In-P, Ga-As, Cd-Te, Al-Sb, and Cd-Se.

Solar cells have a number of advantages over the other solar-energy conversion systems. They are simple, compact, and have a very high power-to-weight ratio. This makes them very attractive for space applications. Solar cells have no moving parts and probably yield the highest overall conversion of solar energy into electricity. Theoretically, the solar cell has unlimited life although, in practice, it is found that they actually suffer from radiation damage, particularly by the bombardment of high-energy charged particles such as electrons in the Van Allen radiation belt around the earth. The radiation damage in a silicon solar cell can be reduced by using an n-on-p configuration.

The major problem associated with solar cells is the high cost associated with the system fabrication. Actually, the principle of fabrication is relatively simple and the raw material is inexpensive, but the actual production techniques are quite expensive. Recently, there have been some significant breakthroughs in fabrication methods that promise a marked reduction in cost. Despite this lower cost, however, the cost must be lowered much more if these systems are to become economically competitive with other solar conversion systems.

A problem associated with any terrestrial solar conversion system is that the system must be integrated with either an energy-storage system or with another type of conversion system to supply energy at night and on days when it is cloudy. Orbiting spacecraft also normally require an energy-storage system to supply energy when the spacecraft is in the shadow of the earth. Storage batteries are commonly employed with low-powered solar-cell systems.

8.4.2 Solar-Cell Performance

The operation of the photoelectric cell is such that part of the current generated by the photoelectric effect J_s is shunted through the internal cell resistance if there is any load on the cell at all. That portion of the current density in the cell that goes through the external load is J_L, and is given by the following equation:

$$J_L = J_s - J \tag{8.67}$$

Substituting Eq. (8.66) into Eq. (8.67) for J gives

$$J_L = J_s - J_0 \left(\exp \frac{ev_L}{kT} - 1 \right) \tag{8.68}$$

If v_L is zero, the short-circuit condition, the exponential term in the last quantity approaches unity and $J_L = J_s$, the short-circuit current density.

The output power from the photovoltaic cell is

$$P = v_L J_L A \tag{8.69}$$

where A is the surface area of the cell. Substituting Eq. (8.68) into Eq. (8.69) gives

$$P = Av_L J_s - Av_L J_0 \left(\exp \frac{ev_L}{kT} - 1 \right) \tag{8.70}$$

Differentiating the above equation with respect to v_L and setting the derivative equal to zero gives the value of the external load voltage $v_{L, \max P}$ that gives the maximum cell output power. This gives the following relationship:

$$\exp \frac{ev_{L, \max P}}{kT} = \frac{1 + J_s/J_0}{1 + ev_{L, \max P}/kT} \tag{8.71}$$

If the short-circuit current density J_s and the reverse-saturation current density J_0 are known, the value of $v_{L, \max P}$ can be evaluated from (8.71) by trial and error. The maximum power output of the cell is then equal to

$$P_{\max} = \frac{Av_{L, \max P}(J_0 + J_s)}{1 + kT/ev_{L, \max P}} \tag{8.72}$$

If the energy flux incident on the cell is known, P_{in}/A, the conversion efficiency for maximum power becomes

$$\eta_{\max P} = \frac{P_{\max}}{P_{\text{in}}} = \frac{v_{L, \max P}(J_0 + J_s)}{P_{\text{in}}(1 + kT/ev_{L, \max P})} = \eta_{\max} \tag{8.73}$$

Since the input energy flux to the cell is normally constant, the efficiency given by Eq. (8.73) is also the maximum possible conversion efficiency, as indicated. The value of J_s is a function of the incident photon flux.

Example 8.4 At a given intensity on a solar cell, the short-circuit current density is 180 A/m^2 and the reverse-saturation current density is $8 \times 10^{-9} \text{ A/m}^2$. At a temperature of 27°C and for the condition of maximum power, find the effective surface area needed for an output of 10 W and also estimate the conversion efficiency if the radiation intensity is 950 W/m^2.

SOLUTION For the condition of maximum power output:

$$\exp \frac{ev_{L, \max P}}{kT} = \frac{1 + J_s/J_0}{1 + ev_{L, \max P}/kT}$$

$$= \exp \frac{11{,}600 v_{L, \max P}}{300} = \frac{1 + 180/(8 \times 10^{-9})}{1 + 38.67 v_{L, \max P}}$$

$$= \exp 38.67 v_{L, \max P} = \frac{22.5 \times 10^9}{1 + 38.67 v_{L, \max P}}$$

$$\left(1 + 38.67 v_{L, \max P}\right) \exp 38.67 v_{L, \max P} = 22.5 \times 10^9 = X$$

By trial and error:

$v_{L, \max P}$	X	$v_{L, \max P}$	X
0.3	1.376×10^6	0.54	25.64×10^9
0.4	8.596×10^7	0.537	22.71×10^9
0.5	5.073×10^9	0.5369	22.62×10^9
0.55	38.41×10^9	0.5368	22.53×10^9

$$v_{L, \max P} = 0.5368 \text{ V}$$

$$\frac{P_{\max}}{A} = \frac{v_{L, \max P}(J_0 + J_s)}{1 + kT/ev_{L, \max P}}$$

$$= \frac{0.5368(180)}{1 + 300/11{,}600(0.5368)}$$

$$= 92.18 \text{ W/m}^2$$

$$\text{Area of cell} = A = \frac{P_{\text{out}}}{P_{\max}/A} = 10/92.18 = 0.10848 \text{ m}^2 = 1084.8 \text{ cm}^2$$

$$\text{Conversion efficiency} = \frac{P_{\max}/A}{(P/A)_{\text{in}}} = 92.18/950 = 0.097 = 9.7\%$$

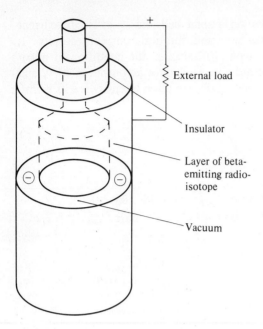

External load

Insulator

Layer of beta-
emitting radio-
isotope

Vacuum

Figure 8.15 The nuclear battery.

8.5 CONVERSION OF NUCLEAR ENERGY INTO ELECTRICITY

There are no large-scale converters that transform nuclear energy directly into electrical energy. The nuclear battery, which is shown schematically in Fig. 8.15, consists of an inner electrode that is coated with a thin layer of an alpha- or a beta-emitting radioisotope. Over half of the alpha particles or electrons emitted by the radioisotope travel across the vacuum gap and are absorbed by the outer case which serves as the other electrode. This system produces very high voltages in the kilovolt range but the current is normally in the micro-microampere range so that the system has very low power. The power output of these systems, like the radioisotope, decays exponentially with time unless the daughter product is also radioactive.

It may be possible to produce electrical energy directly from the fusion reaction by using the interaction between the plasma and the magnetic-field containment system. This is particularly true for the mirror machines.

8.6 CONVERSION OF MECHANICAL ENERGY INTO ELECTRICITY

8.6.1 Electric Generators and Alternators

Almost all of the devices that convert mechanical energy into electrical energy depend on the Faraday effect for the principle of operation. According to the

Faraday effect, a voltage gradient is produced in an electrical conductor that is forced perpendicularly through a magnetic field. The induced voltage gradient, dv/dx, in the conductor is equal to the cross vector product of the conductor velocity $\bar{V_c}$ and the magnetic-field strength $\bar{B}$:

$$\frac{dv}{dx} = \bar{V_c} \times \bar{B} \tag{8.74}$$

where $\bar{V_c}$ is the conductor velocity in meters per second and $\bar{B}$ is the magnetic flux density in gauss. If $\bar{V_c}$ is perpendicular to $\bar{B}$, dv/dx is equal to the product $\bar{V_c}\bar{B}$, but if V_c is parallel to $\bar{B}$, dv/dx is equal to zero.

Almost all of the electrical energy produced in the world is produced in the electrical generator or alternator. These systems obey the same basic laws as the electric motors. They normally have a conversion efficiency ranging from 50 percent for small generators to more than 90 percent for some of the large commercial alternators. If a coil is rotated between the poles of an electromagnet or a permanent magnet, the output from the rotor will be either alternating current or direct current, depending on whether a slip ring (ac) or segmented (dc) commutator is employed. The ac output of an alternator can be converted into direct current with the use of rectifiers, as in the conventional automobile alternator.

The field windings of a conventional dc generator require a dc power supply and it can be "excited" in several ways. Some machines are separately excited from an external power source such as a battery. Other dc generators are either shunt-excited, series-excited, or compound-excited, and these machines are started from the small amount of residual magnetic flux present in the machine. The output voltage of these machines depends on the rotor speed and the number of field poles in the machine. There must be a set of brushes for each set of field poles in the generator.

Alternating-current generators or alternators also require a dc power supply for the field winding. The dc power can be supplied from an external source but in most of these systems, the dc supply is produced by a small dc generator connected to the alternator shaft. This small dc generator is called the "exciter." In a large conventional alternator, the field winding is mounted on the rotor and the power is generated in the stator winding. This arrangement precludes the necessity of having to transfer the high-voltage output through brushes and slip rings. A typical large alternator is shown in Fig. 8.16.

Almost all commercial power generated in the United States is 60-Hz, three-phase current and the angular velocity of the rotor field is

$$N_r = \frac{7200}{p} \tag{8.75}$$

where N_r is the angular velocity of the alternator in revolutions per minute and p is the number of poles on the alternator rotor. Some other countries use systems that generate 50-Hz current. Care must be taken when connecting a generator to an ac power grid that the alternator is in phase with the alternating current in

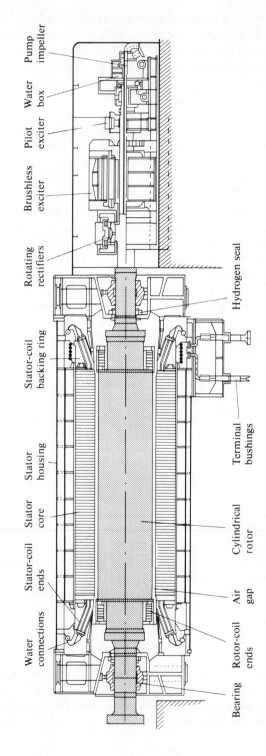

Figure 8.16 A conventional, large, ac generator (alternator). *(Courtesy of Allis-Chalmers Power Systems.)*

the grid at the instant of connection. If the alternator is out of phase with the grid at the instant of connection, the grid will cause immediate acceleration or deceleration of the rotor and severe damage can result.

Almost all large alternators are cooled with hydrogen at pressures above atmospheric pressure. The use of hydrogen reduces the windage losses without losing cooling ability. In the large machines, the hydrogen is pumped through hollow conductors in the stator and armature. Some of the newer machines employ a water coolant in the stator windings. At high powers, the hydrogen pressure may be increased to facilitate the cooling process.

8.6.2 Fluiddynamic Converters

General The fluiddynamic conversion systems transform the kinetic energy or the potential energy of a fluid directly into electrical energy. There are two principle fluiddynamic generators—the magnetohydrodynamic (MHD) converters and the electrogasdynamic (EGD) converters. There are some other minor mechanical-electrical conversion systems such as the electrokinetic generator, the peizoelectric converter, and the fluid-drop, electrostatic generator. The balance of this chapter will deal with the MHD and the EGD conversion systems.

Magnetohydrodynamic generators The magnetohydrodynamic generator or MHD generator depends on the Faraday effect for its mode of operation, just like the conventional mechanical-electrical generators. In the MHD system, an electrically conducting fluid is forced through a perpendicular magnetic field at high velocity, as shown in Fig. 8.17. In most MHD generators, the working fluid is an ionized gas but liquid metals can also be employed in these systems. The

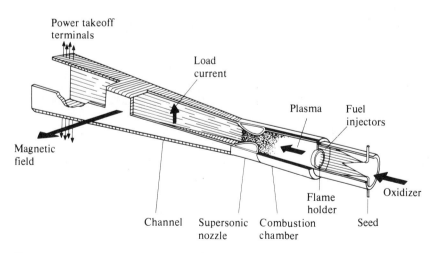

Figure 8.17 The magnetohydrodynamic (MHD) generator. (*From Proceedings of the American Power Conference, vol. 36, p. 585, 1974.*)

MHD converter is usually proposed as a topping cycle or system for a conventional steam-power system to improve the overall conversion efficiency of the system. From Eq. (7.40), it can be shown that the combined conversion efficiency of such a system is

$$\eta_{\text{overall}} = \eta_{\text{MHD}} + \eta_{\text{steam}} - \eta_{\text{MHD}}\eta_{\text{steam}} \tag{8.76}$$

where η_{MHD} and η_{steam} are the individual conversion efficiencies of the MHD and steam systems, respectively.

If the fluid velocity $\bar{V}_c$ and the magnetic-field strength $\bar{B}$ are constant in the interelectrode gap, Eq. (8.74) gives a generated voltage v_g of

$$v_g = \bar{V}_c Bd \tag{8.77}$$

where d is the interelectrode gap width, in meters. If J is the current density, in amperes per square meter, the internal voltage drop is $iR_g = i\rho d/A = J\rho d$ and the external voltage drop from the unit is

$$v_L = v_g - J\rho d \tag{8.78}$$

where ρ is the average electrical resistivity of the MHD fluid in ohm-meters. The ratio of the external load voltage to the generated voltage is defined as the loading factor K of the unit

$$K = \frac{v_L}{v_g} = \frac{v_L}{B\bar{V}_c d} \tag{8.79}$$

The generator current density is equal to the internal cell-voltage drop divided by the product of the interelectrode resistance and the electrode area:

$$J = \frac{i}{A} = \frac{(1/A)(v_g - v_L)}{\rho d/A} = \frac{v_g - v_L}{\rho d} = \frac{\bar{V}_c B}{\rho}(1 - K) \tag{8.80}$$

The output power density (P/V) is the output electrical power of the MHD generator per unit volume of the generator. Thus, the power density is

$$\left(\frac{P}{V}\right)_{\text{out}} = \frac{v_L i}{Ad} = \frac{Kv_g J}{d} = K\bar{V}_c BJ = \frac{\bar{V}_c^2 B^2}{\rho}(K - K^2) \tag{8.81}$$

The maximum power density as a function of the loading factor K can be found by setting $d(P/V)_{\text{out}}/dK = 0$ and then solving for K. This gives the optimum value of K of 0.5, which in turn gives a maximum power density of

$$\left(\frac{P}{V}\right)_{\text{out, max}} = \frac{\bar{V}_c^2 B^2}{4\rho} \tag{8.82}$$

The maximum conversion efficiency of an MHD converter is found by assuming that the only loss in the unit is the joule-heating power loss P_J. The loss per unit volume of converter is P_J/V, or

$$\frac{P_J}{V} = \frac{i^2 R_g}{Ad} = \frac{J^2 A^2(d\rho/A)}{Ad} = J^2\rho = \frac{\bar{V}_c^2 B^2}{\rho}(1 - K)^2 \tag{8.83}$$

and the conversion efficiency for maximum power density becomes

$$\eta = \frac{P_0}{P_0 + P_J} = \frac{1}{1 + P_J/P_0} = \frac{1}{1 + (1 - K)/K} = K \qquad (8.84)$$

and $\qquad \eta_{max} = K_{max} \qquad\qquad\qquad\qquad\qquad\qquad\qquad (8.85)$

For maximum power density, $K_{max} = 0.5$ and, for this condition, the maximum possible conversion efficiency is 50 percent.

Example 8.5 A MHD converter uses a combustion gas that is seeded with 1 percent potassium to increase its electrical resistivity to 0.03 $\Omega \cdot$m. The magnetic field is uniform and perpendicular to the gas velocity and has a flux of 1.5 Wb/m^2. The system is designed for a gas velocity of 900 m/s and a loading factor of 0.55. If the width of the duct d is 0.25 m and the electrode area is 1.2 m^2, determine the output current, voltage, and power. Neglect the Hall effect and the other losses.

SOLUTION

Total generated voltage $= v_g = \bar{V}_c Bd = 900(1.5)(0.25) = 337.5$ V

Voltage drop across
load resistor $= v_L = Kv_g = 0.55(337.5) = 185.6$ V

$$\text{Current density} = J = \frac{v_g - v_L}{\rho d} = \frac{337.5 - 185.6}{0.03(0.25)}$$

$$= 20{,}250 \text{ A/m}^2$$

Output current $= A_e J = 1.2(20{,}250) = 24{,}300$ A

Output power $= iv_L = 24{,}300(185.6) = 4.51 \times 10^6$ W $= 4.51$ MW$_e$

The MHD generator has a number of other losses besides the joule-heating loss. The Hall-effect loss is the loss caused by the Lorentz force acting on the electrons in the interelectrode gap. When any charged particle passes through a perpendicular magnetic field, it is acted on by a force called the Lorentz force F_L. For an electron, this force makes the electron travel in a circular arc and the magnitude of this force is

$$\bar{F}_L = e\bar{V} \times \bar{B} \qquad (8.86)$$

This means that all the electrons try to travel to one end of the collecting electrode and this produces very large currents in the collecting electrode with the accompanying resistance loss. One possible solution to this problem is to use an insulated, segmented collecting electrode, similar to that shown schematically in Fig. 8.18. The problem with the segmented electrode is that each segment must be isolated and loaded independently, as shown in the figure.

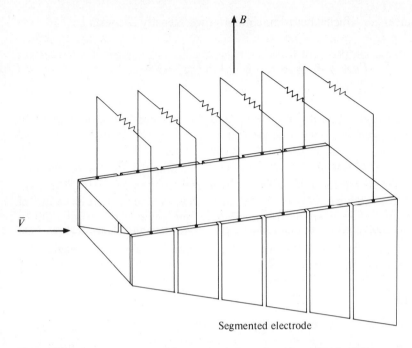

Figure 8.18 A MHD generator with segmented electrodes. *(From Walsh, 1967.)*

The end loss in a MHD generator is a loss that is associated with the reverse flow (short circuit) of electrons through the conducting fluid around the ends of the magnetic field. This loss can be reduced by increasing the aspect ratio (L/d) of the generator, by permitting the magnetic field poles to extend beyond the end of the electrodes, and/or by using insulated vanes in the fluid duct at the inlet and outlet of the generator.

The MHD system also experiences high friction and heat-transfer losses. The fluid flow is highly turbulent through the device and the friction loss may be as high as 12 percent of the input. The high fluid turbulence also increases the convection heat-transfer rate from the gas or liquid metal to the containment walls.

Another loss that can be very important in an ionized-gas or plasma converter is the electrode loss. These systems must operate at extremely high temperatures in order to obtain a high electrical conductivity. Since the electrodes must be relatively cool, the gas in the vicinity of the electrodes is much cooler. This significantly increases the fluid resistivity next to the electrodes resulting in a very large voltage drop across the gas film.

Types of MHD converters The plasma converter employs an ionized gas as the working fluid. The major problem associated with this system is that it is difficult to find a working fluid that can be ionized at a reasonable temperature.

Gases at 2000 K (4060°F) have too high a value of electrical resistivity ρ to serve as a satisfactory working fluid in a MHD converter. Even at the maximum adiabatic flame temperature from combustion of 3000 K (5860°F), the normal combustion products still have too high an electrical resistivity for practical MHD operation.

Most of the research on the plasma converter has been concerned with the use of an additive or "seed" material to the combustion products which will ionize easily. The seed material should have a low ionization potential and should be easily recovered from the exhaust gases. The addition of 1% potassium by volume decreases the resistivity of the gas from 5 $\Omega \cdot$m at 2200 K to about 1 $\Omega \cdot$m and the resistivity decreases further to 0.11 $\Omega \cdot$m at 3000 K.

It may also be possible to decrease the gas resistivity by using a glow discharge or by subjecting the gas to ionizing radiation. There is also research being conducted into the possibility of passing the hypersonic gas stream through a series of shock waves in the converter to increase the conductivity of the gas. A highly conducting region of ionized gas normally occurs in the region immediately behind the shock wave.

A schematic diagram of a possible power system employing a plasma converter is shown in Fig. 8.19. As mentioned earlier, the MHD converter is normally designed to be a topping power system to a conventional steam-power system. It has been estimated that such a topping system could increase the overall conversion efficiency from around 40 to about 55 percent. The system shown in Fig. 8.19 is classed as an open system because the main working fluid is not recirculated. Environmental and economic considerations dictate that the seed material must be recovered. If coal is used as the source of thermal energy in this device, the electrodes may be short circuited by the molten ash. Consequently, oil and natural gas are considered to be much better fuels for this system.

Liquid-metal MHD converters are closed systems that must operate at a much lower temperature than the plasma converters. High temperatures are not required in these systems to achieve a low value of electrical resistivity and these systems are normally designed to operate below 1400 K. This temperature is low enough that the energy can be supplied by a nuclear reactor as well as a fossil-fueled system. A possible liquid-metal system, using either potassium or cesium as the working fluid, is shown in Fig. 8.20.

The main problem associated with the liquid conversion systems is that it is much more difficult to achieve a high fluid velocity in the converter. In the system shown in Fig. 8.20, superheated metallic vapor is expanded through a supersonic nozzle into the drift tube or mixer. Atomized subcooled liquid droplets are accelerated by the vapor. The vapor also condenses on the liquid droplets so that the fluid entering the generator is essentially a liquid. The resulting fluid velocity is in excess of 150 m/s.

At temperatures below 1400 K, it is anticipated that this system could increase the thermal efficiency of a steam-power system by 6 percent. The maximum overall thermal efficiencies are expected to be around 55 percent.

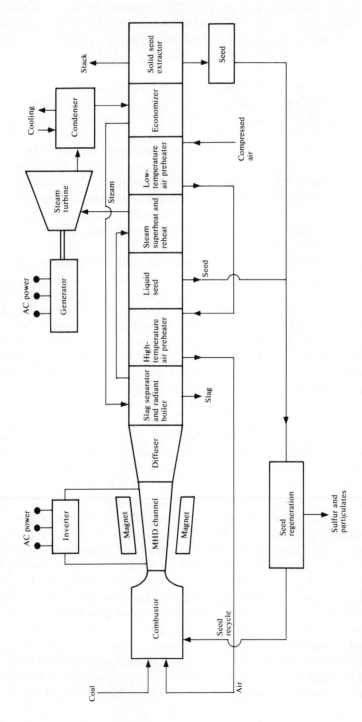

Figure 8.19 A proposed open-cycle MHD-steam binary power system. (*From Angrist, 1976.*)

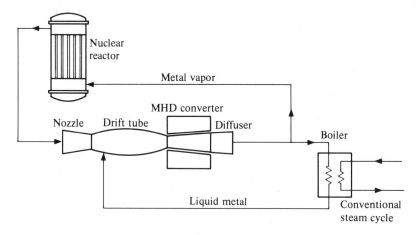

Figure 8.20 The proposed liquid-metal MHD-steam binary power system. *(From Walsh, 1967.)*

The liquid-metal MHD converter has several advantages over the plasma MHD system. This system can utilize nuclear energy since it does not have the high-temperature requirements of the plasma converter. The liquid-metal converter can easily produce ac power directly, while this is almost impossible to do in a plasma system. The power density of the liquid-metal system is around an order of magnitude higher than that of a plasma generator and this significantly reduces the size of the system, including that of the magnets. The magnet cost is a major cost associated with these systems.

The primary advantage of the plasma converter is that it is much easier to achieve a high fluid velocity using a gas and a nozzle. Because of the high gas resistivity, the plasma converter can produce relatively high dc voltages, which is difficult to do in the liquid system. It is generally agreed that the plasma generator can produce higher overall conversion efficiencies than the liquid-metal systems.

Electrostatic mechanical generators The electrostatic mechanical generators convert mechanical energy, usually mechanical potential energy, of a fluid directly to electrical energy. The principal of operation is the same as that for the high-voltage electrostatic generator, commonly called the Van de Graff accelerator. In this device, the electric charge is transferred from one electrode to another by an insulated belt. The operation of the fluiddynamic machines are similar except that the charge is transferred by the fluid instead of a belt. All of these devices are characterized by fairly low currents and very high voltages.

The liquid-drop, electrostatic generator converts the gravitational potential energy of water droplets directly into electrical energy. A schematic diagram of this system is presented in Fig. 8.21. While this device makes an interesting laboratory demonstration, it is not used for the large-scale production of electricity as it cannot compete with the conventional hydroelectric generator.

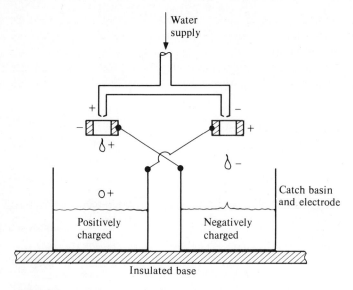

Figure 8.21 The liquid-drop, electrostatic generator.

The electrogasdynamic (EGD) converter is about the only system of this general type that shows much promise for the large-scale production of electrical energy. The EGD generator uses the potential energy of a high-pressure gas to carry electrons from a low-potential electrode to a high-potential electrode, thereby doing work against an electric field. Thus, the potential energy is converted directly into electrostatic electrical energy. A typical EGD duct is shown in Fig. 8.22.

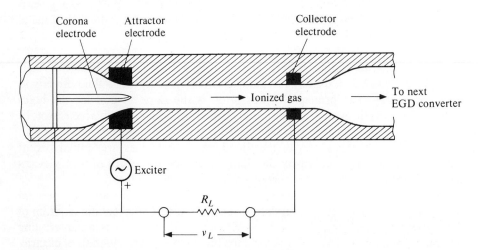

Figure 8.22 A typical gas duct in an electrogasdynamic (EGD) converter. *(From Walsh, 1967.)*

Electrons are generated at the entrance of the EGD duct by some means such as by a corona electrode. The ionized-gas particles are carried down the duct along with the neutral atoms and the ionized particles are neutralized by a collector electrode at the end of the insulated duct. The working fluid for these systems is commonly either combustion gases produced by burning fuel at high pressure or it is a pressurized reactor gas coolant.

EGD performance In the ideal EGD generator, the only major pressure loss is the friction pressure drop Δp_f:

$$\Delta p_f = \frac{C_f \bar{V}^2 L}{2vd} \tag{8.87}$$

where $\bar{V}$ is the gas velocity in meters per second, L is the duct length in meters, d is the duct diameter in meters, v is the specific volume of the gas in cubic meters per kilogram, and C_f is the friction factor. The pressure drop due to the voltage gradient Δp_e is

$$\Delta p_e = E_x n_p eL \tag{8.88}$$

where e is the charge of an electron in coulombs, n_p is the average ion density in the duct in ions per cubic meter, and E_x is the constant electric field in the axial direction in volts per meter. The product of $n_p e$ is equal to $4E_r \varepsilon_0 /d$, where E_r is the radial electric field in volts per meter, and ε_0 is the permitivity of free space. ε_0 has a numerical value of 8.85×10^{-12} C/N. The maximum value of Δp_e occurs when $E_x = E_r = E_b$, where E_b is the limiting or breakdown voltage of the gas:

$$\Delta p_{e,\,\text{max}} = \frac{4\varepsilon_0 E_b^2 L}{d} \tag{8.89}$$

The conversion efficiency of the EGD converter is approximately equal to the ratio of the pressure drops. The output power is proportional to Δp_e, while the total input power is proportional to the total pressure drop $(\Delta p_e + \Delta p_f)$. Thus the maximum conversion efficiency is

$$\eta_{\text{max}} = \frac{\Delta p_{e,\,\text{max}}}{\Delta p_{e,\,\text{max}} + \Delta p_f} = \frac{1}{1 + \Delta p_f / \Delta p_{e,\,\text{max}}}$$

$$= \frac{1}{1 + C_f \bar{V}^2 / 8\varepsilon_0 E_b^2 v} \tag{8.90}$$

The maximum power density in the gas volume $(P/V)_{\text{max}}$ in the EGD generator is

$$\left(\frac{P}{V}\right)_{\text{max}} = \frac{i_{\text{max}} v_{L,\,\text{max}}}{AL} \tag{8.91}$$

where A is the cross-sectional area of the gas duct. The current flow i in the system is equal to the rate of charge flow in the system:

$$i = n_{p,\,\text{max}} e \bar{V} A = \pi \varepsilon_0 E_{r,\,\text{max}} \bar{V} d = \pi \varepsilon_0 E_b \bar{V} d \tag{8.92}$$

The maximum external load voltage is

$$v_{L, \text{max}} = E_{x, \text{max}} L = E_b L \tag{8.93}$$

Most EGD converters have relatively low maximum power outputs of 10 to 30 W per channel. This system usually has several hundred channels connected in series and parallel combinations. These systems are normally very high-voltage systems with output voltages of 100,000 to 200,000 V. Because of the high generated voltages, this machine needs reliable high-voltage insulation for both the ducts and the electrode leads. The material commonly proposed for this system is beryllium oxide (BeO).

Although the EGD systems are still in the early developmental stage, it is anticipated that the overall conversion efficiency of the EGD system can approach that for the MHD-steam combined systems. For an inlet temperature of 2500 K, the predicted conversion efficiency is about 55 percent. At 1100 K, the predicted conversion efficiency is about 35 percent.

The EGD system has a number of advantages over MHD systems. The EGD system operates at relatively low temperatures and there is no need for the injection and collection of seed material from the working fluid. The EGD system is also self-contained in that it does not require a steam plant as a bottoming cycle. It may be possible to extract energy from the gas until it reaches standard stack temperatures. This system does not require large quantities of condenser cooling water.

Of all the so-called direct-energy converters, including thermoelectric, thermionic, solar-cell, and fuel-cell converters, the MHD and the EGD conversion systems appear to offer the best hope or solution in the search for a high-efficiency, large-capacity system for the production of electricity.

PROBLEMS

8.1 The properties for an n-p semiconductor thermoelectric generator are as follows:

Material	n-type	p-type
Seebeck coefficient S, μV/K	$S_n = -104 - 0.35T$	$S_p = 440 - 0.55T$
Electrical resistivity ρ, $\Omega \cdot$m	$\rho_n = 8 \times 10^{-6}$	$\rho_p = 3.5 \times 10^{-5}$
Thermal conductivity k, W/m$\cdot$K	$k_n = 1.32$	$k_p = 1.55$

where the temperature T is in kelvin. If the converter operates between 127 and 427°C, find the average Seebeck coefficient, the average Peltier coefficient, and the average Thomson coefficient. Also find the maximum figure of merit for the converter.

8.2 For the system described in Prob. 8.1, find:

(a) The value of the resistance ratio M_{opt} that gives the maximum thermal efficiency

(b) The optimum value of the ratio of A/L for the n material if the same ratio for the p material is 1.2 cm

(c) The value of the electrical resistance R_g and the thermal conductance K_g for the generator

(d) The number of element-pairs needed to produce 100 W of dc power at the maximum conversion efficiency

(e) The maximum power output of the converter from part (d)

(f) The thermal efficiencies for the condition of (d) and (e)

8.3 A thermoelectric generator operates between 250 and 550°C. The average value of the Seebeck coefficient is 400×10^{-6} V/K, the generator resistance is 0.004 Ω and the thermal conductance is 0.035 W/K·m. Find the open-circuit voltage, the maximum power output, and the thermal efficiency for maximum power output.

8.4 Using the Richardson-Dushman equation, determine the current from a 2-cm square electrode in a vacuum at 2000 K if the electrode is constructed of strontium oxide (SrO), and the space–charge-barrier potential is zero. Also evaluate the current if there is a space–charge-barrier potential of 0.4 eV.

8.5 A thoriated-tungsten emitter is supposed to supply 8 A/cm² when the space–charge-barrier energy is 0.12 V. Determine the emitter temperature in degrees Celsius.

8.6 Consider a vacuum diode with a cathode and anode having a surface area of 10 cm². The emitter is constructed of strontium oxide (SrO) and the collector is constructed of barium oxide (BaO). The space charge barrier for the emitter is 0.18 V and the collector (anode) barrier energy is 0.11 V. The emitter is maintained at a temperature of 2000 K. Find the output current and power of the device and estimate the thermal efficiency if the emissivities of all the surfaces are 0.15. Neglect the conduction loss through the lead-in wire.

8.7 Calculate the higher and lower heating values (HHV and LHV) in British thermal units per pound-mass for the combustion of hydrogen at 1 atm and 25°C, using the enthalpies of formation from Table 8.3.

8.8 Calculate the maximum conversion efficiency of a fuel cell operating at 1 atm and 25°C that is fueled with

(a) Hydrogen and oxygen

(b) Carbon monoxide and oxygen

8.9 Determine the maximum internal cell voltage of a hydrogen-oxygen fuel cell and of a carbon monoxide-oxygen fuel cell operating at 400°C. Assume that the oxygen is supplied from atmospheric air and that all the reactant gases are supplied at a pressure of 1.2 bar.

8.10 A 2-cm square solar cell at 30°C has an output voltage of 0.4 V at maximum power. If the reverse-saturation current density is 10^{-7} A/cm², find the current and the output power of the cell.

8.11 The reverse-saturation current i_0 of a solar cell at 35°C is 1.4×10^{-7} A and the short-circuit current is 6 A when it is exposed to sunlight. Calculate the power output of the cell.

8.12 If the conversion efficiency of a MHD converter is 30 percent and the thermal efficiency of the bottoming steam plant is 35 percent, find the overall conversion efficiency of the binary system.

8.13 A MHD converter uses a seeded combustion gas with an electrical resistivity of 0.028 Ω·m and a velocity of 1000 m/s. The spacing between two 0.75 m² electrodes is 0.35 m and the magnetic field density is 2.2 Wb/m². If the external load has a resistance of 0.015 Ω, find the output voltage, current, and power.

8.14 The following parameters are picked for a MHD plasma converter:

$$\bar{B} = 1.8 \text{ Wb/m}^2$$

$$\bar{V} = 1200 \text{ m/s}$$

$$A = 0.6 \text{ m}^2$$

$$d = 0.25 \text{ m}$$

$$\rho = 0.028 \text{ Ω·m}$$

For an output voltage of 250 V, find the output current and power, the value of the external load resistance, and the conversion efficiency.

REFERENCES

Angrist, S. W.: "Direct Energy Conversion," 3d ed., Allyn and Bacon, Inc., Boston, Mass., 1976.

Berger, C. (ed.): "Handbook of Fuel Cell Technology," Prentice-Hall, Inc., Englewood Cliffs, N.J., 1968.

Cowan, P. L., M. C. Gourdine, and D. H. Malcolm: "EGD Power Generation," Institute of Electrical and Electronics Engineers, Power Group, February, 1968.

Currin, C. G., et al.: Feasibility of Low-cost Silicon Solar Cells, *Conference Record of the Ninth IEEE Photovoltaic Specialists Conference*, 1972, pp. 363–369, IEEE, New York.

"Encyclopedia of Energy," McGraw-Hill Book Company, New York, 1976.

Gietzen, A. J., and W. G. Hohmeyer: Thermionic Reactor Power Systems, *Proceedings Seventh Intersociety Energy Conversion Engineering Conference*, 1972, American Chemical Society, Washington, D.C.

Harris, C. J.: MHD Augmented Shock Tunnel Experiments with Unseeded High Density Air Flows, *AIAA Journal*, vol. 13, no. 2, pp. 229–231, February, 1975.

Harris, L. P., and G. E. Moore: "Combustion-MHD Power Generation for Central Stations," General Electric Report No. 70-C-388, November, 1970.

Kettani, M. A.: "Direct Energy Conversion," Addison-Wesley Publishing Company, Reading, Mass., 1970.

Needrach, L. W., W. T. Grubb, and A. Fickett: "The Current Status of Ion Exchange Membrane Fuel Cells," General Electric Report No. 71-C-303, October, 1971.

Russell, C. R.: "Elements of Energy Conversion," Pergamon Press, Inc., Long Island City, N.Y., 1967.

Smith, A.: Status of Photovoltaic Power Technology, *Journal of Engineering for Power, Transactions of the American Society of Mechanical Engineering*, series A, vol. 91, pp. 1–12, 1969.

Soo, S. L.: "Direct Energy Conversion," Prentice-Hall Inc., Englewood Cliffs, N.J., 1968.

Walsh, E. M.: "Energy Conversion," The Ronald Press Company, New York, 1967.

Weller, A. E., and W. T. Reid: "The Economic Position of MHD for Central Power," ASME Paper 64 WA/ENER-1, September, 1964.

Westinghouse Electric Corporation: "1970 Final Report Project Fuel Cell: Research and Development Report 57," U.S. Government Printing Office, 1970.

Wood, B. D.: "Applications of Thermodynamics," Addison-Wesley Publishing Company, Reading, Mass., 1969.

NINE

ENERGY STORAGE

9.1 INTRODUCTION

All six of the major energy classifications can be stored in some form of that general energy classification except electromagnetic energy which is purely a transitional energy form. Mechanical energy can be stored as either kinetic energy or as potential energy. Electrical energy can be stored as either inductive-field energy or as electrostatic-field energy. Chemical energy and nuclear energy, two of the six major energy classifications, are actually pure forms of stored energy. Thermal energy can be stored as either latent heat and/or as sensible heat.

Energy storage is very important and even essential in many power-generating systems. Terrestrial solar-energy power systems require either an energy-storage system or an alternate source of energy to supply energy when there is insufficient sunlight. Energy-storage systems are also very useful to companies that generate electricity providing the stored energy can be easily and efficiently converted back to electrical energy. With such a storage system, it becomes feasible to produce and store electrical energy during times of low power demand, such as late at night and on weekends. The stored energy is then recovered during times of peak power demand. This stored energy can be produced from the coal- and nuclear-powered, base-loaded units with their low fuel costs. This not only improves the maximum-to-average power ratio for the base-loaded systems, reducing their capital costs, but it also reduces the need for the peaking units with their corresponding high fuel costs.

Table 9.1 Specific energy storage of various materials and systems (all values in kJ/kg)

Deuterium (D-D fusion reaction)	3.5×10^{11}	Silver oxide-zinc battery	437
Uranium-235 (fission reaction)	7.0×10^{10}	Lead-acid battery	119
Heavy water (fusion reaction)	3.5×10^{10}	Flywheel (uniformly	
Reactor fuel (2.5% enriched UO_2)	1.5×10^{9}	stressed disc)	79
Natural uranium	5.0×10^{8}	Compressed gas (spherical	
95% Po-210 (radioactive decay)	2.5×10^{6}	container)	71
80% Pu-238 (radioactive decay)	1.8×10^{6}	Flywheel (cylindrical)	56
Hydrogen (LHV)	1.2×10^{5}	Organic elastomer	20
Methane (LHV)	5.0×10^{4}	Flywheel (rim-arm)	7
Gasoline (LHV)	4.4×10^{4}	Torsion spring	0.24
Lithium hydride (at 700°C)	3.8×10^{3}	Coil spring	0.16
Falling water ($\Delta z = 100$ m)	9.8×10^{2}	Capacitor	0.016

Several things must be considered in the selection, design, and operation of any energy-storage system. First, what is the overall efficiency of the system? This includes the charging process, the storage loss, and the recovery process. Second, what is the energy storage density in kilojoules per cubic meter or British thermal units per cubic foot, and/or the specific energy storage in kilojoules per kilogram or British thermal units per pound-mass? Third, what are the maximum allowable charge and discharge rates? Fourth, what are the economics of the storage system? This includes both the capital and operating costs. Fifth, what are the environmental problems associated with the storage units? Finally, how many times can the system be cycled and how long will it last?

The specific energy storage in kilojoules per kilogram for a number of different materials and systems is given in Table 9.1.

9.2 STORAGE OF MECHANICAL ENERGY

9.2.1 Kinetic-Energy Storage

Mechanical energy can be stored as either kinetic energy or potential energy. Kinetic energy is mechanical energy that is associated with the movement of one mass relative to another. For a linear velocity, the kinetic energy of a given body in joules is

$$KE = \frac{m\bar{V}^2}{2} \tag{9.1}$$

where m is the mass in kilograms and $\bar{V}$ is the linear velocity in meters per second. Kinetic energy can also be stored in a rotating wheel, commonly called a flywheel. The kinetic energy associated with a wheel rim of mass m kilograms, rotating at an angular velocity of ω rad/s at a radius of R m, is

$$KE = \frac{mR^2\omega^2}{2} = 2\pi^2 mR^2 n^2 \tag{9.2}$$

where n is the angular velocity in revolutions per second. For a given flywheel, $\int_0^R r^2 \, dm$ is called the moment of inertia I, and Eq. (9.2) reduces to the more general form of

$$KE = \frac{I\omega^2}{2} = 2\pi^2 n^2 I \tag{9.3}$$

For a thin-rimmed flywheel with all the mass in the rim, the relationship between the tangential stress σ in the wheel, the angular velocity ω, and the radius R is

$$\sigma = \omega^2 R^2 \rho \tag{9.4}$$

where ρ is the density of the material in kilograms per cubic meter. Substituting this equation into Eq. (9.2) gives

$$KE = \frac{m\sigma}{2\rho} \tag{9.5}$$

The specific energy storage in the flywheel is the energy per unit mass. For a thin-rimmed flywheel, the specific energy is

$$\text{Specific energy} = \frac{KE}{m} = \frac{\sigma}{2\rho} \tag{9.6}$$

This equation is commonly written as

$$\text{Specific energy} = K_w \frac{\sigma}{\rho} \tag{9.7}$$

where K_w is the flywheel weight factor and ranges from approximately 1.0 for constant-stress discs to 0.5 for a thin-rimmed flywheel.

In order to compare flywheels, it is common practice to evaluate the maximum stored kinetic energy in the wheel per unit volume of a uniform disc having axial and radial dimensions equal to the maximum width and radius of the given wheel. This energy ratio is usually expressed as

$$\left(\frac{KE}{V}\right)_{\text{max}} = K_v \sigma \tag{9.8}$$

where K_v is equal to the product of K_w and the fraction of the uniform disc occupied by the given flywheel. The values of K_w and K_v for some high-performance flywheels are shown in Fig. 9.1.

It can be seen from Eq. (9.5) that in order to achieve high energy storage in a flywheel, the materials should have a high mechanical strength and a low density. The low density permits operation at high angular velocities or with a larger radius before the centrifugal force tears the wheel apart. Fiber composites, which were developed originally for the aerospace program, appear to have much higher storage capability than the conventional steel systems. Moreover, these materials are less expensive than steel. Some possible materials for flywheels are listed in Table 9.2.

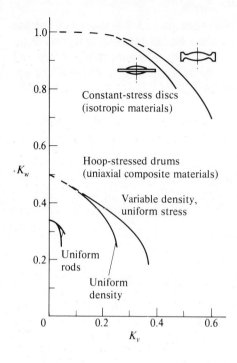

Figure 9.1 The relationship between K_w and K_v for some high-performance flywheels. *(From Fullman, 1975.)*

Table 9.2 Properties of flywheel materials

Materials	Density, kg/m³	Relative cost, $/kg	Cycles	Working stress, 10^6 Pa	Stress density, m²/s²	Stress Density ÷ cost m²·kg/$·s²
Maraging steel (18 Ni-No. 250)	8000	6.60	10^4	7.00	875	132
			10^5	3.38	423	64
4340 steel	7833	1.30	10^4	4.34	554	426
			10^5	2.83	361	278
Ti-6-4	4429	6.60	10^4	6.27	1416	214
			10^5	2.76	623	94
Al 2024-T3	2768	1.10	10^4	2.27	820	745
			10^5	1.17	423	384
60 v/o S-glass/ epoxy	1965	1.80	10^4	10.00	5089	2827
			10^5	7.58	3858	2143
60 v/o E-glass/ epoxy	1993	1.10	10^4	8.27	4150	3772
			10^5	6.21	3116	2833
62 v/o graphite/ epoxy	1689	33.00	$\leq 10^6$	8.27	4896	148
63 v/o Kevlar/ epoxy	1356	7.70	$\leq 10^6$	10.00	7375	958

Flywheels have been proposed as an alternative energy system for the mechanical-energy storage systems of electrical utilities and also as a propulsion system for automobiles and buses. The flywheel systems can normally accept and give up power at much higher rates than some of the more conventional energy-storage systems, such as the storage battery.

9.2.2 Potential-Energy Storage

Potential-energy storage systems are among some of the oldest forms of energy storage. They include springs, torsion bars, and weight systems as well as compressed fluids. Most of these systems have relatively low-storage capability and are used to power clocks, watches, toys, and other systems where small, compact storage systems are required. On the other hand, some of the hydroelectric systems and compressed-air systems employing pumped storage have tremendous power capability.

One general type of potential-energy storage systems includes those systems in which the energy is stored as elastic strain energy, such as springs, torsion bars, etc. The energy stored in a spring is equal to $\int_0^x Kx' \, dx'$, where K is the spring constant. If the spring is a "linear spring," the spring constant is indeed a constant and the stored potential energy, in joules, is

$$PE = \frac{Kx^2}{2} \tag{9.9}$$

Springs can also be wound in a spiral and loaded with a torsional force.

For a bar loaded in tension, compression, and/or bending, the energy stored in the bar is

$$PE = \frac{\sigma A x}{a} \tag{9.10}$$

where σ is the actual stress in pascals, A is the cross-sectional area in square meters, and x is the deformation in meters. The denominator, a, has a value of 2 for pure tension or compression but has a value greater than 2 for bending modes.

For a bar in elastic torsion, the stored energy is a function of the length of the bar L, the torque at the end of the bar T, the polar moment of inertia J, and the torsional modulus of elasticity G. For round bars, the stored potential energy is

$$PE = \frac{T^2 L}{2JG} \tag{9.11}$$

The above equation is valid only for systems that exhibit axial symmetry and is only approximate for other geometries because the angular distortion changes the value of J.

Mass-weight energy-storage systems are relatively simple and consist of raising a weight against a gravitational force field. The amount of energy that can be stored in this manner is

$$PE = mg \, \Delta z \tag{9.12}$$

where m is the mass in kilograms, g is the acceleration of gravity, 9.81 m/s^2, and Δz is the change in elevation in meters. In order to store 1 kW·h of energy with a weight system, a 1000-kg mass would have to be raised 367 m or one of 100 lbm would have to be raised approximately 5 mi into the air.

Despite the small amount of energy that can be obtained from a unit mass, this system is used to store large amounts of energy by moving large quantities of water through reasonable distances in the so-called pumped-storage energy system. In this system, a reversible hydroelectric generator-pump system is used to pump water from a river or lake into a reservoir at higher elevation during periods of low power demand from utilities. During times of peak demand, the system is reversed to recover most of the stored energy. This scheme effectively raises the load factor for the utility and permits the utilization of the low-cost energy sources instead of the high-cost peaking units.

There are a number of pumped-storage systems in use today by various utilities. In 1970, a total of 3600 MW$_e$ had been installed and it is estimated that a total capacity of 27,000 MW$_e$ will be installed by 1980. The storage efficiency of these systems includes the pumping efficiency, the evaporative and leakage losses from the upper reservoir, and the efficiency of the hydraulic turbine and generator when the flow is reversed. The overall storage efficiency of most of these systems is about 65 percent. A typical pumped-storage system is shown in Fig. 9.2, and a number of the pumped-storage systems in the United States are listed in Table 9.3.

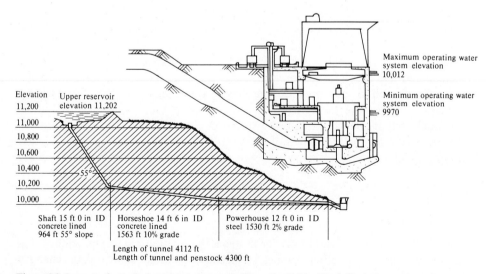

Figure 9.2 Layout of a typical pumped-storage facility—the Cabin Creek Project.

Table 9.3 List of U.S.-designed pump-turbine units for pumped-storage systems

Plant	Year ordered	Manufacturer	Number of units	Head, ft	Output, MW/unit	Speed, rev/min
Flatiron	1953	A-C	1	290	9.0	300/257
Hiwassee	1955	A-C	1	190	62.0	106
Lewiston	1957	A-C	12	75	20.9	113
Provvidenze	1959	A-C	1	850	52.2	375
Hatanagi No. 1	1959	A-C	1	335	51.8	200/167
Taum Sauk	1960	A-C	2	790	220.0	200
Smith Mountain	1960	A-C	2	181	65.0	106
Yards Creek	1960	BLH	3	656	110.0	240
Yagisawa	1961	A-C	2	318	87.5	150
Kisenyama	1962	A-C	1	721	240.0	225
Muma II	1962	A-C	1	88	2.4	400
San Louis	1962	H	8	292	24.0	150/120
Oroville	1963	A-C	3	595	89.7	190
Cabin Creek	1963	A-C	2	1190	166.0	360
Azumi	1964	A-C	4	442	109.0	188
Muddy Run	1964	BLH	8	301	113.0	180
Thermalito	1964	A-C	3	990	25.8	113
De Gray	1964	NN	1	150	33.2	129
Seneca	1965	NN	2	646	162.0	225
Salina	1965	A-C	3	245	44.8	171
Castaic	1965	H	6	1062	200.0	257
Coo-Trois Ponts	1966	A-C	3	895	146.0	300
Grand Coulee	1966	NN	2	272	47.4	200
Longwood Valley	1967	BLH	3	328	45.0	277
Salina	1967	A-C	3	245	46.0	171
Jocassee	1967	A-C	4	293	170.0	120
Northfield	1967	BLH	4	744	257.0	257
Mormon Flat	1968	A-C	1	134	41.8	138.5
Clarence Cannon	1968	A-C	1	57	32.0	75
Blenheim Gilboa	1968	H	4	1081	250.0	257
Carters Dam	1968	A-C	2	344	129.0	150
Horse Mesa	1969	BLH	1	259	100.0	100
Ludington	1969	H	6	361	343.0	100
Bear Swamp	1970	H	2	770	300.0	257
Raccoon Mt.	1970	A-C	4	1040	384.0	300
Mt. Elbert I	1971	A-C	1	406	103.0	180
Wallace	1971	A-C	4	983	54.0	85.8
Fairfield	1972	A-C	8	158	61.9	150
Cornwall	1974	A-C	8	1050	257.0	257
Bath County	1974	......	6	1260	457.0	257
Helms	1974	......	3	1624	358.0	360

Key: A-C   Allis-Chalmers Corporation
BLH   Baldwin-Lima-Hamilton
H   Hitachi
NN   Newport News

Unfortunately, the pumped-storage system requires a specific type of topography that severely limits the application of these systems. In addition to the water system described above and shown in Fig. 9.2, a number of studies have also considered underground water systems that use abandoned mines or natural caverns.

Recently a different kind of pumped-storage system has been proposed for pumped storage by utilities. This system uses compressed air instead of water. The compressed-air, pumped-storage system uses an underground cavity, such as an abandoned mine, abandoned oil and gas fields, a sealed aquifer, a natural cavern, an excavated cavity, or a leached salt deposit to contain the compressed air. Air is pumped into the chamber during times of low demand and then is withdrawn, mixed with fuel and burned, and the hot, compressed exhaust gases are then expanded through a gas turbine to produce power during times of peak demand. Since approximately half of the output of a conventional gas turbine is used to drive the compressor, this system significantly increases the available power from the unit.

The air-storage volume required for this system is directly proportional to the amount of energy to be stored and is inversely proportional to the air pressure to be contained in the cavern. It is desirable to maintain the gas pressure as nearly constant as possible in order to achieve high compressor and turbine efficiencies. This can be accomplished by using water from a nearby lake or reservoir to flood the cavern. The compressed air is then pumped into the cavern and this forces the water back up the supply line into the lake. The resulting hydraulic pressure keeps the gas pressure essentially constant as air is added or extracted from the chamber.

A compressed-air energy-storage system is under construction in Huntorf, West Germany. This system will have a 2-h output capacity of 290 MW_e and the air is stored at a pressure of 68 atm (1000 lb/in^2 abs). The compressed air is extracted, the oil heated and fed to the high-pressure turbine at a pressure of 44.2 atm (650 lb/in^2 abs). This system has both a high-pressure and a low-pressure gas turbine with reheat between the two turbines.

9.3 STORAGE OF ELECTRICAL ENERGY

9.3.1 Introduction

Electrical energy can be stored as electrical energy in either electrostatic fields or in inductive energy fields. The battery is commonly considered to be a device that stores electrical energy and, while it does store electrical energy, it actually converts the electrical energy into chemical energy in an endothermic reaction that is reversible. When the battery is discharged, the reactants in the battery combine in an exothermic chemical reaction that directly produces electricity.

9.3.2 Electrostatic- and Inductive-Field Storage

The storage of electrical energy in the form of electrical energy can be accomplished in either capacitors in which the energy is stored in electrostatic fields, or in the magnetic field established by the flow of electrons through large inductors such as electromagnets. Capacitors are commonly used as storage devices in dc electrical circuits and large banks of capacitors have also been used for the storage of power to improve the lagging power factor of an ac system and also where large bursts of dc energy are desired, such as in some of the fusion experiments. The electrical energy stored in a capacitor is equal to

$$E = \frac{Cv^2}{2} \tag{9.13}$$

where C is the capacitance of the unit in farads and v is the final voltage across the capacitor. Capacitor storage systems have the advantage of rapid charge and discharge without markedly affecting the efficiency or operation of the system.

The electromagnet essentially stores energy in the magnetic field established by the flow of electrons or current through the coils. The amount of energy that is stored in the magnetic field is

$$E = \frac{Li^2}{2} \tag{9.14}$$

where L is the inductance of the coil in henrys and i is the current in the coil in amperes. This system is not commonly used as an energy-storage device because it requires a current flow through the coil to maintain the inductive magnetic field. There is some interest, however, in using a superconducting electromagnet as a possible energy-storage device. In this system, the magnetic field is established by passing the proper current through the superconducting coil and then the coil is short circuited, producing an all superconducting circuit. Since the superconductor has absolutely no electrical resistance, the current will continue to flow through the coil until the energy is needed.

9.3.3 Batteries

Electrochemical batteries are commonly used to store electrical energy although, in reality, the energy is stored as chemical energy. Batteries are normally divided into two general categories—primary batteries and secondary batteries. Primary batteries normally cannot be recharged and consequently only the secondary batteries can be employed for the cyclic storage of electrical energy. The performance of most secondary cells is strongly dependent on the rate that the battery is charged and discharged.

The common battery is composed of two electrodes separated by an electrolyte solution similar to the fuel cells. Unlike the fuel cells, however, all of the chemical reactants are contained within the battery. During the charging process, an endothermic chemical reaction converts the electrical energy into chemical energy. During discharge, an exothermic chemical reaction converts the chemical energy into electricity.

Secondary batteries usually employ an aqueous electrolyte. Lead-acid batteries are commonly used in automobiles and while they can withstand many shallow charge-discharge cycles, they cannot take many deep discharges. Nickel-iron batteries have been used for a long time and they are both relatively low in cost and very long lived. The nickel-iron batteries can withstand deep discharges and are used in battery-powered forklifts and other low-powered vehicles. A number of other promising aqueous electrolyte systems include the nickel-zinc battery, the zinc-bromine battery, the nickel-cadmium battery, the nickel-hydrogen battery, and the zinc-chlorine battery.

Organic electrolyte batteries, such as the sodium-bromine, the lithium-sulfur dioxide, and the lithium-bromine batteries are under investigation because of the relatively low cost of these systems. These systems have high specific energy, in kilojoules per kilogram, but low specific power because of the low conductivity of the organic electrolyte, such as propylene carbonate. The low electrolyte conductivity and the accompanying joule-heating loss yields a low charge-discharge efficiency for these systems.

The metal-air batteries include the zinc-air, the aluminium-air, and the iron-air batteries, although almost any metal can be employed. In these systems, a metal is used to form the negative electrode and a gas electrode, using air as the oxidant, forms the positive electrode. These systems have very high theoretical specific energies with values ranging from 64,620 kJ/kg for beryllium to 873 kJ/kg for lead. Most of the research on metal-air systems has centered on the zinc-air battery.

Recently, a lot of attention and research has been focused on high-temperature batteries using fused-salt or solid electrolytes. The high electrical conductivity of the fused-salt electrolyte means that high power capability is possible. The lithium-sulfur battery employs a eutectic salt of lithium chloride-potassium chloride as an electrolyte. This system has a high theoretical specific energy and they normally operate between 375 and 425°C.

Table 9.4 Characteristics of common secondary batteries

Type	Average voltage	Open-circuit voltage	Charge loss per month, %	Charge-discharge cycles	Specific power, W·h/kg	Power density, kW·h/m³
Nickel-iron	1.2	1.34	30	2000	24	54.9
Lead-acid	2.0	2.14	25	300	33	79.3
Nickel-cadmium	1.2	1.34	2	2000	26	54.9
Silver oxide-cadmium	1.1	1.34	3	2000	53	146.4
Silver oxide-zinc sealed	1.45	1.86	3	100	44–110	79–189
Silver oxide-zinc primary	1.45	1.86	...	.....	121	220.0

The most popular high-temperature battery is the sodium-sulfur cell. This system operates at between 300 and 350°C and employs a solid electrolyte, called beta-alumina, which has very high conductivity. Another promising battery system that operates at high temperature is the lithium-tellurium tetrachloride battery.

The performance characteristics of some of the commonly used secondary batteries are listed in Table 9.4.

9.4 STORAGE OF CHEMICAL ENERGY

Chemical energy is actually stored energy and is one of the more compact forms of stored energy, as can be seen from the values in Table 9.1. Except for batteries, about the only major form of stored chemical energy that is anticipated is the production of elemental or molecular hydrogen. A number of people believe that with the demise of fossil fuels, the human race will shift to a hydrogen or hydrogen-electric economy. Hydrogen makes an excellent fuel as the major combustion product is water and it can then be extracted again from water using another energy source.

Hydrogen can be stored in several ways and the systems used for its production and storage have been well developed. It can be stored as a high-pressure gas, as a liquid at cryogenic temperatures, or it can be stored in metal hydrides. Hydrogen gas has a very low density and consequently either a very high pressure or a very large volume is required to contain a reasonable supply of energy. Liquifaction of hydrogen requires more energy and adds about 30 percent to the cost of the hydrogen product. Storage of hydrogen in metal hydrides results in severe weight and heat-loss penalties as well as the problem of oxygen and/or water contamination of the hydrides that result in a significantly reduced energy-storage capacity.

Hydrogen can be produced from a large number of different reactions. The most familiar reaction is probably that of electrolysis where direct current is passed through a conducting aqueous solution producing hydrogen at one electrode and oxygen at the other. The electrolysis process has an actual conversion efficiency of about 85 percent, but since the input energy for this process is electricity the overall yield of the process in going from thermal to mechanical to electrical to chemical energy is a maximum of about 35 percent. Since hydrogen can be produced more easily and more efficiently from fossil fuels, electrolysis production processes are used only when very pure hydrogen is desired or where inexpensive non-fossil-fuel energy is employed.

Most of the hydrogen produced today is made from methane in the steam-reforming process, which takes place at 900°C. This system uses methane gas in the following reaction:

$$CH_4 + H_2O \longrightarrow 3H_2 + CO$$

This reaction is an endothermic reaction requiring about 230 kJ/mol of CH_4.

The carbon monoxide formed in the preceding reaction is subsequently used to produce additional hydrogen in the following reaction at 400°C:

$$CO + H_2O \longrightarrow H_2 + CO_2$$

The carbon dioxide is then stripped from the fuel gas employing alkali or amines and then discarded.

Other methods of producing hydrogen include the reacting of steam with hot naptha, heavy fuel oil, or even coke, coal, and/or coal char. In an oil refinery, large amounts of hydrogen are generated by converting the common feedstock, naptha (C_8H_{16}), into aromatic compounds. Consider the production of hydrogen during the formation of xylene (C_8H_{10}):

$$C_8H_{16} \longrightarrow C_8H_{10} + 3H_2$$

Hydrogen can also be produced in the steam-iron process where steam is reacted with a hot (900°C) bed of ferrous oxide (FeO). Magnetic iron oxide and hydrogen are produced in this reaction along with about 71 kJ/mol of H_2:

$$H_2O + 3FeO \longrightarrow Fe_3O_4 + H_2$$

The bed is regenerated by passing the exhaust gas obtained by burning carbon with insufficient air. The carbon monoxide in the exhaust reduces the Fe_3O_4 to FeO.

A considerable amount of research has been carried out on the possibility of generating hydrogen in the so-called thermochemical water-splitting process. In this process, the water is split into hydrogen and oxygen in a series of closed-cycle chemical reactions operating at different temperatures. Unless there is a major breakthrough in this process, it is unlikely that it will become competitive until the carbonaceous fossil fuels become scarce and expensive.

Water is dissociated into hydrogen and oxygen by the ionizing radiation as is found in the core of a nuclear reactor. This method is not used in the generation of hydrogen, however, as it is very inefficient. In fact, this reaction presents a problem in the operation of light-water reactors and is discouraged as much as possible with the addition of various inhibitors.

9.5 STORAGE OF NUCLEAR ENERGY

Nuclear energy, like chemical energy, exists only as a stored form of energy and has the highest value of the specific energy storage of any form of stored energy. One kilogram of uranium-235 has a theoretical specific energy of 7×10^{10} kJ/kg in the fission process and 0.6 kg of H-3 reacting with 0.4 kg of H-2 yields a theoretical specific energy storage of 3×10^{11} kJ/kg in the fusion reaction. Radioisotopes are also sources of stored nuclear energy but they have much lower values of specific energy, as listed in Table 9.1.

Stored nuclear energy can be produced by generating radioisotopes like polonium-210 or cobalt-60 from the stable isotopes, bismuth-209 and cobalt-59, respectively, in a nuclear reactor or a particle accelerator. The manufactured fissionable isotopes, uranium-233 and plutonium-239, are also normally generated in a nuclear reactor from the fertile isotopes thorium-232 and uranium-238, respectively. If these isotopes are generated in a breeder reactor, an excess of fuel will be produced, since these reactors produce more fuel than they consume.

Another proposed nuclear-energy storage system actually stores the energy as thermal energy but it uses nuclear energy as the energy source. In this scheme, small thermonuclear (fusion) bombs would be detonated within large salt deposits in the earth's crust. The resulting thermal energy would be used to generate high-pressure steam to drive a conventional turbine-generator system. A private organization, working closely with the Los Alamos Scientific Laboratory in New Mexico, is developing such a system. This is project PACER and is trying to take advantage of geothermal research along with the research carried out under the government's underground testing of nuclear devices.

9.6 STORAGE OF THERMAL ENERGY

Thermal-energy storage (TES) systems utilize essentially three basic modes of thermal-energy storage. These phenomena include sensible-heat storage, latent-heat storage, and quasi-latent-heat storage. This latter system is actually not purely thermal energy as the system involves a chemical reaction.

In the sensible-heat storage systems, the heat is simply stored by increasing the temperature of a solid or liquid. If the specific heat of the material is constant, the amount of energy stored in the system is directly proportional to the temperature rise of the substance.

The storage of thermal energy as latent heat occurs in an isothermal process and occurs as the material undergoes a phase change, usually from a solid to a liquid. Such a phase change is accompanied by the absorption (charging) or the release (discharging) of relatively large amounts of thermal energy. The latent-heat TES systems have much higher energy-storage densities, on the average, than the sensible-heat TES systems.

The quasi-latent-heat TES system operates in essentially the same manner and is essentially indistinguishable from the latent-heat TES system. In this system, thermal energy is converted into chemical energy in a reversible endothermic reaction that takes place at constant temperature. In order to reverse the process, the equilibrium constant is changed by changing the concentration or the pressure of the reactants and/or by changing the temperature. In this last case, the system acts like a sensible-heat TES system with a high specific heat.

Thermal-energy storage systems range from the relatively simple systems where water or air is pumped through the storage system to the very complex system where thermal energy is converted into chemical energy in a reversible endothermic reaction that occurs at constant temperature. In order to reverse the maximum rate of energy addition or removal, the operating temperature range, corrosion and materials compatibility, the environmental risks, and the system economics must be considered. In a two-phase latent-heat system, additional consideration must be given to the problem of melting, expansion, solidification, thermal stability, and containment of the storage media.

The two most important design conditions that affect the design of TES systems are the heat-loss rate and the integrated energy-storage density. The heat-loss rate depends on the surface area of the storage system and the effectiveness of the thermal insulation employed around the containment system. The integrated energy density is the energy stored per unit volume and is usually expressed in units of British thermal units per cubic foot or joules per cubic meter in terms of the integrated intrinsic energy-storage density. The intrinsic value is commonly reported in most tables rather than the practical integrated energy-storage density. This latter value is lower because it includes the volume for heat transfer, controls, etc. In general, the large-capacity TES systems usually have the lowest area-to-volume ratio and some may not require any insulation at all.

TES systems are commonly classed as either low-temperature systems or as high-temperature systems. The low-temperature TES systems normally operate at temperatures below 150°C (300°F) and are commonly composed of either sensible heat storage in water, rocks, and scrap iron or latent heat storage in ice, glauber's salt ($Na_2SO_4 \cdot 10H_2O$), disodium hydrogen phosphate dodecahydrate ($Na_2HPO_4 \cdot 12H_2O$), paraffin waxes, and fatty acids. Some of these latent-heat substances experience deterioration as they are cycled and further developmental work is required before some of them find widespread application.

One application of such a system for home heating-cooling is called ACES (an acronym for *annual cyclic energy storage*). This system uses a water-to-water heat pump and extracts heat from water storage in the winter, freezing it, and melts it during the summer for cooling. Some possible TES materials for low-temperature thermal storage are listed in Table 9.5.

Medium- and high-temperature TES systems have been used extensively in homes and industry. This includes the heating of rocks, brick, bulk iron, and

Table 9.5 Materials for thermal-energy storage

(*a*) Sensible heat storage

Material	Formula	Specific heat, kJ/kg·C°	Volumetric heat capacity, kJ/m³·C°
Water	H_2O	4.18	4191
Isobutyl alcohol	C_4H_9OH	3.01	2381
Ethyl alcohol	C_2H_5OH	2.85	2226
Beryllium	Be	2.81	5231
Limestone	$CaCO_3$	0.91	2548
Sand	SiO_2	0.80	1341
Iron	Fe	0.47	3688

(b) Latent heat storage (solid to liquid)

Material	Formula	Melting point, °C	Heat of fusion	
			kJ/kg	kJ/m³
Ethylene glycol	$C_2H_6O_2$	−13	146.5	162,800
Water	H_2O	0	334.9	305,500
Eastman 1-decanol	$C_{10}H_{22}O$	6	206.1	171,000
Sodium hydroxide $3\frac{1}{2}$ hydrate	$2NaOH \cdot 7H_2O$	15	223.5	364,000
Glycerol	$C_3H_8O_3$	18	200.5	250,000
Sodium sulfate decahydrate	$Na_2SO_4 \cdot 10H_2O$	32	152.8	350,200
Calcium nitrate tetrahydrate	$Ca(NO_3)_2 \cdot 4H_2O$	42	152.8	288,700
Sodium hydroxide octahydrate	$NaOH \cdot 8\ H_2O$	64	272.1	472,000
Barium hydroxide octahydrate	$Ba(OH)_2 \cdot 8H_2O$	78	300.0	655,700
Sodium	Na	98	116.3	117,400
Benzoic acid	$C_2H_6O_2$	122	141.6	179,300
Ammonium thiocyanate	NH_4CNS	146	260.5	337,200
Lithium nitrate	$LiNO_3$	252	367.5	875,500
Sodium hydroxide	NaOH	300	225.6	465,700
Potassium perchlorate	$KClO_4$	527	1251.3	3,148,000
Magnesium	Mg	651	293.1	499,200
Aluminuum	Al	660	395.4	1,655,000
Lithium hydride	LiH	685	3780.0	3,100,000
Sodium chloride	NaCl	810	493.1	763,800
Sodium sulfate	Na_2SO_4	884	162.8	439,600
Magnesium fluoride	MgF_2	1266	930.3	2,273,000
Iron	Fe	1535	151.2	1,192,000

(c) Latent heat storage (solid to solid transition)

Material	Formula	Transition temperature, °C	Heat of transition, kJ/m³
Vanadium tetraoxide	V_2O_4	72	208,600
Silver selenide	Ag_2Se	133	193,700
Ferrous sulfide	FeS	138	231,000

other ceramics. In general, these systems have not received as much attention as may be warranted because of their relatively high cost, operational difficulties, and the current low cost of primary energy supplies. Some of the materials that may be used for the high-temperature TES systems are also listed in Table 9.5. The complexity, expense, and operational problems will probably limit the application of these systems to large industrial power consumers.

REFERENCES

Angrist, S. W.: "Direct Energy Conversion," 3d ed., Allyn and Bacon, Inc., Boston, Mass., 1976.

Brown, J. T., and J. H. Cronin: Battery Systems for Peaking Power Generation, *Ninth Intersociety Energy Conversion Engineering Conference*, 1974, American Society of Mechanical Engineers, New York.

Bundy, F. P., C. S. Herris, and P. G. Kosky: "The Status of Thermal Energy Storage," Power Systems Laboratory, G.E. CRD Report No. 76CRD041, April, 1976.

Fullman, R. L.: "Energy Storage by Flywheels," General Electric Report No. 75CRD051, April, 1975.

Hamlen, R. P., and H. A. Christopher: "Battery Power and Energy Density Requirements for Vehicle Propulsion," General Electric Report No. 71-C-306, October, 1971.

Housz, W.: "Hydrogen Energy and the Environment," General Electric Report No. P-681, TEMPO—Center for Advanced Studies, The General Electric Company, Santa Barbara, Calif., February, 1975.

Russell, C. R.: "Elements of Energy Conversion," Pergamon Press Ltd., Long Island City, N.Y., 1967.

Wentorf, R. H., Jr.: "Hydrogen Generation," General Electric Report No. 75CRD119, May, 1975.

PHYSICAL CONSTANTS

Avagadro's number $= Av = 6.023 \times 10^{26}$ molecules (or atoms)/kg·mol
$= 2.732 \times 10^{26}$ molecules (or atoms)/lbm·mol

Boltzmann's constant $= k = 1.381 \times 10^{-23}$ J/K
$= 8.617 \times 10^{-5}$ eV/°R
$= 1.551 \times 10^{-4}$ eV/K

Electron charge $= e = 1.602 \times 10^{-19}$ C
$= 1.602 \times 10^{-19}$ J/V

Faraday's constant $= F_y = 9.649 \times 10^{7}$ C/kg·mol of electrons

Mass-energy conversion: 1 amu $= 931.5$ MeV
$= 4.147 \times 10^{-17}$ kW·h

Permittivity of free space $= \varepsilon_0 = 8.854 \times 10^{-12}$ F/m

Planck's constant $= h = 6.625 \times 10^{-34}$ J·s
$= 4.136 \times 10^{-15}$ eV·s

Rest masses: Electron: $m_e = 0.0005486$ amu $= 9.1086 \times 10^{-31}$ kg
Neutron: $m_n = 1.0086654$ amu $= 1.6748 \times 10^{-27}$ kg
Proton: $m_p = 1.0072766$ amu $= 1.6725 \times 10^{-27}$ kg

Standard atmospheric pressure $= 1.013 \times 10^{5}$ Pa
$= 1.013$ bar
$= 14.696$ lbf/in^2

Standard gravitational constant $= g = 9.806$ m/s^2
$= 32.17$ ft/s^2

Stefan-Boltzmann's constant $= \sigma = 5.670 \times 10^{-8}$ W/m$^2 \cdot$K^4
$= 0.1714 \times 10^{-8}$ Btu/(h$\cdot$ft$^2 \cdot$°R^4)

Universal gas constant $= R_u = 8.314$ kJ/(kg$\cdot$mol$\cdot$°K)
$= 0.08314$ bar$\cdot$m^3/(kg$\cdot$mol$\cdot$K)
$= 1545$ ft$\cdot$lbf/(lbm$\cdot$mol$\cdot$°R)
$= 1.9857$ Btu/(lbm$\cdot$mol$\cdot$°R)

Velocity of light $= c = 2.998 \times 10^8$ m/s
$= 9.836 \times 10^8$ ft/s

CONVERSION FACTORS

Mass: 1 kilogram = 1 kg = 1000 grams
$\qquad\qquad\qquad$ = 2.205 pounds-mass
$\qquad\qquad\qquad$ = 6.023×10^{26} amu
$\qquad\qquad\qquad$ = 0.001 metric tons
$\qquad\qquad\qquad$ = 0.001102 short tons

Length: 1 meter = 1 m = 10^{10} Angstroms
$\qquad\qquad\qquad$ = 10^6 micrometers
$\qquad\qquad\qquad$ = 1000 millimeters
$\qquad\qquad\qquad$ = 100 centimeters
$\qquad\qquad\qquad$ = 39.37 inches
$\qquad\qquad\qquad$ = 3.281 feet

Area: 1 square meter = 1 m^2 = 10^{28} barns
$\qquad\qquad\qquad$ = 10^4 cm^2
$\qquad\qquad\qquad$ = 1550 in^2
$\qquad\qquad\qquad$ = 10.76 ft^2
$\qquad\qquad\qquad$ = 2.471×10^{-4} acres
$\qquad\qquad\qquad$ = 3.861×10^{-7} square miles

Volume: 1 cubic meter = 1 m^3 = 10^6 cm^3
$\qquad\qquad\qquad$ = 10^3 liters
$\qquad\qquad\qquad$ = 264.2 U.S. gallons
$\qquad\qquad\qquad$ = 35.31 ft^3
$\qquad\qquad\qquad$ = 1.308 yd^3

Density: 1 kilogram/cubic meter = 1 kg/m^3 = 10^{-3} gm/cm^3
$$= 0.008345 \text{ lbm/U.S. gallon}$$
$$= 0.06243 \text{ lbm/ft}^3$$

Viscosity: 1 poise = 100 cP
$$= 0.1 \text{ kg/s} \cdot \text{m}$$
$$= 241.9 \text{ lbm/ft} \cdot \text{h}$$
$$= 0.002089 \text{ lbf} \cdot \text{s/ft}^2$$

Pressure: 1 pascal = 1 Pa = 1 newton/square meter = 1 $kg/m \cdot s^2$
$$= 10^{-5} \text{ bar}$$
$$= 0.9867 \times 10^{-5} \text{ atm}$$
$$= 1.450 \times 10^{-4} \text{ lbf/in}^2$$
$$= 2.953 \times 10^{-4} \text{ in of Hg}$$
$$= 0.004018 \text{ in of H}_2\text{O}$$
$$= 0.007502 \text{ torr} = 0.007502 \text{ mm of Hg}$$

Temperature: 1 degree Celsius = 1°C
$$= 1 \text{ Kelvin degree} = 1 \text{ K}$$
$$= 1.8 \text{ Fahrenheit degrees} = 1.8\text{F}°$$
$$= 1.8 \text{ Rankine degrees} = 1\text{R}°$$

$$°R = °F + 459.67 \qquad K = °C + 273.16$$
$$°F = 32 + 1.8(°C) \qquad °C = 5(°F - 32)/9$$

Thermal conductivity:
$$1 \text{ W/m} \cdot \text{C}° = 1 \text{ J/(s} \cdot \text{m} \cdot °\text{C)} = 1 \text{ N/(s} \cdot °\text{C)} = 1 \text{ kg} \cdot \text{m/(s}^3 \cdot °\text{C)}$$
$$= 0.2388 \text{ cal/(s} \cdot \text{m} \cdot °\text{C)}$$
$$= 0.5778 \text{ Btu/(h} \cdot \text{ft}^2 \cdot \text{F}°/\text{ft)}$$

Energy: 1 joule = 1 J = 1 W $\cdot$ s = 1 N $\cdot$ m = 1 $kg \cdot m^2/s^2$
$$= 6.242 \times 10^{18} \text{ eV}$$
$$= 6.242 \times 10^{12} \text{ MeV}$$
$$= 10^7 \text{ ergs}$$
$$= 0.7376 \text{ ft} \cdot \text{lbf}$$
$$= 0.2388 \text{ cal}$$
$$= 9.478 \times 10^{-4} \text{ Btu}$$
$$= 3.725 \times 10^{-7} \text{ hp} \cdot \text{h}$$
$$= 2.778 \times 10^{-7} \text{ kW} \cdot \text{h}$$

Power: 1 watt = 1 W = 1 J/s = 1 $kg \cdot m^2/s^3$
$$= 0.001 \text{ kW}$$
$$= 3413 \text{ Btu/h}$$
$$= 0.001341 \text{ hp}$$
$$= 6.242 \times 10^{18} \text{ eV/s}$$

Power density and volumetric heat generation rate:
$$1 \text{ W/m}^3 = 1 \text{ kg/m} \cdot \text{s}^3$$
$$= 0.0966 \text{ Btu/h} \cdot \text{ft}^3$$

Specific power and mass heating values:
$$1 \text{ kJ/kg} = 1 \text{ J/gm} = 1000 \text{ m}^2/\text{s}^2$$
$$= 0.430 \text{ Btu/lbm}$$

Heat flux: $1 \text{ W/m}^2 = 1 \text{ kg/s}^3$
$$= 0.3170 \text{ Btu/h} \cdot \text{ft}^2$$

Electrical and magnetic units:
$$1 \text{ ampere} = 1 \text{ watt/volt}$$
$$= 1 \text{ coulomb/second}$$
$$1 \text{ volt} = 1 \text{ watt/ampere} = 1 \text{ joule/coulomb} = 1 \text{ ampere-ohm}$$
$$1 \text{ ohm} = 1 \text{ volt/ampere}$$
$$1 \text{ farad} = 1 \text{ ampere} \cdot \text{second/volt}$$
$$1 \text{ henry} = 1 \text{ volt} \cdot \text{second/ampere}$$
$$1 \text{ weber} = 1 \text{ volt-second}$$
$$1 \text{ weber/m}^2 = 10^4 \text{ gauss}$$
$$= 1 \text{ newton/ampere-meter}$$
$$= 1 \text{ tesla}$$

Decimal multiples and submultiples:

Number	Power of ten	Prefix	Symbol
0.000 000 000 000 000 001	10^{-18}	atto	a
0.000 000 000 000 001	10^{-15}	femto	f
0.000 000 000 001	10^{-12}	pico	p
0.000 000 001	10^{-9}	nano	n
0.000 001	10^{-6}	micro	μ
0.001	10^{-3}	milli	m
1	10^0		
1,000	10^3	kilo	k
1,000,000	10^6	mega	M
1,000,000,000	10^9	giga	G
1,000,000,000,000	10^{12}	tera	T
1,000,000,000,000,000	10^{15}	peta	P
1,000,000,000,000,000,000	10^{18}	exa	E

Standard U.S. fuel energy values:

Coal:
Anthracite: $HHV = 12,700 \text{ Btu/lbm} = 29,540 \text{ kJ/kg}$
$$= 25.4 \times 10^6 \text{ Btu/short ton}$$
Bituminous: $HHV = 11,750 \text{ Btu/lbm} = 27,330 \text{ kJ/kg}$
$$= 23.5 \times 10^6 \text{ Btu/short ton}$$
Lignite: $HHV = 11,400 \text{ Btu/lbm} = 26,515 \text{ kJ/kg}$
$$= 22.8 \times 10^6 \text{ Btu/short ton}$$

Crude Oil: HHV = 18,100 Btu/lbm = 42,100 kJ/kg
= 138,100 Btu/gallon
= 5,800,000 Btu/bbl

Natural gas (dry): HHV = 24,700 Btu/lbm = 57,450 kJ/kg
= 1021 Btu/ft^3

Fuel-energy equivalents:

One barrel (42 gallons) of oil = 460 lbm of coal
= 5680 ft^3 of natural gas
= 1700 kW·h of electricity

One short ton of coal = 4.345 bbl of crude oil
= 24,682 ft^3 of natural gas
= 7386 kW·h of electricity

1000 ft^3 of natural gas = 0.176 bbl of crude oil
= 81.0 lbm of coal
= 300 kW·h of electricity

AVERAGE ANALYSES OF COALS OF THE UNITED STATES

(Compiled from U.S. Bureau of Mines Bulletins)

| | Moisture and ash free | | | | | | | | As received | |
States and counties	VM	FC	C	H_2	O_2	N_2	S	kJ/kg	M	A
ALABAMA										
Jefferson, Tuscaloosa	27.7	72.3	88.1	5.2	4.2	1.7	0.8	36,340	2–5	3–10
Jefferson, Tuscaloosa	32.9	67.1	86.7	5.3	5.0	1.8	1.2	35,875	2–5	2–10
Jefferson, St. Clair	35.4	64.6	85.2	5.4	5.8	1.8	1.8	35,470	2–5	4–12
Walker, Bibb, Shelby	38.0	62.0	84.3	5.4	7.6	1.7	1.0	35,120	1–5	2–14
ARKANSAS										
Sebastian, Logan	19.0	81.0	89.3	4.4	2.0	1.8	2.5	36,040	2–4	6–12
Franklin, Johnson	16.1	83.9	89.4	4.2	2.1	1.8	2.5	35,935	2–4	6–12
COLORADO										
Las Animas	36.5	63.5	84.9	5.5	7.4	1.5	0.7	35,525	1–7	6–20
Huerfano, Gunnison, Garfield	40.7	59.3	80.6	5.5	11.6	1.6	0.7	33,575	3–10	3–12
Weld, Boulder	42.5	57.5	75.0	5.1	17.9	1.5	0.5	30,085	17–30	3–6
Routt, Fremont	43.0	57.0	76.6	5.2	16.0	1.3	0.9	31,515	10–18	3–10
El Paso	46.4	53.6	71.5	5.0	21.8	1.1	0.6	28,095	20–35	5–14
ILLINOIS										
Franklin, Williamson	39.1	60.9	81.3	5.3	9.8	1.7	1.9	33,725	8–12	8–12
Saline, Perry	39.6	60.4	80.6	5.4	10.3	1.7	2.0	33,770	6–11	7–12
Macoupin, Sangamon	46.3	53.7	77.5	5.4	10.2	1.4	5.5	32,970	12–16	8–16
Madison, St. Clair	46.7	53.3	77.0	5.4	10.6	1.3	5.7	32,865	9–16	11–20

continued

	Moisture and ash free								As received	
States and counties	VM	FC	C	H_2	O_2	N_2	S	kJ/kg	M	A
INDIANA										
Sullivan, Greene	45.2	54.8	80.9	5.6	9.7	1.8	2.0	33,725	10–14	6–12
Vigo, Vermilion, Knox	48.0	52.0	79.6	5.6	8.8	1.5	4.5	33,750	6–12	7–12
IOWA										
Appanoose, Polk, Lucas	47.9	52.1	77.0	5.5	9.7	1.5	6.3	33,155	13–19	7–15
KANSAS										
Cherokee, Crawford	39.5	60.5	83.0	5.4	6.1	1.5	4.0	34,955	2–8	9–12
Leavenworth	47.2	52.8	79.6	5.4	7.8	1.5	5.7	33,560	10–12	12–16
KENTUCKY										
Letcher, Pike	37.8	62.2	85.2	5.4	7.0	1.6	0.8	35,320	1–4	2–7
Harlan, Perry, Bell, Knox	40.2	59.8	83.5	5.6	7.9	1.9	1.1	34,785	2–6	2–8
Muhlenburg, Hopkins	44.7	55.3	80.3	5.4	8.6	1.7	4.0	33,595	4–10	5–12
Webster, Union, Butler	45.0	55.0	81.0	5.5	7.7	1.7	4.1	33,945	4–10	5–12
MARYLAND										
Allegheny, Garrett	18.2	81.8	89.3	4.7	2.7	1.7	1.6	33,400	2–4	6–14
MISSOURI										
Barton	38.0	62.0	83.0	5.5	5.1	1.4	5.0	35,165	5–7	8–13
Bates	42.3	57.7	81.0	5.5	7.5	1.5	4.5	34,190	7–12	11–15
Ray, Lafayette, Linn, Adair	46.3	53.7	78.6	5.6	9.3	1.3	5.2	33,315	11–16	7–15
Henry, Macon, Randolph, Clay	45.4	54.6	79.2	5.5	8.4	1.3	5.6	33,420	9–16	8–16
MONTANA										
Rosebud, Custer	44.4	55.6	71.7	4.4	22.0	1.1	0.8	27,980	27–30	3–15
Musselshell, Gallatin	42.8	57.2	78.5	5.4	13.6	1.4	1.1	32,365	4–14	6–30
Carbon, Fergus	40.6	59.4	76.0	5.1	14.2	1.4	3.3	31,025	8–24	8–18
Cascade	35.0	65.0	77.8	4.8	12.6	1.0	3.8	31,445	3–12	14–25
Valley	47.4	52.6	69.0	4.6	24.1	1.2	1.1	26,490	32–45	4–9
NEW MEXICO										
Colfax, Lincoln	43.1	56.9	82.6	5.7	9.3	1.5	0.9	34,560	2–5	9–17
McKinley	47.2	52.8	79.2	5.5	13.0	1.4	0.9	32,420	10–16	3–10
NORTH DAKOTA										
McLean, Morton, Stark	54.0	46.0	72.4	4.7	18.6	1.5	2.8	28,920	35–43	5–12
Williams, Ward	47.6	52.4	72.1	4.9	21.0	1.1	0.9	27,910	35–43	5–12
OHIO										
Belmont, Guernsey	46.0	54.0	80.3	5.6	8.0	1.5	4.6	34,120	3–6	9–16
Tuscarawas, Noble, Jackson	46.2	53.8	79.2	5.6	9.2	1.5	4.5	33,680	3–9	7–15
Athens, Hocking, Meigs, Perry	41.8	58.2	79.7	5.5	10.8	1.4	2.6	33,455	6–10	4–12
Jefferson, Guernsey	41.5	58.5	82.2	5.5	7.7	1.7	2.9	34,560	3–7	5–12
OKLAHOMA										
Pittsburgh, Latimer	40.0	60.0	84.0	5.5	7.4	2.0	1.1	34,990	2–10	4–8
Okmulgee, Tulsa	41.9	58.1	82.4	5.5	7.0	2.0	3.1	34,655	2–8	4–10
Coal	46.6	53.4	77.8	5.2	10.4	1.8	4.8	32,420	5–7	9–12
Haskell	23.6	76.4	89.1	4.8	3.2	1.9	1.0	36,165	2–4	3–8
LeFlore	18.2	81.8	90.4	4.6	2.2	1.9	0.9	36,190	1–3	6–12

States and counties	Moisture and ash free								As received	
	VM	FC	C	H$_2$	O$_2$	N$_2$	S	kJ/kg	M	A
PENNSYLVANIA										
Fayette	31.4	68.6	86.0	5.2	6.3	1.5	1.0	36,095	3–5	7–9
Fayette, Washington, Elk	37.8	62.2	84.8	5.5	5.7	1.6	2.4	35,525	2–4	6–13
Allegheny, Butler	39.6	60.4	83.7	5.5	7.1	1.7	2.0	35,050	2–5	4–13
Cambria, Center, Clearfield	24.6	75.4	88.2	5.1	3.4	1.4	1.9	36,320	1–5	5–12
Somerset, Tioga	23.5	76.5	88.6	4.8	3.1	1.6	1.9	36,145	1–5	5–12
Westmoreland, Indiana	26.5	73.5	87.6	5.2	3.3	1.4	2.5	36,355	1–5	5–12
Westmoreland, Jefferson	35.4	64.6	85.0	5.4	5.8	1.7	2.1	35,550	2–4	7–15
Cambria, Bedford	19.8	80.2	89.4	4.8	2.4	1.5	1.9	36,435	1–6	5–12
Huntingdon, Somerset	17.1	82.9	90.3	4.6	2.3	1.4	1.4	36,715	1–6	5–12
Sullivan	10.6	89.4	91.6	3.8	2.5	1.2	0.9	35,925	3–4	10–15
Lackawanna, Luzerne	7.3	92.7	93.5	2.6	2.3	0.9	0.7	35,120	2–6	6–16
Schuylkill	2.0	98.0	93.9	2.1	2.3	0.8	0.9	34,585	2–3	8–13
TENNESSEE										
Campbell, Anderson	40.3	59.7	83.1	5.5	7.4	2.1	1.9	34,610	2–6	2–8
Claiborne, Overton, Scott	40.5	59.5	83.5	5.6	6.5	2.0	2.4	34,770	2–6	2–11
Morgan, Fentress, White	41.6	58.4	83.0	5.7	6.1	1.7	3.5	35,095	2–5	4–12
Marion, Hamilton	31.1	68.9	87.3	5.4	4.2	1.6	1.5	35,875	3–5	3–11
Rhea, Roane, Grundy	34.5	65.5	85.7	5.3	6.1	1.6	1.3	35,260	2–4	9–15
TEXAS										
Houston, Milam, Wood	53.3	46.7	72.8	5.3	19.3	1.4	1.2	29,550	29–37	6–13
UTAH										
Carbon, Emery, Grand	45.8	54.2	80.3	5.7	11.7	1.6	0.7	33,420	3–10	4–20
Summit, Uintah, Iron	45.3	54.7	76.2	5.4	13.7	1.2	3.5	31,491	5–17	3–13
VIRGINIA										
Wise, Russell	36.9	63.1	86.3	5.5	5.7	1.6	0.9	35,840	2–5	4–9
Tazewell	17.8	82.2	90.4	4.8	2.9	1.2	0.7	36,960	2–6	2–6
Lee, Scott	37.8	62.2	83.6	5.4	8.3	1.6	1.1	35,085	3–6	3–9
Dickenson, Buchanan, Henico	31.2	68.8	87.3	5.2	4.3	1.8	1.4	35,855	2–4	3–18
Montgomery	13.6	86.4	90.7	4.2	3.3	1.0	0.8	35,715	1–5	16–25
WASHINGTON										
Kittitas, King	43.3	56.7	80.5	6.0	10.8	1.9	0.8	34,005	3–9	10–20
King, Pierce	39.4	60.6	82.5	5.9	8.6	2.0	1.0	34,655	2–6	8–20
Lewis, Thurston	49.3	50.7	71.7	5.8	19.5	1.3	1.7	29,305	16–30	6–23
King	44.8	55.2	76.6	5.7	15.1	1.8	0.8	31,865	7–18	5–16
Pierce	25.4	74.6	87.5	5.4	4.0	2.5	0.6	36,120	2–5	9–15
WEST VIRGINIA										
Logan, Fayette	36.2	63.8	86.3	5.5	5.4	1.7	1.1	35,680	1–3	4–8
McDowell, Fayette, Wyoming	17.7	82.3	90.4	4.8	2.7	1.3	0.8	36,540	2–4	3–6
Raleigh, Mercer, Fayette	18.5	81.5	90.1	4.8	3.0	1.4	0.7	36,560	2–4	3–6
Monongalia, Greenbrier	30.2	69.8	87.2	5.2	5.1	1.6	0.9	35,910	2–4	4–12
Preston, Fayette, Randolph	31.0	69.0	87.5	5.3	4.2	1.5	1.5	36,050	2–4	4–12
Marion, Kanawha, Harrison	40.9	59.1	83.9	5.6	7.4	1.6	1.5	35,075	2–4	4–10

continued

States and counties	Moisture and ash free								As received	
	VM	FC	C	H_2	O_2	N_2	S	kJ/kg	M	A
WYOMING										
Sweetwater, Hot Springs	41.8	58.2	77.3	5.3	14.8	1.5	1.1	31,920	8–16	3–8
Sweetwater, Carbon	39.5	60.5	76.2	5.1	16.3	1.6	0.8	30,840	15–25	4–7
Lincoln, Uinta, Sweetwater	43.4	56.6	79.8	5.4	12.2	1.5	1.1	33,075	6–18	4–12
Sheridan, Campbell Carbon	45.3	54.7	74.1	5.1	18.7	1.3	0.8	29,795	18–30	5–15
Albany, Converse, Freemont	45.1	54.9	73.6	5.2	18.8	1.3	1.1	29,610	17–27	4–15

CONVERTING PROXIMATE TO ULTIMATE ANALYSIS

This method gives good results for the great bulk of coals used in steam generation. Its accuracy falls off with anthracite, cannel, and coals with unusual amounts of resins and waxes.

Start with the as-received proximate analysis:

Volatile matter	= 38.0%	Sulfur = 2.4%
Fixed carbon	= 52.3%	HHV = 13,770 Btu/lbm
Moisture	= 1.8%	
Ash	= 7.9%	
	100.0%	

Convert the as-received proximate analysis to a dry, ash-free basis by dividing the values above by $(1 - M - A)$ or $(1 - 0.018 - 0.079) = (0.903)$:

Volatile matter	= 42.1%	Sulfur = 2.7%
Fixed carbon	= 57.9%	HHV = 15,250 Btu/lbm
	100.0%	

Locate the intersection of dry, ash-free volatile matter (42.1%) and British thermal unit (15,250) lines on the graph. This corresponds to 84.9% carbon and 5.9% hydrogen. Assuming a value of 1.5% nitrogen for the dry, ash-free condition, build up the following ultimate analysis for the dry, ash-free basis.

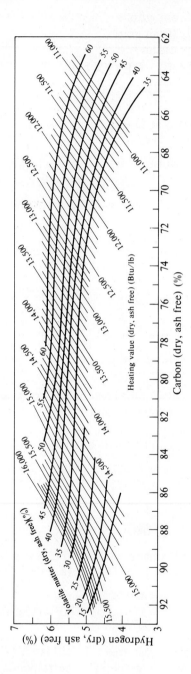

Dry, ash-free ultimate analysis (estimated)

$$\%C = \%C \text{ (from graph)} - \%S/4 = 84.9 - 0.7 = 84.2$$
$$\%H_2 = \%H_2 \text{ (from graph)} \qquad\qquad\qquad = 5.9$$
$$\%N_2 \text{ (always assumed to be } 1.5\%) \qquad\quad = 1.5$$
$$\%S \text{ (from the proximate analysis)} \qquad\quad = 2.7$$
$$\%O_2 = 100 - \text{(sum of other values)} \qquad\quad = 5.7$$
$$\overline{\qquad\qquad\qquad\qquad\qquad\qquad\qquad\qquad 100.0\%}$$

Convert the above analysis to an as-received, ultimate analysis by multiplying the values by $(1 - M - A)$ or 0.903:

$C = 76.0\%$	Moisture may be distributed to H_2 and O_2
$H_2 = 5.3\%$	$C = 76.0\%$
$N_2 = 1.4\%$	$H_2 = 5.5\%$
$S = 2.4\%$	$N_2 = 1.4\%$
$O_2 = 5.1\%$	$S = 2.4\%$
$M = 1.8\%$	$O_2 = 6.7\%$
$A = 7.9\%$	$A = 7.9\%$
$\overline{99.9\%}$	$\overline{99.9\%}$

This method has been checked against laboratory analyses for many coals; here are some results[†]:

Producing area	Proximate analysis, as-received						Ultimate analysis, as-received				
	Moist	VM	FC	Ash	Sulfur	Btu		C	H_2	O_2	N_2
Penn.(1)	2.2	16.2	72.1	9.5	2.1	13,710	Actual	79.2	4.4	3.4	1.4
							Calculated	79.4	4.5	3.2	1.3
Penn.(2)	3.3	34.4	54.1	8.2	1.1	13,380	Actual	74.5	5.3	9.4	1.5
							Calculated	75.0	5.3	9.0	1.4
W. Va.(1)	3.3	16.0	76.1	4.6	0.6	14,450	Actual	83.9	4.6	5.2	1.1
							Calculated	83.7	4.7	4.9	1.5
W. Va.(2)	1.8	38.0	52.3	7.9	2.4	13,770	Actual	76.0	5.2	7.1	1.4
							Calculated	76.0	5.4	6.9	1.4
Ky.(1)	3.6	37.2	56.5	2.7	0.7	14,220	Actual	79.6	5.5	9.9	1.6
							Calculated	79.6	5.7	9.8	1.5
Ky.(2)	3.4	37.4	54.8	4.4	1.0	13,800	Actual	77.4	5.5	10.2	1.5
							Calculated	77.6	5.6	9.9	1.5
Ill.	9.2	33.8	48.6	8.4	0.9	11,930	Actual	67.3	5.5	16.3	1.5
							Calculated	67.1	5.6	16.7	1.3
Colo.	14.4	38.4	41.2	6.0	0.8	10,600	Actual	60.6	5.8	25.6	1.2
							Calculated	60.2	6.1	25.6	1.3
Tenn.	5.1	36.3	55.8	2.8	1.0	13,570	Actual	76.5	5.6	12.2	1.9
							Calculated	76.4	5.6	12.7	1.5
Iowa	16.6	36.5	37.4	9.5	4.3	10,480	Actual	57.4	6.0	21.8	1.0
							Calculated	57.6	6.1	21.3	1.2
Alaska	10.7	30.4	44.0	14.9	0.7	9,640	Actual	55.3	5.0	23.6	0.6
							Calculated	55.5	5.2	22.5	1.2

[†] E. O. Smith, *Asst. Manager, Inspection Dept., Eastern Gas and Fuel Associates.*

LIQUID FUELS

Commercial fuels	Molecular weight	Specific gravity, °API	Flash point, °F	Higher heating value, kJ/kg	Relative cost, Cost / Unit energy
Propane (LPG)†	44	112.5	....	50,400	122
Butane (LPG)†	58	103.1	....	49,590	118
Gasolene	113	70.0	0	47,590	167
Gasolene	126	60.0	0	47,120	144
Alcohol (denatured)	....	48.0	170	29,770	
Aviation jet fuel	....	45.0	110	46,050	
Aviation jet fuel	....	50.0	....	46,010	
Kerosene	154	40.0	130	45,940	116
Diesel oil (1-D)	170	30.2	100	44,750	
Diesel oil (2-D)	184	22.2	125	44,450	
Diesel oil (4-D)	198	15.8	130	43,800	
No. 1 fuel oil	....	42.0	100	46,070	118
No. 2 fuel oil	....	34.0	100	45,260	100
No. 4 fuel oil	....	22.5	130	43,820	76
No. 5 fuel oil	....	18.0	130	43,170	60
No. 6 fuel oil	....	14.5	150	42,330	51

† Under pressure.

Crude oils	Mass fractions				Specific gravity, °API	HHV, kJ/kg
	C	H_2	$O_2 + N_2$	S		
Texas No. 1	84.60	10.90	2.87	1.63	21.68	44,140
Texas No. 2	83.26	12.41	3.83	0.50	21.37	45,710
Pennsylvania	84.90	13.70	1.40	0.00	28.20	44,680
California	81.52	11.51	6.42	0.55	14.98	43,420

F

FUEL-GAS ANALYSES

Gases	Heating value,† kJ/liter	Composition in percentage by volume								
		CH_4	C_2H_4	C_2H_6	H_2	CO	O_2	N_2	CO_2	H_2O
Natural gases:										
Alabama	35.97	97.6						2.1	0.3	
Arkansas	36.60	99.2						0.6	0.2	
California-A	38.86	77.5		16.0					6.5	
California-B	40.84	83.4		15.8				0.5	0.7	
Illinois	35.24	95.6						3.9	0.5	
Indiana-A	42.85	75.4		23.4				1.2		
Kansas	36.16	98.0						0.8	1.2	
Kentucky	43.11	75.0		24.0				1.0		
Louisiana-A	35.13	78.8	9.5				0.3	11.3	0.1	
Louisiana-B	36.66	90.0		5.0				5.0		
Missouri	35.31	84.1		6.7				8.4	0.8	
New York	40.65	84.0		15.0				1.0		
Ohio-A	35.28	93.3	0.3		1.8	0.5	0.3	3.4	0.2	0.2
Ohio-B	36.96	93.4	0.4		1.6	0.4	0.4	3.4	0.4	
Oklahoma-A	38.93	73.5		18.4				8.1		
Oklahoma-B	35.64	84.1		6.7				8.4	0.8	
Pennsylvania-A	38.97	90.0		9.0				0.8	0.2	
Pennsylvania-B	41.31	83.4		15.8				0.8		
West Virginia	42.78	76.8		22.5				0.7		

Gases	Heating value,† kJ/liter	Composition in percentage by volume								
		CH_4	C_2H_4	C_2H_6	H_2	CO	O_2	N_2	CO_2	C_6H_6
Artificial gases:										
Producer gas										
Anthracite	4.961				15.5	22.7	0.3	56.0	5.5	
Bituminous	5.525	3.7		0.1	11.6	24.4	0.6	54.8	4.8	
Blast-furnace gas	3.597	0.2			3.6	26.5		57.0	12.7	
B-F gas (lean)	3.066	0.1			2.5	24.1		58.4	14.9	
Coke-oven gas	21.59	33.9	5.2		47.9	6.1	0.6	3.7	2.6	
Illuminating gas	18.44	23.6		10.5	11.7	13.7	0.7	32.6	7.2	
Water gas (Carb.)	19.54	15.5	4.7		34.0	32.0	0.7	6.5	4.3	2.3

† All gas values are corrected to 1 atm and 20°C (68°F).

Courtesy of Babcock and Wilcox.

G

REACTANT PROPERTIES

Substance	Chemical formula	Molecular weight	Density,† kg/liter	Higher heating value		Lower heating value	
				kJ/liter†	kJ/kg	kJ/liter†	kJ/kg
FUELS:							
Hydrogen	H_2	2.016	0.0000838	11.908	142,097	10.062	120,067
Carbon	C	12.011	....	....	32,778	....	32,778
Sulfur	S	32.064	....	....	9,257	....	9,257
Hydrogen sulfide	H_2S	34.080	0.001437	23.720	16,506	21.848	15,204
Carbon monoxide	CO	28.006	0.001167	11.802	10,110	11.802	10,110
Methane	CH_4	16.043	0.000670	37.204	55,529	33.496	49,994
Methyl alcohol	CH_3OH	32.043	0.001334	31.827	23,858	28.129	21,086
Ethane	C_2H_6	30.071	0.001267	65.782	51,920	60.168	47,489
Ethylene	C_2H_4	28.055	0.001170	58.877	50,322	55.173	47,156
Acetylene	C_2H_2	26.039	0.001099	54.961	50,010	53.087	48,305
Ethyl alcohol	C_2H_5OH	46.071	0.001918	58.714	30,610	53.164	27,717
Propane	C_3H_8	44.099	0.001887	95.103	50,399	87.501	46,370
Propylene	C_3H_6	42.083	0.001751	85.714	48,954	80.172	45,789
n-Butane	C_4H_{10}	58.126	0.002495	123.725	49,589	114.191	45,768
Isobutane	C_4H_{10}	58.126	0.002495	123.435	49,472	113.901	45,652
n-Butene	C_4H_8	56.110	0.002335	113.255	48,503	105.863	45,338
Isobutene	C_4H_8	56.110	0.002335	112.619	48,231	105.228	45,065
n-Pentane	C_5H_{12}	72.153	0.003003	147.337	49,064	136.246	45,370
Isopentane	C_5H_{12}	72.153	0.003003	147.003	48,952	135.911	45,258

Substance	Chemical formula	Molecular weight	Density,† kg/liter	Higher heating value		Lower heating value	
				kJ/liter†	kJ/kg	kJ/liter†	kJ/kg
Neopentane	C_5H_{12}	72.153	0.003003	146.521	48,791	135.429	45,098
n-Pentene	C_5H_{10}	70.137	0.002919	140.670	48,191	131.431	45,026
n-Hexane	C_6H_{14}	86.181	0.003586	174.866	48,764	161.930	45,156
Benzene	C_6H_6	78.117	0.003249	137.410	42,293	131.871	40,588
Toluene	C_7H_8	92.169	0.003835	165.021	43,030	157.626	41,102
Xylene	C_8H_{10}	106.172	0.004421	191.769	43,377	182.618	41,307
Naphthalene	$C_{10}H_8$	128.179	0.005338	214.818	40,244	207.431	38,860
Ammonia	NH_3	17.031	0.000719	16.166	22,484	13.353	18,572
NONFUELS:							
Oxygen	O_2	31.999	0.001334				
Nitrogen	N_2	28.013	0.001164				
Air	....	28.97	0.001208				
Carbon Dioxide	CO_2	44.010	0.001845				
Sulfur Dioxide	SO_2	64.063	0.002733				

† All gas values corrected to 1 atm and 20°C (68°F).

ALPHABETICAL LIST OF THE ELEMENTS

Element	Symbol	Atomic number Z	Element	Symbol	Atomic number Z
Actinium	Ac	89	Einsteinium	Es	99
Aluminum	Al	13	Erbium	Er	68
Americium	Am	95	Europium	Eu	63
Antimony	Sb	51	Fermium	Fm	100
Argon	Ar	18	Fluorine	F	9
Arsenic	As	33	Francium	Fr	87
Astatine	At	85	Gadolinium	Gd	64
Barium	Ba	56	Gallium	Ga	31
Berkelium	Bk	97	Germanium	Ge	32
Beryllium	Be	4	Gold	Au	79
Bismuth	Bi	83	Hafnium	Hf	72
Boron	B	5	Hahnium	—	105
Bromine	Br	35	Helium	He	2
Cadmium	Cd	48	Holmium	Ho	67
Calcium	Ca	20	Hydrogen	H	1
Californium	Cf	98	Indium	In	49
Carbon	C	6	Iodine	I	53
Cerium	Ce	58	Iridium	Ir	77
Cesium	Cs	55	Iron	Fe	26
Chlorine	Cl	17	Krypton	Kr	36
Chromium	Cr	24	Lanthanum	La	57
Cobalt	Co	27	Lawrencium	Lw	103
Copper	Cu	29	Lead	Pb	82
Curium	Cm	96	Lithium	Li	3
Dysprosium	Dy	66	Lutecium	Lu	71

Element	Symbol	Atomic number Z	Element	Symbol	Atomic number Z
Magnesium	Mg	12	Ruthenium	Ru	44
Manganese	Mn	25	Rutherfordium	—	104
Mendelevium	Md	101	Samarium	Sm	62
Mercury	Hg	80	Scandium	Sc	21
Molybdenum	Mo	42	Selenium	Se	34
Neodymium	Nd	60	Silicon	Si	14
Neon	Ne	10	Silver	Ag	47
Neptunium	Np	93	Sodium	Na	11
Nickel	Ni	28	Strontium	Sr	38
Niobium	Nb	41	Sulfur	S	16
Nitrogen	N	7	Tantalum	Ta	73
Nobelium	No	102	Technetium	Tc	43
Osmium	Os	76	Tellurium	Te	52
Oxygen	O	8	Terbium	Tb	65
Palladium	Pd	46	Thallium	Tl	81
Phosphorus	P	15	Thorium	Th	90
Platinum	Pt	78	Thulium	Tm	69
Plutonium	Pu	94	Tin	Sn	50
Polonium	Po	84	Titanium	Ti	22
Potasium	K	19	Tungsten (Wolfram)	W	74
Praseodymium	Pr	59	Uranium	U	92
Promethium	Pm	61	Vanadium	V	23
Protactinium	Pa	91	Xenon	Xe	54
Radium	Ra	88	Ytterbium	Yb	70
Radon	Rn	86	Yttrium	Y	39
Rhenium	Re	75	Zinc	Zn	30
Rhodium	Rh	45	Zirconium	Zr	40
Rubidium	Rb	37			

PARTIAL LIST OF THE ISOTOPES

Element Symbol Atomic weight	Atomic number Z	Atomic mass number A	Isotopic mass, amu	Natural abundance, %	Half-life	Type of decay
Electron	−1	0	0.000549			
Neutron	0	1	1.008665		11.7 m	β^-
Hydrogen	1	1	1.007825	99.985		
H	1	2	2.01410	0.015		
1.00797	1	3	3.01605		12.26 y	β^-
Helium	2	3	3.01603	0.00013		
He	2	4	4.00260	99.99987		
4.0026	2	5	5.01230		2.0×10^{-21} s	α
	2	6	6.01888		0.8 s	β^-
	2	8	8.0375			β^-
Lithium	3	5	5.0125		$\sim 10^{-21}$ s	α
Li	3	6	6.01512	7.42		
6.939	3	7	7.01600	92.58		
	3	8	8.02247		0.85 s	β^-
Beryllium	4	6	6.0197		4.0×10^{-21} s	β^+
Be	4	7	7.0169		56.3 d	K
9.0122	4	8	8.0053		10^{-16} s	2α
	4	9	9.01218	100.00		
	4	10	10.0135		2.0×10^6 y	β^-
	4	11	11.0216		13.6 s	β^-

Element Symbol Atomic weight	Atomic number Z	Atomic mass number A	Isotopic mass, amu	Natural abundance, %	Half-life	Type of decay
Boron	5	8	8.0246		0.77 s	β^+
B	5	9	9.01333		8.0×10^{-19} s	p/2α
10.811	5	10	10.01294	19.78		
	5	11	11.00931	80.22		
	5	12	12.0143		0.02 s	β^-
	5	13	13.0178		0.019 s	β^-
Carbon	6	10			19.0 s	β^+
C	6	11	11.01141		20.3 m	β^+
12.011	6	12	12.00000	98.89		
	6	13	13.00335	1.11		
	6	14	14.00323		5730.0 y	β^-
	6	15	15.00939		2.40 s	β^-
Nitrogen	7	12	12.01895		0.011 s	β^+
N	7	13	13.00572		9.96 m	β^+
14.0067	7	14	14.00307	99.63		
	7	15	15.00011	0.37		
	7	16	16.00656		7.20 s	β^-
	7	17	17.00862		4.16 s	β^-
Oxygen	8	14	14.00856		73.0 s	β^+
O	8	15	15.0030		0.122 s	β^+
15.9994	8	16	15.99491	99.759		
	8	17	16.99914	0.037		
	8	18	17.99915	0.204		
	8	19	19.00344		29.0 s	β^-
Fluorine	9	16	16.01171		10^{-19} s	β^+
F	9	17	17.00210		66.0 s	β^+
18.9984	9	18	18.00094		109.7 s	β^+
	9	19	18.99841	100.00		
	9	20	19.99999		11.4 s	β^-
Neon	10	18	18.00546		1.67 s	β^+
Ne	10	19	19.00187		17.5 s	β^+
20.179	10	20	19.99244	90.52		
	10	21	20.99395	0.26		
	10	22	21.99138	9.22		
	10	23	22.99437		37.6 s	β^-
Sodium	11	20	20.00887		0.4 s	β^+
Na	11	21	20.99760		23.0 s	β^+
22.9898	11	22	21.99432		2.62 y	β^+
	11	23	22.98977	100.00		
	11	24	23.99102		15.0 h	β^-
	11	25	24.98984		60.0 s	β^-

continued

Element Symbol Atomic weight	Atomic number Z	Atomic mass number A	Isotopic mass, amu	Natural abundance, %	Half-life	Type of decay
Magnesium	12	23	22.99380		11.3 s	β^+
Mg	12	24	23.98504	78.99		
24.305	12	25	24.98504	10.00		
	12	26	25.98259	11.01		
	12	27	26.98436		9.5 m	β^-
	12	28	27.98381		21.3 h	β^-
Aluminum	13	24	24.00006		2.10 s	β^+
Al	13	25	24.99036		7.20 s	β^+
26.98153	13	26	25.98793		7.4×10^5 y	β^+
	13	27	26.98153	100.00		
	13	28	27.98193		2.31 m	β^-
	13	29	28.98053		6.60 s	β^-
Silicon	14	27	26.98667		4.20 s	β^+
Si	14	28	27.97693	92.21		
28.086	14	29	28.97649	4.70		
	14	30	29.97376	3.09		
	14	31	30.97536		2.62 h	β^-
	14	32	31.97396		280 y	β^-
Phosphorus	15	28	27.99168		0.28 s	β^+
P	15	29	28.98178		4.40 s	β^+
30.9738	15	30	29.97863		2.50 m	β^+
	15	31	30.97376	100.00		
	15	32	31.97392		14.3 d	β^-
	15	33	32.97168		25.0 d	β^-
	15	34	33.97331		12.4 s	β^-
Sulfur	16	31	30.97901		2.60 s	β^+
S	16	32	31.97207	95.00		
32.064	16	33	32.97146	0.76		
	16	34	33.96786	4.22		
	16	35	34.96923		86.7 d	β^-
	16	36	35.96709	0.014		
	16	37	36.97029		5.10 m	β^-
Chlorine	17	32	31.98601		0.31 s	β^+
Cl	17	33	32.99725		2.50 s	β^+
35.453	17	34	33.97376		1.56 s	β^+
	17	35	34.96885	75.77		
	17	36	35.96852		3.1×10^5 y	β^-
	17	37	36.96590	24.23		
	17	38	37.96797		37.3 m	β^-
	17	39	38.96742		55.5 m	β^-
Argon	18	35	34.97459		1.83 s	β^+
Ar	18	36	35.96755	0.337		
39.948	18	37	36.96674		35.0 d	K
	18	38	37.96272	0.063		
	18	39	38.96428		265.0 y	β^-
	18	40	39.96238	99.60		
	18	41	40.96454		1.83 h	β^-

Element Symbol Atomic weight	Atomic number Z	Atomic mass number A	Isotopic mass, amu	Natural abundance, %	Half-life	Type of decay
Potassium	19	37	36.97324		1.20 s	β^+
K	19	38	37.96905		7.70 m	β^+
39.0983	19	39	38.96371	93.26		
	19	40	39.9740	0.012	1.28×10^9 y	K/β^-
	19	41	40.96184	6.73		
	19	42	41.96352		12.4 h	β^-
	19	43	42.96066		22.4 h	β^-
	19	44	43.96192		22.0 m	β^-
Calcium	20	39	38.97100		0.87 s	β^+
Ca	20	40	39.96259	96.97		
40.08	20	41	40.96228		8.0×10^4 y	K
	20	42	41.95863	0.647		
	20	43	42.95878	0.135		
	20	44	43.95549	2.09		
	20	45			165.0 d	β^-
	20	46	45.95367	0.0033		
	20	48	47.95253	0.187		
	20	49	48.95559		8.80 m	β^-
Scandium	21	40	39.97753		0.18 s	β^+
Sc	21	41	40.96860		0.87 s	β^+
44.956	21	43	42.96106		3.89 h	β^+
	21	44	43.95928		3.92 h	β^+
	21	45	44.95592	100.00		
	21	46	45.95487		83.8 d	β^-
	21	47	46.95230		3.40 d	β^-
	21	48	47.95216		44.0 h	β^-
	21	49	48.94997		57.5 m	β^-
Titanium	22	45	44.95797		3.08 h	β^+
Ti	22	46	45.95263	8.25		
47.90	22	47	46.95177	7.45		
	22	48	47.94795	73.70		
	22	49	48.94787	5.40		
	22	50	49.94479	5.20		
	22	51	50.94645		5.80 m	β^-
Vanadium	23	46	45.96028		0.40 s	β^+
V	23	47	46.95469		32.0 m	β^+
50.942	23	48	47.95220		16.1 d	β^+
	23	49	48.94847		330.0 d	K
	23	50	49.9472	0.24	6.0×10^{15} y	K
	23	51	50.9440	99.76		
	23	52	51.94418		3.76 m	β^-

continued

Element Symbol Atomic weight	Atomic number Z	Atomic mass number A	Isotopic mass, amu	Natural abundance, %	Half-life	Type of decay
Chromium	24	49	48.95122		42.0 m	β^+
Cr	24	50	49.94605	4.31		
51.996	24	51	50.94418		27.8 d	K
	24	52	51.9405	83.76		
	24	53	52.94065	9.55		
	24	54	53.9389	2.38		
	24	55	54.94095		3.50 m	β^-
Manganese	25	50	49.95411		0.29 s	β^+
Mn	25	51	50.94809		45.0 m	β^+
54.9380	25	52	51.94618		5.60 d	β^+
	25	53	52.94126		3.8×10^6 y	K
	25	54	53.9404		303.0 d	K
	25	55	54.93805	100.00		
	25	56	55.93904		2.58 h	β^-
Iron	26	52	51.94769		8.00 h	K
Fe	26	53	52.94541		9.00 m	β^+
55.847	26	54	53.9396	5.82		
	26	55	54.93856		2.60 y	K
	26	56	55.9349	91.66		
	26	57	56.9354	2.19		
	26	58	57.9333	0.33		
	26	59	58.9349		45.1 d	β^-
Cobalt	27	54	53.94904		0.18 s	β^+
Co	27	55	54.94188		18.0 h	β^+
58.9332	27	56	55.93982		77.3 d	K/β^+
	27	57	56.93587		270.0 d	K
	27	58	57.93520		71.3 d	K
	27	59	58.9332	100.00		
	27	60	59.93344		5.26 y	β^-
	27	61	60.93199		99.0 m	β^-
	27	62	61.93324		13.9 m	β^-
Nickel	28	57	56.9394		36.1 h	K
Ni	28	58	57.9353	68.30		
58.71	28	59	58.9342		8.0×10^4 y	K
	28	60	59.9308	26.23		
	28	61	60.9310	1.19		
	28	62	61.9283	3.66		
	28	63	62.9286		100.0 y	β^-
	28	64	63.9280	1.08		
	28	65	64.9291		2.56 h	β^-
Copper	29	58	57.9456		3.30 s	β^+
Cu	29	60	59.9375		24.0 m	β^+
63.546	29	61	60.9327		3.30 h	β^+
	29	62	61.9316		9.80 m	β^+
	29	63	62.9298	69.20		
	29	64	63.9288		12.9 h	$\beta^+/K/\beta^-$

Element Symbol Atomic weight	Atomic number Z	Atomic mass number A	Isotopic mass, amu	Natural abundance, %	Half-life	Type of decay
	29	65	64.9278	30.80		
	29	66	65.9288		5.10 m	β^-
	29	67	66.9278		61.0 h	β^-
Zinc	30	62	61.9339		9.30 h	β^+
Zn	30	63	62.9330		38.8 m	β^+
65.37	30	64	63.9291	48.60		
	30	65	64.9283		243.6 d	K/β^+
	30	66	65.9260	27.90		
	30	67	66.9271	4.10		
	30	68	67.9249	18.80		
	30	69	68.9257		58.0 m	β^-
	30	70	69.9253	0.60		
	30	71	70.9273		2.20 m	β^-
Gallium	31	64	63.9368		2.60 m	β^+
Ga	31	65	64.9325		15.0 m	β^+
69.72	31	66	65.9315		9.50 h	β^+
	31	67	66.9283		78.0 h	K
	31	68	67.9270		68.3 m	β^+
	31	69	68.9257	60.10		
	31	70	69.9259		21.0 m	β^-
	31	71	70.9249	39.90		
	31	72	71.9245		14.1 h	β^-
	31	73	72.9248		4.80 h	β^-
Germanium	32	67	66.9330		19.0 m	β^+
Ge	32	69	68.9280		40.0 h	K
72.59	32	70	69.9243	20.52		
	32	71	70.9251		11.0 d	K
	32	72	71.9221	27.43		
	32	73	72.9234	7.76		
	32	74	73.9212	36.54		
	32	75	74.9228		82.8 m	β^-
	32	76	75.9214	7.76		
	32	77	76.9215		11.3 h	β^-
Arsenic	33	71	70.9271		62.0 h	K/β^+
As	33	72	71.9264		26.0 h	β^+
74.9216	33	73	72.9237		80.3 d	K
	33	74	73.9217		17.9 d	K
	33	75	74.9216	100.00		
	33	76	75.9201		26.5 h	β^-
	33	77	76.9206		39.0 h	β^-
	33	78	77.9217		91.0 m	β^-
	33	79	78.9209		9.0 m	β^-
Selenium	34	73	72.9266		7.10 h	β^+
Se	34	74	73.9225	0.087		
78.96	34	75	74.9225		120.4 d	K

continued

Element Symbol Atomic weight	Atomic number Z	Atomic mass number A	Isotopic mass, amu	Natural abundance, %	Half-life	Type of decay
	34	76	75.9192	9.02		
	34	77	76.9199	7.58		
	34	78	77.9173	23.52		
	34	79	78.9185		6.5×10^4 y	β^-
	34	80	79.9165	49.82		
	34	81	80.9185		18.6 m	β^-
	34	82	81.9167	9.19		
Bromine	35	78	77.9211		6.40 m	β^+
Br	35	79	78.9183	50.69		
79.904	35	80	79.9172		17.6 m	β^-
	35	81	80.9163	49.31		
	35	82	81.9158		35.5 h	β^-
Krypton	36	78	77.9204	0.35		
Kr	36	79	78.9200		34.9 h	K
83.8	36	80	79.9164	2.27		
	36	81	80.9165		2.1×10^5 y	K
	36	82	81.9135	11.56		
	36	83	82.9141	11.55		
	36	84	83.9116	56.90		
	36	85	84.9126		10.76 y	β^-
	36	86	85.9109	17.37		
	36	87	86.9136		76.0 m	β^-
Rubidium	37	84	83.9142		33.0 d	K
Rb	37	85	84.9117	72.15		
85.47	37	86	85.9100		18.66 d	β^-
	37	87	86.9092	27.85	5.0×10^{10} y	β^-
	37	88	87.9113		17.7 m	β^-
Strontium	38	84	83.9134	0.56		
Sr	38	85	84.9095		64.0 d	K
87.62	38	86	85.9094	9.86		
	38	87	86.9089	7.02		
	38	88	87.9056	82.56		
	38	89	88.9057		52.0 d	β^-
	38	90	89.9072		28.1 y	β^-
	38	91	90.9097		9.67 h	β^-
Yttrium	39	88	87.9096		106.6 d	K
Y	39	89	88.9054	100.00		
88.906	39	90	89.9066		64.0 h	β^-
	39	91	90.9069		58.8 d	β^-
	39	92	91.9083		3.54 h	β^-
Zirconium	40	89	88.9086		78.4 h	K
Zr	40	90	89.9047	51.46		
91.22	40	91	90.9056	11.23		
	40	92	91.9050	17.11		
	40	93	92.9063		1.5×10^6 y	β^-
	40	94	93.9061	17.40		

Element Symbol Atomic weight	Atomic number Z	Atomic mass number A	Isotopic mass, amu	Natural abundance, %	Half-life	Type of decay
	40	95	94.9072		65.0 d	β^-
	40	96	95.9082	2.80	3.6×10^{17} y	β^-
	40	97	96.91104		17.0 h	β^-
Niobium	41	92	91.9062		10.13 d	K/β^+
Nb	41	93	92.9064	100.00		
92.906	41	94	93.9063		2.0×10^4 y	β^-
	41	95	94.9060		3.50 d	β^-
Molybedenum	42	92	91.9068	14.80		
Mo	42	93	92.9057		3500 y	K
95.940	42	94	93.9047	9.30		
	42	95	94.9046	15.90		
	42	96	95.9046	16.70		
	42	97	96.9058	9.60		
	42	98	97.9055	24.10		
	42	99	98.9069		66.7 h	β^-
	42	100	99.9076	9.60		
Technetium	43	95	94.9073		20.0 h	K
Tc	43	97	96.9068		2.6×10^6 y	K
	43	98			4.2×10^6 y	β^-
	43	99	98.9054		2.12×10^5 y	β^-
Ruthenium	44	95	94.9095		1.70 h	K
Ru	44	96	95.9076	5.51		
101.07	44	97			2.90 d	K
	44	98	97.9055	1.87		
	44	99	98.9061	12.72		
	44	100	99.9042	12.62		
	44	101	100.9056	17.07		
	44	102	101.9043	31.61		
	44	103	102.9058		39.6 d	β^-
	44	104	103.9055	18.58		
	44	105	104.9075		4.44 h	β^-
	44	106	105.9073		367.0 d	β^-
Rhodium	45	102	101.9064		206.0 d	$K/\beta^+/\beta^-$
Rh	45	103	102.9055	100.00		
102.905	45	104	103.9064		43.0 s	β^-
Palladium	46	102	101.9056	0.96		
Pd	46	103	102.9058		17.0 d	K
106.4	46	104	103.9040	10.97		
	46	105	104.9051	22.23		
	46	106	105.9032	27.33		
	46	107	106.9049		7.0×10^6 y	β^-
	46	108	107.9039	26.71		
	46	109	108.9059		13.47 h	β^-
	46	110	109.9052	11.81		
	46	111	110.9076		22.0 m	β^-

continued

Element Symbol Atomic weight	Atomic number Z	Atomic mass number A	Isotopic mass, amu	Natural abundance, %	Half-life	Type of decay
Silver	47	106	105.9061		8.4 d	K
Ag	47	107	106.9051	51.82		
107.87	47	108	107.9059		2.20 m	β^-
	47	109	108.9047	48.18		
	47	110	109.9072		24.4 s	β^-
Cadmium	48	106	105.9061	1.22		
Cd	48	107	106.9064		6.50 h	K
112.40	48	108	107.9040	0.88		
	48	109	108.9048		450.0 d	K
	48	110	109.9030	12.39		
	48	111	110.9042	12.75		
	48	112	111.9028	24.07		
	48	113	112.9046	12.26		
	48	114	113.9036	28.86		
	48	115	114.9070		53.5 h	β^-
	48	116	115.9050	7.58		
	48	117	116.9076		2.40 h	β^-
Indium	49	113	112.9043	4.28		
In	49	114	113.9070		72.0 s	β^-
114.82	49	115	114.9041	95.72	6.0×10^{14} y	β^-
	49	116	115.9071		14.0 s	β^-
Tin	50	112	111.9048	0.95		
Sn	50	113			115.0 d	K
118.69	50	114	113.9030	0.65		
	50	115	114.9035	0.35		
	50	116	115.9021	14.30		
	50	117	116.9031	7.61		
	50	118	117.9018	24.03		
	50	119	118.9034	8.58		
	50	120	119.9022	32.85		
	50	121	120.9025		27.0 h	β^-
	50	122	121.9034	4.72		
	50	123	122.9037		42.0 m	β^-
	50	124	123.9052	5.94		
Antimony	51	120	119.9040		15.9 m	K
Sb	51	121	120.9038	57.25		
121.75	51	122	121.9035		2.80 d	β^-
	51	123	122.9041	42.75		
Tellurium	52	120	119.9032	0.089		
Te	52	122	121.9030	2.46		
127.60	52	123	122.9042	0.87		
	52	124	123.9028	4.61		
	52	125	124.9044	6.99		
	52	126	125.9032	18.71		
	52	127	126.9053		9.40 h	β^-
	52	128	127.9047	31.79		
	52	129	128.9066		69.0 m	β^-

Element Symbol Atomic weight	Atomic number Z	Atomic mass number A	Isotopic mass, amu	Natural abundance, %	Half-life	Type of decay
	52	130	129.9062	34.48		
	52	131	130.9084		25.0 m	β^-
Iodine	53	126	125.9053		13.0 d	$K/\beta^+/\beta^-$
I	53	127	126.9044	100.00		
126.9044	53	128	127.9060		25.08 m	β^-
	53	129	128.9047		1.7×10^7 y	β^-
	53	130	129.9065		12.3 h	β^-
	53	131	130.9060		8.07 d	β^-
Xenon	54	124	123.9061	0.096		
Xe	54	125			17.0 h	K
131.30	54	126	125.9042	0.09		
	54	127	126.9055		36.41 d	K
	54	128	127.9035	1.92		
	54	129	128.9048	26.44		
	54	130	129.9035	4.08		
	54	131	130.9051	21.18		
	54	132	131.9042	26.89		
	54	133	132.9054		5.27 d	β^-
	54	134	133.9054	10.44		
	54	135			9.20 h	β^-
	54	136	135.9072	8.87		
Cesium	55	132	131.9060		6.50 d	K
Cs	55	133	132.9051	100.00		
132.905	55	134	133.9064		2.05 y	β^-
	55	137	136.9073		33.0 y	β^-
Barium	56	130	129.9062	0.101		
Ba	56	132	131.9050	0.097		
137.34	56	134	133.9043	2.42		
	56	135	134.9056	6.59		
	56	136	135.9044	7.81		
	56	137	136.9058	11.32		
	56	138	137.9050	71.66		
	56	139	138.9079		82.9 m	β^-
	56	140	139.9099		12.8 d	β^-
Lanthanum	57	138	137.9071	0.089	1.1×10^{11} y	$K/\beta^+/\beta^-$
La	57	139	138.9064	99.911		
138.91	57	140	139.9085		40.22 h	β^-
	57	141	140.9095		3.90 h	β^-
Cerium	58	136	135.9071	0.19		
Ce	58	137			9.00 h	K
140.12	58	138	137.9060	0.25		
	58	139	138.9054		140.0 d	K
	58	140	139.9053	88.48		
	58	141	140.9069		33.0 d	β^-
	58	142	141.9093	11.08	5.0×10^{16} y	β^-
	58	143	142.9111		33.0 h	β^-
	58	144	143.9127		285.0 d	β^-

continued

Element Symbol Atomic weight	Atomic number Z	Atomic mass number A	Isotopic mass, amu	Natural abundance, %	Half-life	Type of decay
Praseodymium	59	140	139.9079		3.39 m	K
Pr	59	141	140.9077	100.00		
140.907	59	142	141.9087		19.2 h	β^-
	59	143	142.9096		13.7 d	β^-
Neodymium	60	142	141.9075	27.11		
Nd	60	143	142.9096	12.17		
144.24	60	144	143.9099	23.85		
	60	145	144.9122	8.30		
	60	146	145.9131	17.22		
	60	147			11.1 d	β^-
	60	148	147.9169	5.73		
	60	149	148.9169		1.73 h	β^-
	60	150	149.9207	5.62		
Promethium	61	146	145.9125		5.52 y	K/β^-
Pm	61	147	146.9138		2.50 y	β^-
	61	148	147.9171		5.39 d	β^-
	61	149	148.9175		53.1 h	β^-
Samarium	62	144	143.9120	3.09		
Sm	62	145			340.0 d	K
150.35	62	146	145.9129		1.03×10^8 y	α
	62	147	146.9149	14.97	1.06×10^{11} y	α
	62	148	147.9146	11.24	8×10^{15} y	α
	62	149	148.9171	13.83	10^{16} y	α
	62	150	149.9173	7.44		
	62	151			93.0 y	β^-
	62	152	151.9195	26.72		
	62	153			46.8 h	β^-
	62	154	153.9220	22.71		
	62	155	154.9242		22.0 m	β^-
Europium	63	148	147.9182		54.5 d	K
Eu	63	151	150.9199	47.82		
151.96	63	153	152.9212	52.18		
	63	154	153.9240		8.2 y	β^-
	63	155	154.9219		4.76 y	β^-
Gadolinium	64	148	147.9181		93.0 y	α
Gd	64	149	148.9193		9.0 d	K
157.24	64	150	149.9185		1.8×10^6 y	α
	64	152	151.9198	0.20	1.1×10^{14} y	α
	64	154	153.9207	2.15		
	64	155	154.9226	14.73		
	64	156	155.9221	20.47		
	64	157	156.9239	15.68		
	64	158	157.9241	24.87		
	64	159			18.0 h	β^-
	64	160	159.9271	21.90		
	64	161			3.70 m	β^-

Element Symbol Atomic weight	Atomic number Z	Atomic mass number A	Isotopic mass, amu	Natural abundance, %	Half-life	Type of decay
Terbium	65	158			150 y	K
Tb	65	159	158.9253	100.00		
158.925	65	160	159.9269		73.0 d	β^-
Dysprosium	66	156	155.9243	0.052		
Dy	66	157			8.1 h	K
162.50	66	158	157.9243	0.090		
	66	160	159.9252	2.29		
	66	161	160.9269	18.88		
	66	162	161.9268	25.53		
	66	163	162.9287	24.97		
	66	164	163.9291	28.18		
	66	165	164.9303		2.3 h	β^-
Holmium	67	164	163.9306		37.0 m	β^-/K
Ho	67	165	164.9303	100.00		
164.93	67	166			26.9 h	β^-
Erbium	68	162	161.9288	0.136		
Er	68	164	163.9293	1.56		
167.26	68	165			10.3 h	K
	68	166	165.9304	33.41		
	68	167	166.9320	22.94		
	68	168	167.9324	27.07		
	68	170	169.9355	14.88		
	68	171			7.5 h	β^-
Thulium	69	168			93.1 d	K
Tm	69	169	168.9344	100.00		
168.934	69	170			128.0 d	β^-
Ytterbium	70	168	167.9339	0.135		
Yb	70	170	169.9349	3.03		
173.04	70	171	170.9365	14.31		
	70	172	171.9366	21.82		
	70	173	172.9383	16.13		
	70	174	173.9390	31.84		
	70	175			101.0 h	β^-
	70	176	175.9427	12.73		
Lutecium	71	173			1.37 y	K
Lu	71	175	174.9409	97.41		
174.97	71	176	175.9427	2.59	3.0×10^{10} y	β^-
	71	177			6.7 d	β^-
Hafnium	72	174	173.9400	0.18	2.0×10^{15} y	
Hf	72	175			70.0 d	K
178.49	72	176	175.9414	5.20		
	72	177	176.9435	18.50		
	72	178	177.9439	27.14		
	72	179	178.9460	13.75		
	72	180	179.9468	35.24		
	72	181			42.4 d	β^-

continued

Element Symbol Atomic weight	Atomic number Z	Atomic mass number A	Isotopic mass, amu	Natural abundance, %	Half-life	Type of decay
Tantalum	73	180	179.9475	0.0123		
Ta	73	181	180.9480	99.988		
180.948	73	182	181.9475		115.0 d	β^-
Tungsten	74	180	179.9467	0.14		
W	74	181			121.0 d	K
183.85	74	182	181.9483	26.41		
	74	183	182.9503	14.40		
	74	184	183.9510	30.64		
	74	185			75.8 d	β^-
	74	186	185.9543	28.41		
	74	187	186.9530		24.0 h	β^-
Rhenium	75	185	184.9530	37.07		
Re	75	186	185.9515		90.0 h	β^-/K
186.207	75	187	186.9560	62.93	5.0×10^{10} y	β^-
	75	188	187.9565		16.7 h	β^-
Osmium	76	184	183.9526	0.018		
Os	76	185			94.0 d	K
190.20	76	186	185.9539	1.59	2.0×10^{15} y	α
	76	187	186.9560	1.64		
	76	188	187.9560	13.30		
	76	189	188.9582	16.10		
	76	190	189.9586	26.40		
	76	191	190.9607		15.0 d	β^-
	76	192	191.9615	41.00		
	76	193	192.9650		31.0 h	β^-
Iridium	77	191	190.9606	37.30		
Ir	77	192	191.9636		74.0 d	β^-
192.22	77	193	192.9629	62.70		
	77	194	193.9647		19.2 h	β^-
Platinum	78	190	189.9600	0.0127	6.0×10^{11} y	α
Pt	78	191			3.0 d	K
195.09	78	192	191.9611	0.786	1.0×10^{15} y	α
	78	193	192.9640		50.0 y	K
	78	194	193.9628	32.90		
	78	195	194.9648	33.80		
	78	196	195.9650	25.30		
	78	197	196.9666		18.0 h	β^-
	78	198	197.9679	7.21		
	78	199			30.0 m	β^-
Gold	79	196	195.9658		6.18 d	K/β^-
Au	79	197	196.9666	100.00		
196.967	79	198	197.9675		2.693 d	β^-
	79	199	198.9677		3.15 d	β^-

Element Symbol Atomic weight	Atomic number Z	Atomic mass number A	Isotopic mass, amu	Natural abundance, %	Half-life	Type of decay
Mercury	80	196	195.9658	0.146		
Hg	80	197			65.0 h	K
200.59	80	198	197.9668	10.01		
	80	199	198.9683	16.84		
	80	200	199.9683	23.13		
	80	201	200.9703	13.22		
	80	202	201.9706	29.80		
	80	203	202.9719		46.57 d	β^-
	80	204	203.9735	6.85		
	80	205	204.9751		5.50 m	β^-
Thallium	81	203	202.9723	29.50		
Tl	81	204	203.9721		3.80 y	β^-
204.37	81	205	204.9745	70.50		
	81	206	205.9747		4.19 m	β^-
Lead	82	204	203.9730	1.48		
Pb	82	205	204.9731		1.4×10^7 y	K
207.18	82	206	205.9745	24.10		
	82	207	206.9759	22.10		
	82	208	207.9766	52.30		
	82	209	208.9798		3.30 h	β^-
	82	210	209.9828		21.0 y	β^-
Bismuth	83	208	207.9784		3.7×10^5 y	K
Bi	83	209	208.9804	100.00	2.0×10^{18} y	α
208.98	83	210	209.9841		5.01 d	β^-/α
	83	211	210.9873		2.15 m	α/β^-
Polonium	84	206	205.9805		8.8 d	K/α
Po	84	207	206.9816		5.7 h	K/α
	84	208	207.9813		2.93 y	α/K
	84	209	208.9825		103.0 y	α/K
	84	210	209.9829		138.4 d	α
	84	211	210.9866		0.52 s	α
	84	212	211.9889		0.304 μs	α
	84	213	212.9928		4.120 μs	α
	84	214	213.9952		0.000162 s	α
	84	215	214.9995		0.00178 s	α
	84	216	216.0019		0.15 s	α
	84	218	218.0089		3.05 m	α
Astantine	85	210	209.9871		8.30 h	K/α
As	85	211	210.9875		7.20 h	K/α
	85	212	211.9907		0.30 s	α
	85	213	212.9929		0.10 μs	α
	85	214	213.9963		2.00 μs	α
	85	215	214.9987		0.0001 s	α
	85	216	216.0024		0.0003 s	α
	85	217	217.0046		0.032 s	α
	85	218	218.0086		2.0 s	α
	85	219	219.0114		0.9 m	α

continued

Element Symbol Atomic weight	Atomic number Z	Atomic mass number A	Isotopic mass, amu	Natural abundance, %	Half-life	Type of decay
Radon	86	219	219.0095		4.0 s	α
Rn	86	220	220.0114		55.0 s	α
	86	221	221.0154		25.0 m	β^-/α
	86	222	222.0175		3.823 d	α
Francium	87	220	220.0123		27.55 s	α
Fr	87	221	221.0142		4.80 m	α
	87	222	222.0161		14.8 m	β^-
	87	223	223.0198		22.0 m	β^-
Radium	88	220	220.0110		0.023 s	α
Ra	88	221	221.0139		30.0 s	α
	88	222	222.0154		38.0 s	α
	88	223	223.0186		11.43 d	α
	88	224	224.0202		3.65 d	α
	88	225	225.0219		14.8 d	β^-
	88	226	226.0254		1600.0 y	α
	88	227	227.0276		41.2 m	β^-
	88	228	228.0296		5.75 y	β^-
Actinium	89	225	225.0231		10.0 d	α
Ac	89	226	226.0261		29.0 h	β^-/α
	89	227	227.0278		21.6 y	β^-
	89	228	228.0295		6.13 h	β^-
	89	229	229.0308		66.0 m	β^-
Thorium	90	228	228.0287		1.913 y	α
Th	90	229	229.0316		7340 y	α
232.038	90	230	230.0331		80,000 y	α
	90	231	231.0347		25.5 h	β^-
	90	232	232.0382	100.00	1.41×10^{10} y	α
	90	233	233.0387		22.2 m	β^-
Protactinium	91	226	226.0278		1.80 m	α
Pa	91	227	227.0289		38.3 m	α/K
	91	228	228.0310		22.0 h	K
	91	229	229.0321		1.50 d	K
	91	230	230.0345		17.4 d	K/β^-
	91	231	231.0359		33,500 y	α
	91	232	232.0371		1.31 d	β^-
	91	233	233.0384		27.0 d	β^-
	91	234	234.0414		6.75 h	β^-
	91	235	235.0438		23.7 m	β^-
Uranium	92	227	227.0309		1.3 m	α
U	92	228	228.0313		9.3 m	α/K
238.029	92	229	229.0335		58.0 m	K/α
	92	230	230.0339		20.8 d	α
	92	231	231.0363		4.3 d	K
	92	232	232.0372		73.6 y	α
	92	233	233.0395		1.65×10^5 y	α
	92	234	234.0409	0.006	2.47×10^5 y	α

Element Symbol Atomic weight	Atomic number Z	Atomic mass number A	Isotopic mass, amu	Natural abundance, %	Half-life	Type of decay
	92	235	235.0439	0.720	7.1×10^8 y	α
	92	236	236.0457		2.39×10^7 y	α
	92	237	237.0469		6.75 d	β^-
	92	238	238.0508	99.274	4.51×10^9 y	α
	92	239	239.0526		23.5 m	β^-
	92	240	240.0546		14.1 h	β^-
Neptunium	93	231	231.0383		50.0 m	α
Np	93	233	233.0406		35.0 m	K
	93	234	234.0419		4.40 d	K
	93	235	235.0441		410.0 d	K
	93	236	236.0466		22.0 h	β^-/K
	93	237	237.0480		2.14×10^6 y	α
	93	238	238.0494		2.10 d	β^-
	93	239	239.0513		2.35 d	β^-
	93	240	240.0537		63.0 m	β^-
	93	241	241.0558		16.0 m	β^-
Plutonium	94	232	232.0411		36.0 m	K/α
Pu	94	234	234.0433		9.0 h	K
	94	235	235.0453		26.0 m	K
	94	236	236.0461		2.85 y	α
	94	237	237.0483		45.6 d	K
	94	238	238.0495		86.0 y	α
	94	239	239.0522		24.400 y	α
	94	240	240.0540		6580 y	α
	94	241	241.0568		14.7 y	β^-
	94	242	242.0587		3.79×10^5 y	α
	94	243	243.0601		5.0 h	β^-
	94	244	244.0642		8.0×10^7 y	α
Americium	95	240	240.0539		51.0 h	K
Am	95	241	241.0567		458.0 y	α
	95	242	242.0574		16.0 h	β^-/K
	95	243	243.0614		7370 y	α
	95	244	244.0625		10.1 h	β^-
	95	245	245.0648		2.1 h	β^-
Curium	96	238	238.0530		2.5 h	K/α
Cm	96	240	240.0555		26.8 d	α
	96	241	241.0577		32.8 d	K
	96	242	242.0558		163.0 d	α
	96	243	243.0614		28.5 y	α
	96	244	244.0629		17.6 y	α
	96	245	245.0653		8500 y	α
	96	246	246.0674		4730 y	α
Berkelium	97	245	245.0664		4.98 d	K
Bk	97	246	246.0672		1.80 d	K
	97	247	247.0702		1400 y	α
	97	248	248.0730		9.0 y	β^-/K
	97	249	249.0750		314.0 d	β^-

continued

Element Symbol Atomic weight	Atomic number Z	Atomic mass number A	Isotopic mass, amu	Natural abundance, %	Half-life	Type of decay
Californium	98	244	244.0659		20.0 m	α
Cf	98	245	245.0666		44.0 m	K/α
	98	246	246.0688		36.0 h	α
	98	247	247.0690		2.5 h	K
	98	248	248.0724		350.0 d	α
	98	249	249.0748		360.0 y	α
	98	250	250.0766		13.0 y	α
Einsteinium	99	251	251.0800		1.5 d	K
Es	99	252	252.0829		471.0 d	α
	99	253	253.0847		20.47 d	α
	99	254	254.0881		276.0 d	α/β^-
	99	255	255.0900		38.3 d	β^-
Fermium	100	250	250.0795		30.0 m	α
Fm	100	252	252.0827		23.0 h	α
	100	253	253.0837		3.0 d	K/α
	100	254	254.0870		3.2 h	α
	100	255	255.0899		20.1 h	α
Mendelevium	101	255	255.0911		27.0 m	K/α
Mv	101	256	256.0939		76.0 m	K
Nobelium	102	253	253.0905		95.0 s	α
No	102	254	254.0910		55.0 s	α
Lawrencium Lw	103	256	256.0986		31.0 s	α
Rutherfordium	104	259	259.1057		0.0003 s	SF

Abbreviations used in this table.

    For half-lives: s = seconds, m = minutes, h = hours, d = days, y = years.

    For type of radioactive decay: α = alpha decay, β^- = beta decay,

                  β^+ = positron decay, K = K-capture decay,

                  SF = spontaneous fission.

RADIOISOTOPE FUELS

Radioisotope and half-life	Active material	Mass fraction of radioisotope in material	Melting point, °C	Specific power, kW/kg	Power density, kW/liter
$^{90}_{38}Sr$	Metal	55% Sr-90	772	0.50	1.28
	$SrTiO_3$	24.5% Sr-90	1910	0.23	1.17
$T_{1/2} = 28.1$ y	SrO	44.0% Sr-90	2457	0.42	1.94
	SrF_2	36.0% Sr-90	1463	0.34	1.44
	$SrTiO_4$	31.5% Sr-90	1860	0.30	1.48
$^{137}_{55}Cs$	CsCl	28.9% Cs-137	645	0.12	0.37
$T_{1/2} = 33.0$ y	Cs_2SO_4	26.9% Cs-137	1019	0.11	0.46
$^{144}_{58}Ce$	Ce_2O_3	10.8% Ce-144	2190	2.76	19.0
$T_{1/2} = 285.0$ d	Ce_2O_2S	10.3% Ce-144	1890	2.64	15.8
	Ce_2S_3	9.0% Ce-144	1437	2.31	14.3
$^{147}_{61}Pm$	Metal	95% Pm-147	865	0.31	2.30
$T_{1/2} = 2.50$ y	Pm_2O_3	78% Pm-147	2130	0.27	1.87
$^{210}_{84}Po$	Metal	95.0% Po-210	254	144.00	1324.0
$T_{1/2} = 138.4$ d	GdPo-Ta (98%)	1.06% Po-210	1675	1.60	16.6
	GdPo-Ta (91%)	4.95% Po-210	1675	7.50	77.2
$^{238}_{94}Pu$	Metal	80.0% Pu-238	600	0.45	6.80
$T_{1/2} = 86.0$ y	PuO_2	71.0% Pu-238	2150	0.40	2.70
	PuC	74.6% Pu-238	1654	0.42	5.70
	PuN	74.6% Pu-238	2570	0.42	6.20
	PuZr	73.0% Pu-238	730	0.41	5.60
	PuZr	77.0% Pu-238	615	0.44	6.50

continued

Radioisotope and half-life	Active material	Mass fraction of radioisotope in material	Melting point, °C	Specific power, kW/kg	Power density, kW/liter
$^{242}_{96}$Cm $T_{1/2} = 163.0$ d	$Cm_2O_3\text{-}AmO_2$	35.7% Cm-242	2000	42.8	500.00
$^{244}_{96}$Cm $T_{1/2} = 17.6$ y	Metal	95.5% Cm-242	1340	2.67	36.00
	Cm_2O_3	86.9% Cm-242	1950	2.42	26.10
	Cm_2O_2S	84.4% Cm-242	2000	2.35	23.30
	CmF_3	77.5% Cm-242	1406	2.15	21.10

From G. L. Tuve and R. E. Bole, "Handbook of Tables for Applied Engineering Science," The Chemical Rubber Company, 1970, p. 348.

APPENDIX

K

SOLAR POSITION AND IRRADIATION VALUES

(For 40 degrees North latitude; 1.0 clearness factor; 0% ground reflectance)

Date	Solar time		Solar position		Total solar insolation, W/m²†		
	A.M.	P.M.	Altitude β_1	Azimuth α_1	Direct normal	Horizontal	South vertical
Jan 21	8	4	8.1	55.3	448	88	265
	9	3	16.8	44.0	753	262	539
	10	2	23.8	30.9	864	400	703
	11	1	28.4	16.0	911	485	797
	12		30.0	0.0	927	517	829
	Surface daily totals, W·h/m²				6878	2988	5440
Feb 21	7	5	4.8	72.7	217	32	69
	8	4	15.4	62.2	706	230	337
	9	3	25.0	50.2	863	416	526
	10	2	32.8	35.9	930	561	662
	11	1	38.1	18.9	961	649	744
	12		40.0	0.0	971	681	772
	Surface daily totals, W·h/m²				8321	4457	5453

† 1 W/m² = 0.3173 Btu/h·ft².

continued

(Solar position and irradiation values for 40 degrees North latitude)

Date	Solar time A.M.	Solar time P.M.	Altitude β_1	Azimuth α_1	Direct normal	Horizontal	South vertical
Mar 21	7	5	11.4	80.2	539	145	110
	8	4	22.5	69.6	788	359	281
	9	3	32.8	57.3	889	545	435
	10	2	41.6	41.9	936	687	555
	11	1	47.7	22.6	961	779	630
		12	50.0	0.0	968	810	656
	Surface daily totals, W·h/m²				9191	5838	4678
Apr 21	6	6	7.4	98.9	281	63	13
	7	5	18.9	89.5	649	274	38
	8	4	30.3	79.3	794	479	167
	9	3	41.3	67.2	864	652	293
	10	2	51.2	51.4	901	788	397
	11	1	58.7	29.3	920	873	463
		12	61.6	0.0	924	905	485
	Surface daily totals, W·h/m²				9746	7168	3221
May 21	5	7	1.9	114.7	3	0	0
	6	6	12.7	105.6	454	154	28
	7	5	24.0	96.6	681	359	41
	8	4	35.4	87.2	788	552	79
	9	3	46.8	76.0	842	716	189
	10	2	57.5	60.9	873	842	280
	11	1	66.2	37.1	892	924	340
		12	70.0	0.0	895	949	359
	Surface daily totals, W·h/m²				9960	8044	2282
June 21	5	7	4.2	117.3	69	13	3
	6	6	14.8	108.4	489	189	32
	7	5	26.0	99.7	681	388	44
	8	4	37.4	90.7	775	574	50
	9	3	48.8	80.2	829	734	148
	10	2	59.8	65.8	857	857	233
	11	1	69.2	41.9	873	933	290
		12	73.5	0.0	879	958	309
	Surface daily totals, W·h/m²				10,023	8346	1923

† 1 W/m² = 0.3173 Btu/h·ft².

(Solar position and irradiation values for 40 degrees North latitude)

Date	Solar time A.M.	Solar time P.M.	Solar position Altitude β_1	Solar position Azimuth α_1	Total solar insolation, W/m²† Direct normal	Total solar insolation, W/m²† Horizontal	Total solar insolation, W/m²† South vertical
July 21	5	7	2.3	115.2	6	0	0
	6	6	13.1	106.1	435	158	28
	7	5	24.3	97.2	656	359	44
	8	4	35.8	87.8	760	548	76
	9	3	47.2	76.7	816	709	183
	10	2	57.9	61.7	848	835	271
	11	1	66.7	37.9	867	914	328
		12	70.6	0.0	870	939	350
	Surface daily totals, W·h/m²				9651	7987	2213
Aug 21	6	6	7.9	99.5	255	66	16
	7	5	19.3	90.0	602	274	38
	8	4	30.7	79.9	747	473	158
	9	3	41.8	67.9	820	646	281
	10	2	51.7	52.1	857	775	378
	11	1	59.3	29.7	876	860	441
		12	62.3	0.0	883	889	463
	Surface daily totals, W·h/m²				9191	7073	3083
Sep 21	7	5	11.4	80.2	469	136	101
	8	4	22.5	69.6	725	344	265
	9	3	32.8	57.3	829	526	416
	10	2	41.6	41.9	883	665	530
	11	1	47.7	22.6	905	753	605
		12	50.0	0.0	914	785	630
	Surface daily totals, W·h/m²				8536	5636	4463
Oct 21	7	5	4.5	72.3	151	22	50
	8	4	15.0	61.9	643	214	315
	9	3	24.5	49.8	810	397	504
	10	2	32.4	35.6	882	536	640
	11	1	37.6	18.7	917	627	722
		12	39.5	0.0	927	656	750
	Surface daily totals, W·h/m²				7735	4249	5213
Nov 21	8	4	8.2	55.4	429	88	255
	9	3	17.0	44.1	731	258	526
	10	2	24.0	31.0	845	397	690
	11	1	28.6	16.1	892	482	782
		12	30.2	0.0	908	514	813
	Surface daily totals, W·h/m²				6707	2969	5314

† 1 W/m² = 0.3173 Btu/h·ft².

continued

(Solar position and irradiation values for 40 degrees North latitude)

Date	Solar time		Solar position		Total solar insolation, W/m^2†		
	A.M.	P.M.	Altitude β_1	Azimuth α_1	Direct normal	Horizontal	South vertical
Dec 21	8	4	5.5	53.0	281	44	177
	9	3	14.0	41.9	684	205	514
	10	2	20.7	29.4	823	337	697
	11	1	25.0	15.2	883	422	794
		12	26.6	0.0	898	451	829
	Surface daily totals, $W \cdot h/m^2$				6235	2465	5188

† $1\ W/m^2 = 0.3173\ Btu/h \cdot ft^2$.

Reprinted with permission from the 1974 Applications Volume, ASHRAE Handbook and Product Directory.

VARIATION OF SOLAR RADIATION WITH LATITUDE

Date	Degrees latitude	Direct normal insolation, $W \cdot h/m^2$	Total horizontal insolation, $W \cdot h/m^2$	Date	Degrees latitude	Direct normal insolation, $W \cdot h/m^2$	Total horizontal insolation, $W \cdot h/m^2$
Jan 21	24	8718	5113	Apr 21	24	9569	7735
$\delta = -20°$	32	7748	4060	$\delta = +11.9°$	32	9696	7533
	40	6878	2988		40	9746	7168
	48	5390	1879		48	9696	6638
	56	3549	889		56	9532	5964
	64	965	142		64	9399	5182
Feb 21	24	9569	6298	May 21	24	9557	8057
$\delta = -10.6°$	32	9053	5434	$\delta = +20.3°$	32	9809	8138
	40	8321	4457		40	9960	8044
	48	7344	3404		48	10257	7823
	56	6260	2332		56	10528	7483
	64	4514	1261		64	10937	7048
Mar 21	24	9702	7155	June 21	24	9437	8113
$\delta = 0.0°$	32	9494	6569	$\delta = +23.45°$	32	9721	8302
	40	9191	5838		40	10023	8346
	48	8763	4974		48	10439	8277
	56	8151	3997		56	10837	8075
	64	7237	2938		64	11505	7842

continued

Date	Degrees latitude	Direct normal insolation, $W \cdot h/m^2$	Total horizontal insolation, $W \cdot h/m^2$	Date	Degrees latitude	Direct normal insolation, $W \cdot h/m^2$	Total horizontal insolation, $W \cdot h/m^2$
July 21	24	9242	7962	Oct 21	24	9040	6077
$\delta = +20.5°$	32	9494	8063	$\delta = -10.7°$	32	8498	5213
	40	9651	7987		40	7735	4249
	48	9954	7798		48	6789	3221
	56	10212	7477		56	5686	2169
	64	10628	7086		64	3902	1128
Aug 21	24	9027	7590	Nov 21	24	8529	5075
$\delta = +12.1°$	32	9147	7414	$\delta = -19.9°$	32	7584	4035
	40	9191	7073		40	6707	2969
	48	9134	6575		48	5257	1879
	56	8983	5938		56	3448	895
	64	8851	5188		64	952	145
Sep 21	24	9071	6915	Dec 21	24	8271	4646
$\delta = 0.0°$	32	8851	6348	$\delta = -23.45°$	32	7401	3581
	40	8536	5636		40	6235	2465
	48	8094	4797		48	4551	1406
	56	7464	3845		56	2357	492
	64	6537	2811		64	76	6

Reprinted with permission from the 1974 Applications Volume, ASHRAE Handbook and Product Directory.

INDEX